Power to Fuel

Power to Fuel

How to Speed Up a Hydrogen Economy

Edited by

Giuseppe Spazzafumo

Department of Civil and Mechanical Engineering, University of Cassino and Southern Lazio, Cassino, Italy

ELSEVIER

ACADEMIC PRESS

An imprint of Elsevier

Academic Press is an imprint of Elsevier
125 London Wall, London EC2Y 5AS, United Kingdom
525 B Street, Suite 1650, San Diego, CA 92101, United States
50 Hampshire Street, 5th Floor, Cambridge, MA 02139, United States

British Library Cataloguing-in-Publication Data
A catalogue record for this book is available from the British Library

Library of Congress Cataloging-in-Publication Data
A catalog record for this book is available from the Library of Congress

ISBN: 978-0-12-822813-5

For Information on all Academic Press publications
visit our website at https://www.elsevier.com/books-and-journals

Publisher: Joe Hayton
Acquisitions Editor: Lisa Reading
Editorial Project Manager: Alice Grant
Production Project Manager: Nirmala Arumugam
Cover Designer: Matthew Limbert

Typeset by MPS Limited, Chennai, India

Contents

List of contributors

Ramy E. Ali Fraunhofer Institute for Solar Energy Systems ISE, Freiburg, Germany

Eleonora Bargiacchi Department of Energy, Systems, Territory and Constructions Engineering, University of Pisa, Pisa, Italy

Dmitri Bessarabov HySA Infrastructure Centre, North-West University, Potchefstroom, South Africa

Daniele Candelaresi Department of Civil and Mechanical Engineering, University of Cassino and Southern Lazio, Cassino, Italy

Sudipta Chatterjee KAUST Catalysis Centre, King Abdullah University of Science and Technology, Thuwal, Saudi Arabia; Division of Physical Science and Engineering, King Abdullah University of Science and Technology, Thuwal, Saudi Arabia

Indranil Dutta KAUST Catalysis Centre, King Abdullah University of Science and Technology, Thuwal, Saudi Arabia; Division of Physical Science and Engineering, King Abdullah University of Science and Technology, Thuwal, Saudi Arabia

Malte Gierse Fraunhofer Institute for Solar Energy Systems ISE, Freiburg, Germany

Christopher Hebling Fraunhofer Institute for Solar Energy Systems ISE, Freiburg, Germany

Kuo-Wei Huang KAUST Catalysis Centre, King Abdullah University of Science and Technology, Thuwal, Saudi Arabia; Division of Physical Science and Engineering, King Abdullah University of Science and Technology, Thuwal, Saudi Arabia

Meng Ni Building Energy Research Group, Department of Building and Real Estate, The Hong Kong Polytechnic University, Hong Kong, P.R. China; Environmental Energy Research Group, Research Institute for Sustainable Urban Development (RISUD), The Hong Kong Polytechnic University, Hong Kong, P.R. China

Kingsley Onyebuchi Obodo HySA Infrastructure Centre, North-West University, Potchefstroom, South Africa

Mohamed Ouda Fraunhofer Institute for Solar Energy Systems ISE, Freiburg, Germany

Cecil Naphtaly Moro Ouma HySA Infrastructure Centre, North-West University, Potchefstroom, South Africa

András Perl Hanze University of Applied Sciences, Groningen, The Netherlands

Jörg Sauer Institute of Catalysis Research and Technology (IKFT), Karlsruhe Institute of Technology (KIT), Eggenstein-Leopoldshafen, Germany

Achim Schaadt Fraunhofer Institute for Solar Energy Systems ISE, Freiburg, Germany

Giuseppe Spazzafumo Department of Civil and Mechanical Engineering, University of Cassino and Southern Lazio, Cassino, Italy

Haoran Xu State Key Laboratory of Clean Energy Utilization, Zhejiang University, Hangzhou, China; Department of Chemical Engineering, Loughborough University, Loughborough, United Kingdom; Building Energy Research Group, Department of Building and Real Estate, The Hong Kong Polytechnic University, Hong Kong, P.R. China

Preface

The transition from fossil fuels to renewable energy sources is increasingly desired, but a further acceleration will only be possible if the storage problem is solved. In fact, almost all these sources are characterised by intermittent availability and generally do not have their own storage capacity.

The final uses of energy are mainly satisfied by electricity and fuels with an average ratio of 1:4. With the current technologies, the most widespread renewable sources generate almost exclusively electricity. Even in the case of a greater use of electricity in the upcoming years, fuels will remain the predominant energy carrier.

Thanks to its characteristics, hydrogen seems to be the best candidate to help electricity as an energy carrier. It does not contain carbon and can be easily produced from electricity and converted back into it. It can also be produced from biomass and waste materials, and it is better than electricity when it is necessary to store energy in the long run and in a large amount. Therefore, it qualifies as the ideal partner for renewable sources.

However, the development of an energy system based on hydrogen that is able to replace commonly used fuels still faces obstacles. From a technical point of view, on-board storage and distribution still represent a bottleneck.

Hydrogen-powered vehicles are now equipped with tanks that contain compressed gas up to 700 bar. Despite the high pressure, these tanks are still considered too bulky, but the alternatives have proved uncompetitive or are still in the research and development phase.

Hydrogen is now distributed by tank trucks or produced on-site, but for its large-scale use, it will be necessary to arrive at a system similar to that used for natural gas.

From an economic point of view, the required investments are enormous, both by those who will undertake the production and distribution of hydrogen and by those who want to use it. If the end-user faces high prices (e.g., to buy a car with fuel cells), there will be risks to limit the hydrogen demand and no one will undertake to produce an under-demanded good in large quantities, just as no one will spend money to create a distribution infrastructure that will be underused. Incentives may come from the governments of many countries, but they may not be sufficient.

The solution may be to take one step at a time. The first step can be that of production, since without production there can be neither distribution nor use. The key point might be to create a great hydrogen demand which, however, will not be used in pure form, at least in the first moment. In this way, investments will be concentrated on the production plants, while at the same time research will keep developing to overcome technical obstacles and to reduce costs for the end-user.

Of course, hydrogen should no longer be produced from fossil fuels as it is done currently. Indeed, it should replace commonly used fuels. This is precisely the concept of power-to-fuel: converting electricity produced from renewable sources into hydrogen. Furthermore, at least for now, using hydrogen to produce synthetic fuels is capable of replacing traditional fuels without requiring major changes to the distribution infrastructure and end-use equipment.

Such a strategy would allow a significant increase in the hydrogen demand and consequently a significant increase in the use of renewable energy sources. Therefore, power-to-fuel seems to be the most reasonable way towards a fast decarbonisation of the energy system.

This book aims to present the most promising technologies that respond to the concept of power-to-fuel. The synthetic fuels covered in this book can be deployed without major changes to the existing infrastructure and can be used in existing equipment either directly or with minimal changes. Some can already be used in direct fuel cells today without the need for an external reforming process. Moreover, all of them are hydrogen carriers.

Chapter 1 is an introduction to the concept of power-to-fuel. It is divided into three main sections. Initially, the issue of carbon dioxide emissions into the atmosphere and the need for a transition from fossil fuels to renewable energy sources are addressed. This section highlights the existing obstacles to a further spread of the latter and indicates the solution of power-to-hydrogen in the long term. The second section lists the characteristics of hydrogen and briefly describes the production processes starting from the most varied primary sources of energy. The current obstacles to a rapid transformation of the energy system towards the one based on hydrogen are then highlighted, and the path of its use to produce synthetic fuels is indicated as viable. Finally, the main synthetic fuels such as methane, methanol, dimethyl ether, ammonia, urea and formic acid which can be obtained by applying the concept of power-to-fuel are briefly described.

It is not easy to go directly from electricity to a synthetic fuel and therefore the first step of a power-to-fuel process is the power-to-hydrogen carried out through the electrolysis process. Until now, electrolysis has been considered a niche process, for small flow rates or for high purities. Now, however, new horizons are opening up thanks to the desire to decarbonise the energy system and there is talk of multi-megawatt installations and beyond.

Low-temperature electrolysis is discussed in Chapter 2. Fundamentals of water electrolysis and modelling of the reaction mechanism of the oxygen evolution reaction both in acidic and alkaline media are addressed. In addition, a brief overview of recent developments in low-temperature water electrolysis is included. Different modelling techniques applicable to electrocatalysts in the oxygen evolution reaction are presented at macroscopic, mesoscopic and microscopic length scale. A section of the chapter is devoted to iridium and its oxides in crystalline and amorphous phases since Ir-based compounds are typical materials for the oxygen evolution reaction.

Chapter 3 deals with high-temperature electrolysis that is based on a solid oxides electrolyte. In the first section, the principles of the solid oxides electrochemical process are presented in detail. In the next section, the direct production of a syngas is presented. Such a syngas can be produced by means of the co-electrolysis of water and a carbon source. Finally, the last section deals with the production of a synthetic liquid fuel by combining a co-electrolysis process of water and carbon dioxide with a Fischer–Tropsch process.

Direct injection of hydrogen into the natural gas pipelines represents a very interesting option to store a large amount of energy. However, there are technical and commercial limits to the amount of hydrogen that can be introduced. Chapter 4 is then regarded as one of the most promising and efficient pathways to store energy through the power-to-fuel concept. It is the simplest process that allows carbon utilisation to produce a synthetic fuel. Natural gas is already widely used in industry and in the residential sector, and in the transport sector in some countries as well. Therefore the implementation of power-to-substitute of natural gas on a large-scale could allow a vast and fast decarbonisation of the energy sector. Both chemical and biological methanation, including the state-of-the-art and literature studies, are presented in detail with their most important features and operational parameters. A comparison between different available technologies is carried out with an outlook on power-to-methane plant economy. Finally, an insight on substitute natural gas production from biomass gasification and electrolytic hydrogen is proposed.

Another interesting alternative is discussed in Chapter 5. Methanol is a very important raw material in the chemical industry and could be also used as a substitute for liquefied petroleum gas and diesel fuel. Methanol has historically been produced starting from syngas that is mainly composed of carbon monoxide and hydrogen. However, today its production from carbon dioxide and hydrogen seems more interesting. These two processes are described and compared. In addition, two innovative technologies are also mentioned, namely, co-electrolysis and biological oxidation of methane.

Chapter 6 deals with power-to-dimethyl ether (DME) which is another liquid product with a broad spectrum of applications not only as a fuel but also as a compound for the chemical industry and as a chemical building block. Nowadays, DME is produced from natural gas or coal, but it can be produced from carbon dioxide and hydrogen. State-of-the-art and emerging DME production pathways are presented in detail with a particular focus on the influence of a renewable production based on carbon dioxide and hydrogen. The last section also reports economic insights into a power-to-DME process.

Ammonia and urea are examined in Chapter 7. Ammonia is heavily used in agriculture as a fertiliser and in the industry as a raw material for the production of various organic nitrogen compounds. The current industrial ammonia production is dominated by fossil energy sources. However, its high hydrogen content and its established infrastructure for both storage and distribution make ammonia a prominent candidate for storing fluctuating renewable energy. The traditional Haber–Bosch process consisting of a heterogeneous reaction of hydrogen and nitrogen on an iron-based catalyst can become green and carbon-free if renewable electricity is used for hydrogen generation. Basarov reaction is instead used to generate urea, which is another potential hydrogen carrier having the advantage to be safer because solid.

Formic Acid, discussed in Chapter 8, is another promising hydrogen energy carrier thanks to its high volumetric hydrogen capacity and low toxicity and flammability. In the past two decades, great progress has been made in both the use and production of formic acid. Important developments in both hydrogenation and electrochemical reduction of carbon dioxide for formate/formic acid synthesis are presented here. Also, consideration of formic acid cost production and observations on future challenges and opportunities are reported at the end of the chapter.

Chapter 9 reviews 48 pilot and demonstration worldwide power-to-X projects in which the hydrogen is partially or totally produced through electrolysis. The first three sections report the specific features of the most relevant power-to-methane, power-to-methanol and power-to-ammonia plants. The following sections present an analysis of the plant data, which includes the number of projects per technology, the average and total installed power based on the electrolyser size and the project geographical distribution. Regardless of the final product, it is possible to obtain a fair comparison between different power-to-X plants on the basis of the electrolyser size.

In Chapter 10 the authors tried to give a rough idea of the potential penetration of power-to-fuel in the market, from a perspective of complete replacement of fossil fuels with hydrogen and/or synthetic carbon-neutral fuels. Starting from an analysis of world energy consumption, the potential global demand for hydrogen and other synthetic renewable fuels necessary for a complete energy transition was calculated, with a focus on their application in various sectors, such as industrial, transport, building and electricity generation sectors.

To conclude, I express special thanks to all the authors for their excellent contribution. Also, many thanks must be addressed to all the Elsevier staff for the great work in helping us during all the various steps for realising this book in the best possible way.

Giuseppe Spazzafumo

Introduction: the power-to-fuel concept

1

Daniele Candelaresi and Giuseppe Spazzafumo
Department of Civil and Mechanical Engineering, University of Cassino and Southern Lazio, Cassino, Italy

Since the discovery of fire, man has learned to use fuels: by switching from wood to coal, to oil and finally to natural gas and increasing enormously, at the same time, the global fuel consumption. Fig. 1.1 shows as the resulting CO_2 emissions led the atmospheric concentration to more than 400 ppm and this could be the main cause of global warming and climate change. Today, efforts are being made from an increasing number of countries to reduce, and potentially eliminate, the use of fossil fuels through the use of biofuels and hydrogen. Hydrogen is an important feedstock gas in industrial applications (Bessarabov and Millet, 2018). It is generally produced through steam methane reforming (SMR), whereas electrolytic hydrogen production has been considered an option for small flows of hydrogen or for high purity hydrogen required by niche applications.

The renewable energy sources (RESs) with the highest potential are solar and wind, and luckily, the cost of energy generated from these RESs is constantly decreasing. However, they produce electric power which covers only around 20% of the final energy demand.

Electricity, although it was discovered by the Greek Thales of Miletus in the 4th century BCE, has entered into use only in recent times. In the second half of the 19th century, with the invention of the dynamo by the Italian Antonio Pacinotti and with its industrialisation carried out by the Frenchman Zénobe-Théophile Gramme, the conversion from fuel to electricity began. And ever since then the direction of the process has been this.

Today the reversal of the direction of the process, that is from electricity to fuel, is carefully considered. The reason lies on the one hand in the need to replace fossil fuels with renewable and carbon-neutral fuels and on the other hand with the difficulty of storing the electricity obtainable from most RESs.

1.1 Renewable energy sources and energy storage

Unlike fossil resources, RESs are well spread on the planet and this is undoubtedly a great advantage. Unfortunately, the time distribution slows down the spread of exploitation. Almost all RESs are in fact characterised by a random availability accompanied by the lack of intrinsic storage capacity. Actually, apart from large hydropower, only geothermal energy offers constant availability, whereas tidal energy is variable,

Power to Fuel. DOI: https://doi.org/10.1016/B978-0-12-822813-5.00005-9

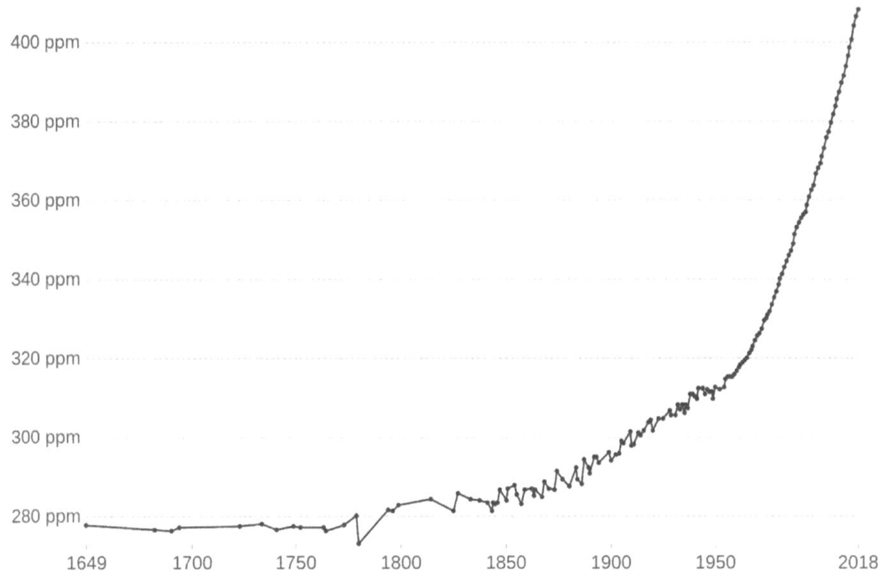

Figure 1.1 Global mean annual concentration of carbon dioxide in atmosphere (Our World in Data, 2020).

but its variability is foreseeable. Biomass can be stored; however, the sources whose exploitation shows the most favourable diffusion trend are solar and wind power, which generate electricity and are characterised by intermittency, with a consequent mismatch between production and demand. Moreover, a local surplus of electric power could not be completely transferred to another region due to probable grid instability problems. This will force to find an energy storage solution to allow the continuous replacing of fossil fuels with renewable and nonclimate-changing fuels.

Therefore electricity from solar and wind can be fed into the grid only if there is an effective demand and if the quality of the energy produced is such as not to jeopardise the stability of the grid itself. Otherwise, it is necessary to accumulate the surplus electricity.

In the event that the problem lies in the lack of electricity demand, it is still possible to transfer the surplus to centralised storage systems, such as pumping stations. If, on the other hand, the inability to transfer electricity to the grid is linked to the poor quality of the energy produced, then it is necessary to increase in some way its quality or to store energy on-site.

Unlike wind power, for solar energy, there is, at least in perspective, the possibility of direct conversion to fuel by means of thermochemical cycles for the production of hydrogen. Unfortunately, this is a prospect thought up about 60 years ago and not yet in view of a practical application. Over 100 different thermochemical cycles have been conceived, but only a few are still being studied, for example the sulphur-iodine cycle (Mathias and Brown, 2003). Initially designed to store the energy produced by nuclear reactors, these cycles are now also studied for solar energy storage, for example the zinc oxide cycle (Weidenkaff et al., 2000).

Electricity can be easily stored in rather small amount and for short periods, whereas the most promising system for storing large quantities of energy over long periods seems to be the conversion of the surplus of electricity into hydrogen by means of the electrolysis process. Hydrogen can also be used in all those applications or sectors where, for technical or economic reasons, electrification is difficult or not convenient, thus allowing to increase the penetration of RES in the market. Furthermore, hydrogen could play an important role to achieve a greater energy system integration, creating new links between different sectors, energy carriers and infrastructures, which today are separate from each other. This would result in a more interconnected, flexible, efficient and cleaner energy system (Fig. 1.2). Finally, electrolytic hydrogen produced from RES power can also be used as a basis to obtain a variety of chemicals and synthetic carbon-neutral fuels, such as those that will be presented in the following sections.

1.2 Power-to-fuel role in the energy transition

The hydrogen atom consists of a single proton and a single electron, which has a mass about 1800 times lower than that of the proton. The atomic weight is 1.00794 amu. The radius of the orbit of the electron is about 100,000 times higher than the radius of the proton and, therefore, the space occupied by the atom is practically empty.

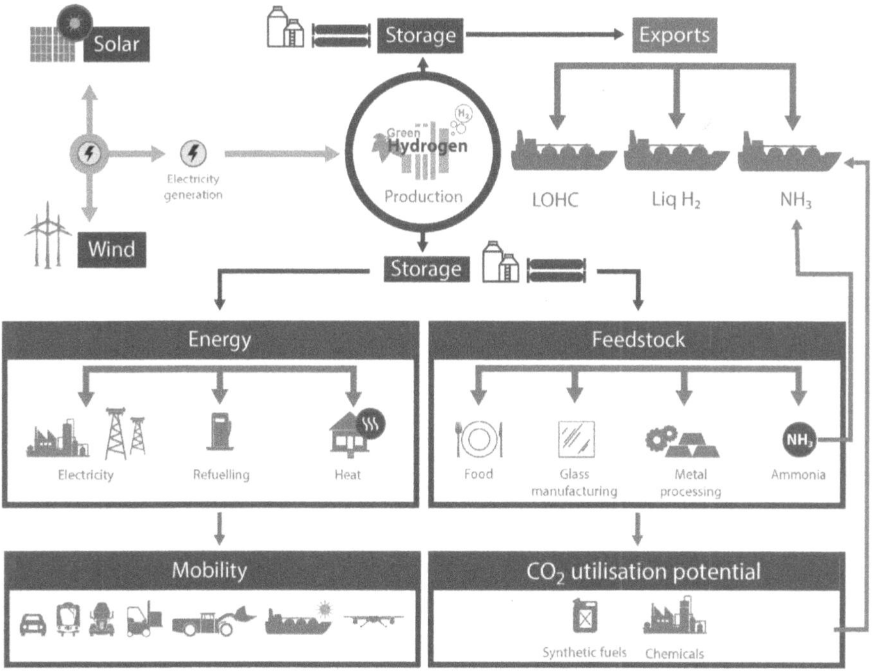

Figure 1.2 Hydrogen as an enabling chemical for sectoral integration: various examples. *Source*: Courtesy D. Bessarabov, HySA.

Its molecule is diatomic with covalent bonding and the protons may have parallel (ortho-hydrogen) or antiparallels (para-hydrogen) spins. The molecular weight is 2.016 amu. At room temperature or above, hydrogen is a mixture (75% ortho-hydrogen + 25% para-hydrogen) and it is called normal-hydrogen. Hydrogen is colourless, odourless, tasteless, practically insoluble in water, highly flammable and nontoxic (simple asphyxiant). Several other characteristics of hydrogen are shown in Table 1.1.

It is the main source of energy in the universe and constitutes 75% of the known matter. Hydrogen is an extremely versatile chemical element. Although on Earth it is present almost exclusively in combined form, it can be produced from any primary source by extracting it from water or other substances that contain it. Table 1.2 shows the main hydrogen production options.

The main current processes for hydrogen production are by far based on reactions between fossil fuels and water: SMR and coal gasification. Neglecting the oxygen required for heat production, the reactions involved are respectively:

$$CH_4 + 2\,H_2O \Rightarrow CO_2 + 4\,H_2 \tag{1.1}$$

and

$$C + 2\,H_2O \Rightarrow CO_2 + 2\,H_2 \tag{1.2}$$

Table 1.1 Hydrogen main characteristics.

Critical temperature	33.25K
Critical pressure	13.2 bar
Density (@15°C, 1 atm)	0.0892 kg m^{-3}
Fusion point	14K
Boiling point	20.3K
Molar volume	$11.42 \cdot 10^{-3}$ m^3 mol^{-1}
Gas constant	4.12 kJ kg^{-1} K^{-1}
Specific heat at $p = cost$	14.89 kJ kg^{-1} K^{-1}
Speed of sound	1270 m s^{-1} at 298.15K
Lower heating value	119.93 MJ kg^{-1}
Higher heating value	141.86 MJ kg^{-1}
Energy density (gas, @15°C, 1 atm)	10.7 MJ Nm^{-3}
Energy density (liquid)	8491 MJ Nm^{-3}
CAS number	1333-74-0
Colour of the cylinder ogive	Red

Table 1.2 Main hydrogen production options.

<table>
<tr><th>Process</th><th>Energy source</th><th>Hydrogen donor</th></tr>
<tr><td>Coal gasification</td><td>Coal</td><td>Water/coal</td></tr>
<tr><td>Hydrocarbons partial oxidation</td><td rowspan="2">Hydrocarbons</td><td rowspan="2">Water/hydrocarbons</td></tr>
<tr><td>Hydrocarbons reforming</td></tr>
<tr><td>Kvaerner process</td><td rowspan="2">Fossil/nuclear/renewable</td><td>Methane</td></tr>
<tr><td>Electrolysis</td><td rowspan="3">Water</td></tr>
<tr><td>Thermolysis</td><td>Nuclear/renewable</td></tr>
<tr><td>Photoelectrolysis</td><td rowspan="4">Sun</td></tr>
<tr><td>Photosynthesis</td><td rowspan="3">Water/biomass</td></tr>
<tr><td>Biomass gasification</td></tr>
<tr><td>Biomass anaerobic digestion</td></tr>
<tr><td>Reaction water/hydride</td><td>Chemical hydride</td><td>Water/hydride</td></tr>
</table>

plus the water gas shift reaction:

$$H_2O + CO \Leftrightarrow H_2 + CO_2 \tag{1.3}$$

which regulates the chemical balance between carbon oxides.

The methane steam reforming process has the advantage to produce a syngas with a high hydrogen concentration since methane is characterised by the highest hydrogen to carbon ratio among the hydrocarbons. Moreover, the process is based on not expensive catalysts based on nickel. Heat is required both for steam production and for the advancement of reaction (1.1), which is endothermic. Heat recovery from syngas cooling allows to reduce the need for external combustion of methane, which is the usual heat source. A complete methane conversion can be obtained at about 900°C. If hydrogen is separated during the process, for example using a membrane, a lower temperature for a complete methane conversion is required and concentrated solar heat could replace the methane combustion reducing the carbon footprint of the process. A high pressure is unfavourable to the reaction advancement because of increasing mole number. However, pressures up to 30 bar are adopted both for reducing the reactor's volume and for reducing the power required for compression when compressed hydrogen is required.

Unless hydrogen is separated during the process, purification is required. Reaction (1.3) is favourable to hydrogen production when the temperature is low. Therefore a two-step water gas shift reaction is usually performed: a high-temperature reactor operates at about 400°C, allowing a good kinetics, whereas a low-temperature reactor works at about 200°C to reach a better chemical equilibrium. Finally, steam is condensed and carbon dioxide is separated, as well as residual carbon monoxide and other impurities.

Hydrogen can be produced also from heavier hydrocarbons by means of partial oxidation or autothermal reforming, both followed by water gas shift reaction.

Coal can be used to produce hydrogen by gasification. Ideally, the process would require only steam, but also reaction (1.2) is endothermic and oxygen is supplied to generate the required heat by internal combustion of coal. Different kinds of gasifiers (fixed bed, fluidised bed, and entrained bed), operating at different temperature and pressure, have been conceived. The product of coal gasification is a syngas mainly composed by hydrogen and carbon monoxide, and shift reactors to convert the latter into hydrogen are used, followed by final purification.

The last process based on fossil fuels is the Kvaerner process, which consists in methane dissociation using an electric arc:

$$CH_4 \Rightarrow C_{(s)} + 2\ H_2 \tag{1.4}$$

Its main advantage is the complete removing of carbon oxide productions and emissions. The main drawback is the very low thermal efficiency because the energy content of carbon is lost. It is a process which can find application when carbon black has a market for not energy applications.

Hydrogen production from RES covers only a minimal share of the current hydrogen production.

Biomass gasification is in principle similar to coal gasification. There are three main differences:

- Biomass has a lower carbon content and, therefore, a lower heating value.
- Biomass is more reactive and, therefore, lower temperature is required.
- Biomass has to be collected from quite large land extension and, therefore, plant size is smaller.

Whereas gasification is suitable for lignocellulosic biomass, anaerobic digestion can be used for the other types of biomass. In general, the final product is a biogas mainly composed by methane and carbon dioxide; however, it is possible to block the activity of methanogenic bacteria, in such a way to obtain a mix of hydrogen and carbon dioxide.

Remaining in the field of bioprocesses, photosynthetic hydrogen production is another option. Hydrogen is spontaneously produced by some microorganisms as a parasitic process to consume the energy surplus generated by an overexposure to sunlight. Responsible for the process are two different enzymes: hydrogenase and nitrogenase. The process can be aerobic (water dissociation) or anaerobic (organic substances dissociation) depending on the microorganism. Current efficiencies are very low, but genetic engineering could give good results in the future.

Another process based on sunlight is photoelectrolysis, in which both photochemical and electrochemical reactions are combined to dissociate water.

Water thermolysis could be a very interesting process, with good promising efficiency. Direct water thermolysis is not practicable due to very high temperature required. However, in the last 60 years, more than one hundreds of different thermochemical cycles involving from 2 to 5 reactions were developed. They consist in water dissociation using other substances which are recycled along the process. In general, the more reactions are used, the lower is the maximum temperature required. Some of the proposed cycles are 'hybrid' because they contain also one electrochemical reaction. Currently, only less than 10 of the proposed cycles are still under study since the others were not enough efficient or showed some drawbacks.

Just to make an example, S–I cycle carries out the water dissociation with three reactions:

$$xI_{2(g)} + SO_{2(g)} + 2\,H_2O_{(l)} \Rightarrow 2HI_{x(aq)} + H_2SO_{4(aq)} \quad 200^{\circ}C \tag{1.5}$$

$$2\,HI_{(g)} \Rightarrow H_{2(g)} + I_{2(g)} \quad 550^{\circ}C \tag{1.6}$$

$$H_2SO_{4(aq)} \Rightarrow H_2O_{(g)} + SO_{2(g)} + \frac{1}{2}O_{2(g)} \quad 900^{\circ}C \tag{1.7}$$

A hydrogen production technology which is often listed among hydrogen storage system is the reaction between water and the so-called chemical hydrides. Actually, these are not technologies for hydrogen storage because for some cases hydrogen is totally supplied by water. They could be considered energy storage technologies since energy is required to produce these hydrides, possibly recycling the waste material generated.

Simple chemical hydrides are composed by hydrogen and one alkaline (e.g. sodium) or alkaline earth metal (e.g. calcium) and react spontaneously with water generating hydrogen and a metal hydroxide or oxide. For example:

$$NaH + H_2O \Rightarrow NaOH + H_2 \tag{1.8}$$

Complex chemical hydrides contain also another element (e.g. boron) and require a catalyst to react with water generating hydrogen and an oxide. For example:

$$NaBH_4 + 2\,H_2O \Rightarrow NaBO_2 + 4\,H_2 \tag{1.9}$$

Also reaction between water and some metals (e.g. aluminium) can generate hydrogen.

All these technologies for hydrogen production have one or more of the following drawbacks:

- have a high carbon footprint;
- are still at laboratory level;
- are still too expensive to compete in the market; and
- generate a waste material.

Another technology to produce 'green hydrogen' is electrolysis, as long as energy from RES is used. In the past, electrolysis was used for the production of extremely pure hydrogen because of the high cost of electricity, but the need for energy storage from RES will change the scenario. Electrolysis will be extensively covered in the next two chapters because it is the basis for power to fuels, whereas the technologies briefly listed above can be explored for example in Sørensen and Spazzafumo (2018; section 2.1).

Hydrogen can also be easily converted into heat or electricity. In this second case, it is possible to use traditional internal combustion engines (Yip et al., 2019) and gas turbines (ENEL, 2009) but also fuel cells (Grove, 1838). Furthermore, it is possible, having pure oxygen available, to directly generate steam (Sternfeld and Heinrich, 1989) to use it as such or for high-performance thermodynamic cycles (Sørensen and Spazzafumo, 2018; section 2.5.4).

Hydrogen is therefore the key to significantly increase the exploitation of RES well beyond the absorption capacity of the electricity grid.

The bottleneck lies in the storage and distribution of hydrogen. Actually, hydrogen has the highest energy density per unit of mass but the lowest energy density per unit of volume (Table 1.3). This means that the storage as a gas, still the most used, is particularly bulky even when compressed at 70 MPa. The main alternatives are liquefied hydrogen, metal hydrides, organic liquids, carbon nanostructures, metal organic framework and chemical hydrides. Unfortunately, they are either not yet mature or have some disadvantages that limit their use.

Distribution is also problematic: although there are recently installed natural gas pipelines ready to accept 100% hydrogen (e.g. H21, 2018), many existing traditional natural gas pipelines are not suitable for the transport of hydrogen and however there is the need to replace pumping stations.

A first simple idea to use hydrogen avoiding storage and distribution problems is to mix it with natural gas. Several papers connected to the EU project NaturalHy are downloadable from the project website (NaturalHy, 2010). The end user could burn the mix of natural gas and hydrogen or could separate hydrogen for a specific application (e.g. low-temperature fuel cells). When burning the mix, it is necessary to take into account that

Table 1.3 Energy density of different fuels.

Fuel	LHV (MJ m^{-3})	Status[a]	HHV (MJ kg^{-1})	LHV (MJ kg^{-1})
Hydrogen	10.78	Gas	141.86	119.93
	8491	Liquid		
Methane	32.56	Gas	55.53	50.02
	20,920	Liquid		
Propane	86.67	Gas	50.36	45.60
	23,489	Liquid		
Gasoline	31,150	Liquid	47.50	44.50
Diesel fuel	31,436	Liquid	44.80	42.50
Methanol	15,800	Liquid	22.88	19.96
Dimethyl ether	59.3	Gas	31.68	28.70
	19,230	Liquid		
Ammonia	14.34	Gas	22.50	18.65
	12,700	Liquid		
Urea	12,103	Solid	10.40	9.10
Formic acid	5606.18	Liquid	5.53	4.59

HHV, Higher heating value; *LHV*, lower heating value; *STP*, standard temperature and pressure.
[a]For volumetric energy density, gas density is considered at STP conditions, liquid density at boiling point temperature and ambient pressure.

some properties change depending on hydrogen content. Recently, SNAM, a leading Italian energy infrastructure operator, tested the distribution of natural gas added with 5% and 10% of hydrogen to a pasta factory and both distribution and utilisation were successful (SNAM, 2020a). Together with Baker Hughes, SNAM also tested a gas turbine fuelled by natural gas blended with 10% of hydrogen (SNAM, 2020b).

But hydrogen is not just a fuel, so much so that its almost exclusive use has always been linked to the chemical industry. Here, then, that its use to produce synthetic fuels can be decisive in accelerating the transition towards a system based mainly on electricity and hydrogen. In fact, among the synthetic fuels that can be produced, there are some that lend themselves very well to replace traditional liquid and gaseous fossil fuels, with the advantage of not requiring substantial changes to the existing distribution system.

In this way, it will be possible to increase the level of replacement of fossil fuels with carbon-neutral fuels and this will require an investment concentrated above all on a greater exploitation of renewable sources and on a greater production of hydrogen, leaving the investment in the pure hydrogen distribution infrastructure at a later date with also a view to hydrogen storage and distribution systems technically better and hopefully less expensive than the current ones.

1.3 Main synthetic fuels

The most interesting synthetic fuels which can be produced starting from hydrogen and a carbon source are methane, methanol, dimethyl ether (DME), urea and formic acid. In addition, also ammonia, which does not contain carbon, can be considered, since it can be produced starting from electrolytic hydrogen and a source of nitrogen. Next sections of this chapter report a brief summary of their characteristics.

The main potential carbon sources are fossil fuels, biomass and carbon dioxide. Obviously, 'green carbon' can be obtained only from biomass or directly captured from the air. Taking into consideration the low CO_2 concentration into the atmosphere, a very large volume of air should be treated to obtain even a single cubic metre of CO_2. However, researchers are working on carbon dioxide capture from air and new technologies are continuously developed (e.g. Monash University, 2020).

During the transition to a carbon-free energy system, we will continue to use fossil fuels, albeit in smaller and smaller quantities. So, there will be availability of carbon dioxide in great concentration at the smokestack of power plants and industrial plants that will make use of fossil fuels. This could be the main short- and medium-term carbon source. Oil industries are also working in this direction (e.g. Repsol, 2020). Obviously, a separation and purification process will be necessary.

The most used technology for carbon dioxide capture is chemical absorption with amines, either monoethanolamine or diethanolamine. Amines are liquid sorbents capable to separate carbon dioxide from a gas stream and regenerable by heating, generally using steam available inside the plant or generated by heat recovery. Other capture technologies are based on solid materials capable to adsorb carbon dioxide on their surface and releasing it by pressure or temperature changes. Also porous or semiporous membranes can be used and the driving force is the difference in pressure between the two sides of the membrane.

As already stated, biomass also can be a source of carbon. When biomass is burned or converted into a syngas or biofuel which is burned in a stationary plant, carbon dioxide can be captured with the same technologies mentioned above. An interesting alternative is the direct treatment with hydrogen of biomass (hydrogasification) or biogas (methanation). These processes will be treated in a subsequent chapter.

Finally, it is interesting to consider the by-product of electrolysis: oxygen. Often electrolytic oxygen is released into the atmosphere if there is no possibility to sell. The use of oxygen for oxycombustion of fossil fuels, biomass or biofuels allows to obtain exhaust gases with a higher carbon dioxide concentration and consequent easier capture process.

1.3.1 Methane

Methane is the lightest hydrocarbon composed by a carbon and four hydrogen atoms located at the top of a regular tetrahedron, in the centre of which there is the carbon atom. It was discovered and isolated by the Italian scientist Alessandro Volta in the second half of the 18th century.

At room temperature and standard pressure, methane is a colourless and odourless gas. It is nontoxic, but it is an asphyxiant since its presence in the air reduces the oxygen concentration. It is extremely flammable and may form explosive mixtures with air in volume concentration between 5% and 15%.

It is by far the main constituent of natural gas, representing 85%–90% by volume, and pure methane can be used for all applications in place of natural gas, including direct use in high-temperature fuel cells. Moreover, large methane reservoirs in the form of clathrates ($4\ CH_4 \cdot 23\ H_2O$) are located on the ocean floor and in the arctic permafrost. These clathrates are not only a significant amount of fuel but also a potential risk for global warming since the global warming potential of methane is 28–36 times and 84–87 times that of carbon dioxide over a period of 100 and 20 years, respectively (Environmental Protection Agency, 2020; IPCC, 2013; section 8.7).

Most of the world's methane emissions are anthropogenic and derive mainly from waste landfills, coal mines, oil industry, agriculture and livestock, and for this reason, its concentration in the atmosphere has more than doubled in the last 250 years, from the preindustrial level of 722 ± 25 to 1803 ± 2 ppb in the year 2011 (IPCC, 2013; section 2.2).

Methane can also be produced starting from biomass and from the humid fraction of waste materials, and in this case, it is called bio-methane. The common process is anaerobic digestion that allows to obtain a raw gas which main components are methane and carbon dioxide. A subsequent purification step proceeds to remove carbon dioxide. In addition to being a renewable fuel, bio-methane contributes to reducing the emissions of climate-changing gases that would be emitted into the atmosphere due to the natural activity of microorganisms on the biomass left outdoors and on the wet waste sent to landfills.

Methane can also be obtained from methanation of carbon oxides with hydrogen (see Chapter 4: Power to Methane):

$$CO + 3\,H_2 \Rightarrow CH_4 + H_2O \qquad (1.10)$$

$$CO_2 + 4\,H_2 \Rightarrow CH_4 + 2\,H_2O \qquad (1.11)$$

and this is a renewable fuel when hydrogen is produced from a RES and carbon comes from biomass or from captured CO_2.

Reactions (1.10) and (1.11) are exothermic reactions favoured at low temperature. The reverse reactions, endothermic and occurring at high temperature, constitute the SMR process and together with the water shift reaction (1.3) allow to generate hydrogen from methane. Therefore methane can be considered also a hydrogen carrier.

1.3.2 Methanol

Methanol is the simplest alcohol and is composed by a methyl group and a hydroxyl group (CH_3-OH). It is a volatile, colourless, flammable liquid (flammability limits 6.7%−36%), toxic with an odour similar to that of ethanol. Methanol is miscible with water and biodegradable; therefore unlike gasoline, it cannot accumulate in water, air or soil. It is used also as a polar solvent.

Methanol is industrially produced by hydrogenation of carbon monoxide:

$$CO + 2\,H_2 \Rightarrow CH_4O \qquad (1.12)$$

usually starting from a syngas obtained from partial oxidation of fossil fuels. A similar process can be accomplished also using lignocellulosic biomass as a feedstock and the final product is renewable and called bio-methanol. Obviously, the hydrogen required could come also from electrolysis using power from RES: this option will be discussed in Chapter 5, Power to Methanol.

Methanol can be also burned or used directly in low-temperature fuel cells (direct methanol fuel cells).

Hydrogen can be obtained from methanol steam reforming:

$$CH_4O + H_2O \Rightarrow 3\,H_2 + CO_2 \qquad (1.13)$$

therefore, combining reaction (1.12) and (1.13), it can be also considered as a hydrogen carrier. The main advantage of synthetic methanol versus hydrogen and methane as a hydrogen carrier lies in its liquid form at ambient conditions, which makes it easier to store and transport, as it already happens with oil derivatives.

1.3.3 Dimethyl ether

DME is the simplest ether (CH_3OCH_3) and is a colourless, not toxic and highly flammable gas in the range 3.4%−27% by volume when mixed with air

(Mitsubishi Gas Chemical, 2020). It can be produced by dehydration of methanol:

$$2\,CH_3OH \Rightarrow (CH_3)_2O + H_2O \tag{1.14}$$

but also from syngas or biogas (see Chapter 6: Power-to-DME: A Cornerstone Towards a Sustainable Energy System).

DME is interesting as an alternative fuel for diesel engines since it requires minor modifications of a diesel engine, is sulphur free and has a cetane number of 55, higher than that of oil-derived diesel fuel.

Also, DME direct fuel cells have been developed and tested. And also hydrogen can be produced from DME, so that also DME can be considered a hydrogen carrier.

1.3.4 Ammonia

Ammonia is the simplest nitrogen hydride, composed by a nitrogen atom and three hydrogen atoms (NH_3). It is a colourless gas, lighter than air, with a characteristic pungent smell. It was first isolated by the Scottish chemist Joseph Black. Its boiling point is −33.3°C, whereas freezing point is −77.7°C. Ammonia can act both as a base, forming salts containing ammonium ion (NH_4^+), and as an acid, forming compounds containing amide ion (NH_2^-). The ammonia molecule is polar and highly miscible with water.

The Haber–Bosch process, developed by Fritz Haber and Carl Bosch in the first decade of the 20th century, is the main process for ammonia production. Hydrogen and nitrogen react at high temperature (300°C–550°C) under high pressure (15–30 MPa). A mixture of iron, potassium oxide and aluminium oxide is used as a catalyst. Ammonia produced is then liquefied at −33.3°C and separated from the residual gaseous mixture of hydrogen and nitrogen. The hydrogen used for the current industrial process is produced starting from fossil fuels, but it would be possible to use hydrogen generated starting from RES (see Chapter 7: Power to Ammonia and Urea) obtaining 'green ammonia' (Fúnez Guerra et al., 2020).

Although it is caustic and hazardous in its concentrated form, ammonia is widely used. It is a common nitrogenous waste, particularly among aquatic organisms, and it contributes significantly to the nutritional needs of terrestrial organisms by serving as a precursor to food and fertilizers. Ammonia is also used for the synthesis of many pharmaceutical products and for many commercial cleaning products.

Ammonia is a fuel, although burns with difficulty in air and only when mixed at a concentration of 15%–27%:

$$4\,NH_3 + 3\,O_2 \Rightarrow 2\,N_2 + 6\,H_2O_{(g)} \tag{1.15}$$

However, it could also be used as a fuel in direct ammonia fuel cells.

Finally, ammonia decomposes in hydrogen and nitrogen when heated at high temperature (850°C–950°C) in the presence of a suitable catalyst (e.g. nickel). Therefore ammonia can be considered also a hydrogen carrier.

1.3.5 Urea

Urea was characterised in the 18th century by the Dutch chemist Herman Boerhaave and is an amide with formula $CO(NH_2)_2$, that is composed by two $-NH_2$ groups and a carbonyl functional group $CO=$. It is a colourless, odourless solid, highly soluble in water, and can irritate skin, eyes, and the respiratory tract.

The German chemist Friedrich Wöhler produced urea artificially in 1828, but the industrial process (Bosch–Meiser) was developed only one century later, in 1922. This process is based on the reaction between ammonia and carbon dioxide:

$$2\,NH_3 + CO_2 \Leftrightarrow NH_4[H_2NCO_2] \tag{1.16}$$

The ammonium carbamate is then decomposed in urea and water by supplying heat:

$$NH_4[H_2NCO_2] \Leftrightarrow CO(NH_2)_2 + H_2O \tag{1.17}$$

It is widely used in fertilizers because has a nitrogen content higher than that of other solid fertilizers and because many bacteria can produce ammonia from urea.

Urea can be used as a hydrogen carrier: electrolysis of urea aqueous solutions requires a voltage of 0.37 V (Boggs, King, and Botte, 2009) that is much lower than the voltage required for water electrolysis. Moreover, urea (and urine) can be directly used in alkaline fuel cells to produce electric power (Lan, Tao, and Irvine, 2010).

1.3.6 Formic acid

Formic (or methanoic) acid is the simplest carboxylic acid (HCOOH) and at room temperature is a colourless liquid having a penetrating odour, with flammability limits 12%–38%. It is slightly toxic, but it is urticant: in nature, it is found in some plants (e.g. nettle) and animals (especially ants), and actually its name comes from 'formica', the Latin word for ant. It was first isolated by the English naturalist John Ray and was first synthesised by the French chemist Joseph Gay-Lussac.

It is miscible in water, methanol, glycerol and ethyl acetate. It is a strong reducing agent.

Formic acid can be obtained from several chemical processes, for example by hydrolysis of methyl formate in presence of a large water excess:

$$HCO_2CH_3 + H_2O \Rightarrow HCOOH + CH_3OH \tag{1.18}$$

in a two-step process which produces the methyl formate from methanol and carbon monoxide:

$$CH_3OH + CO \Rightarrow HCO_2CH_3 \tag{1.19}$$

resulting in the global reaction:

$$CO + H_2O \Rightarrow HCOOH \tag{1.20}$$

Power to formic acid will be discussed in detail in Chapter 8, Power to Formic Acid.

Thermal decomposition of formic acid can produce carbon monoxide and water which can be converted in carbon dioxide and hydrogen (water gas shift reaction):

$$HCOOH \Rightarrow CO + H_2O \Leftrightarrow CO_2 + H_2 \tag{1.21}$$

Direct decomposition to hydrogen and carbon dioxide can be also obtained in presence of catalysts like platinum and ruthenium. Therefore also formic acid can be considered a hydrogen carrier. The hydrogen content of formic acid is 53 g L^{-1} at room temperature and atmospheric pressure.

Also, formic acid fuel cells have been developed and tested.

Nomenclature

DME dimethyl ether
HHV higher heating value
LHV lower heating value
RES renewable energy source
SMR steam methane reforming
STP standard temperature and pressure

References

Bessarabov, D., and Millet, P. (2018). Chapter 2—Brief historical background of water electrolysis. In B. G. Pollet (Ed.), *PEM water electrolysis* (pp. 17–42). Academic Press. Available from https://doi.org/10.1016/b978-0-12-811145-1.00002-2.

Boggs, B. K., King, R. L., and Botte, G. G. (2009). Urea electrolysis: Direct hydrogen production from urine. *Chemical Communications* (32), 4859–4861. Available from https://doi.org/10.1039/b905974a.

ENEL. (2009). *ENEL: First hydrogen-fuelled power now on line in Venice*. Available at https://www.enel.com/media/explore/search-press-releases/press/2009/08/enel-first-hydrogen-fuelled-power-now-on-line-in-venice- (Accessed 3 May 2020).

Environmental Protection Agency. (2020). *Understanding global warming potentials | Greenhouse gas (GHG) emissions | US EPA*. Available at https://www.epa.gov/ghgemissions/understanding-global-warming-potentials (Accessed 22 August 2020).

Fúnez Guerra, C., et al. (2020). Technical-economic analysis for a green ammonia production plant in Chile and its subsequent transport to Japan. *Renewable Energy*, *157*, 404–414. Available from https://doi.org/10.1016/j.renene.2020.05.041.

Grove, W. R. (1838). On a new voltaic combination. *The London, Edinburgh, and Dublin Philosophical Magazine and Journal of Science*, *13*(84), 430–431. Available from https://doi.org/10.1080/14786443808649618.

H21. (2018). *H21 north of England.* Available at https://www.h21.green/wp-content/uploads/2019/01/H21-NoE-PRINT-PDF-FINAL-1.pdf.

IPCC. (2013). In T. F. Stocker, et al. (Eds.), *Climate change 2013: The physical science basis—Contribution of working group I to the fifth assessment report of the Intergovernmental Panel on Climate Change.* Cambridge: Cambridge University Press.

Lan, R., Tao, S., and Irvine, J. T. S. (2010). A direct urea fuel cell—Power from fertiliser and waste. *Energy and Environmental Science*, *3*(4), 438–441. Available from https://doi.org/10.1039/b924786f.

Mathias, P. M., and Brown, L. C. (2003). Thermodynamics of the sulfur-iodine cycle for thermochemical hydrogen production. In *68th Annual meeting of the Society of Chemical Engineers, Japan* (p. 23). Available at http://www.aspentech.com/publication_files/TP51.pdf (Accessed 19 March 2020).

Mitsubishi Gas Chemical. (2020). *Dimethyl ether | products | Mitsubishi Gas Chemical Company, Inc.* Available at https://www.mgc.co.jp/eng/products/nc/dimethyl-ether.html (Accessed 27 July 2020).

Monash University. (2020). *Australian researchers set record for carbon dioxide capture.* Available at https://www.monash.edu/news/articles/australian-researchers-set-record-for-carbon-dioxide-capture2 (Accessed 2 October 2020).

NaturalHy. (2010). *NaturalHy project.* Available at http://66.39.116.196/scientific_papers.htm (Accessed 21 August 2020).

Our World in Data. (2020). *Atmospheric concentrations—Our World in Data.* Available at https://ourworldindata.org/atmospheric-concentrations#how-have-global-co2-concentrations-changed-over-time (Accessed 1 October 2020).

Repsol. (2020). *Repsol to develop two major emissions-reductions projects in Spain.* Available at https://www.repsol.com/en/press-room/press-releases/2020/repsol-to-develop-two-major-emissions-reductions-projects-in-spain.cshtml (Accessed 2 October 2020).

SNAM. (2020a). *Snam: Hydrogen blend doubled to 10% in Contursi trial.* Available at https://www.snam.it/en/Media/news_events/2020/Snam_hydrogen_blend_doubled_in_Contursi_trial.html (Accessed 25 November 2020).

SNAM. (2020b). *Snam and Baker Hughes test world's first hydrogen blend turbine for gas networks.* Available at https://www.snam.it/export/sites/snam-rp/repository/ENG_file/Media/Press_releases/2020/PR_Baker-Hughes-Snam-Hydrogen-Test.pdf (Accessed 1 October 2020).

Sørensen, B., and Spazzafumo, G. (2018). In B. Sørensen, and G. Spazzafumo (Eds.), *Hydrogen and fuel cells: Emerging technologies and applications, hydrogen and fuel cells* (3rd ed.). Academic Press. Available from https://doi.org/10.1016/C2015-0-01635-5.

Sternfeld, H. J., and Heinrich, P. (1989). A demonstration plant for the hydrogen/oxygen spinning reserve. *International Journal of Hydrogen Energy*, *14*(10), 703–716. Available from https://doi.org/10.1016/0360-3199(89)90087-6.

Weidenkaff, A., et al. (2000). Thermogravimetric analysis of the ZnO/Zn water splitting cycle. *Thermochimica Acta*, *359*(1), 69–75. Available from https://doi.org/10.1016/S0040-6031(00)00508-6.

Yip, H. L., et al. (2019). A review of hydrogen direct injection for internal combustion engines: Towards carbon-free combustion. *Applied Sciences (Switzerland)*, *9*((22), 1–30. Available from https://doi.org/10.3390/app9224842.

Low-temperature water electrolysis

2

Kingsley Onyebuchi Obodo, Cecil Naphtaly Moro Ouma and Dmitri Bessarabov
HySA Infrastructure Centre, North-West University, Potchefstroom, South Africa

2.1 Fundamentals: water electrolysis and the oxygen evolution reaction

The variable nature of energy generated using renewable sources has been a key impediment towards its wider deployment and utilisation. Now, recent developments in battery technology and water electrolysis (WE) have created avenues for energy storage during times of low demand or higher generation for use at peak times.

WE can be used to produce hydrogen gas on large scale; this hydrogen can be stored in different forms and subsequently used in various sectors (e.g. the transport sector) (Bessarabov and Millet, 2018; Ma, Dai, and Qiao, 2016). The process of WE is largely considered as a clean way to produce hydrogen, provided that the energy applied for the splitting of water molecules is derived from a renewable source. Globally, there is an increase in efforts to further develop, and deploy, electrolytic hydrogen technology, including the roll-out of hydrogen refuelling stations, increasing the production of renewable hydrogen by means of WE and preparing various industries for the adoption of renewable hydrogen use. Renewable hydrogen can act as an enabler to link together many industries that require energy and chemical feedstock at the same time. Recently, many countries have announced ambitious plans to increase investments in the so-called 'hydrogen economy'.

Electrolysis is considered as the 'clean way to produce hydrogen', when the required electricity is supplied by renewable energy sources. Given the fact that the global water volume is estimated to be around 1.4×10^9 km^3 (Djinović and Schüth, 2015), and hydrogen could be seen as an almost everlasting resource.

Electrolysis involves the flow of electric current through an ionic medium (electrolyte), with a resultant chemical reaction. The decomposition of material takes place at the electrodes, via chemical reaction. This reaction entails the loss (oxidation) and gain (reduction) of electrons. The ions are transferred between the electrodes, via an electrolyte, which can be in solution or a solid medium. To date, the most mature method for the large-scale production of hydrogen without the use of fossil fuel is via WE. The process is simple and efficient, with no moving or mechanical parts. There are various WE technologies available, both on the laboratory scale and the industrial scale—they differ mainly in terms of the type of electrolyte used in the cells (see Figs 2.1, 2.2 and 2.3).

Power to Fuel. DOI: https://doi.org/10.1016/B978-0-12-822813-5.00003-5

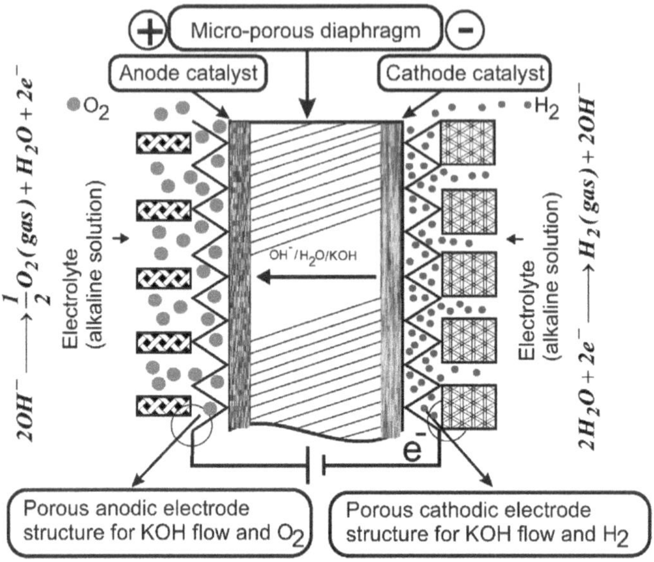

Figure 2.1 Schematic representation of the zero-gap cell design with a microporous separator. *Source*: Inspired by Bessarabov, D., and Millet, P. (2018). PEM water electrolysis. Academic Press. https://doi.org/10.1016/C2016-0-00574-0.

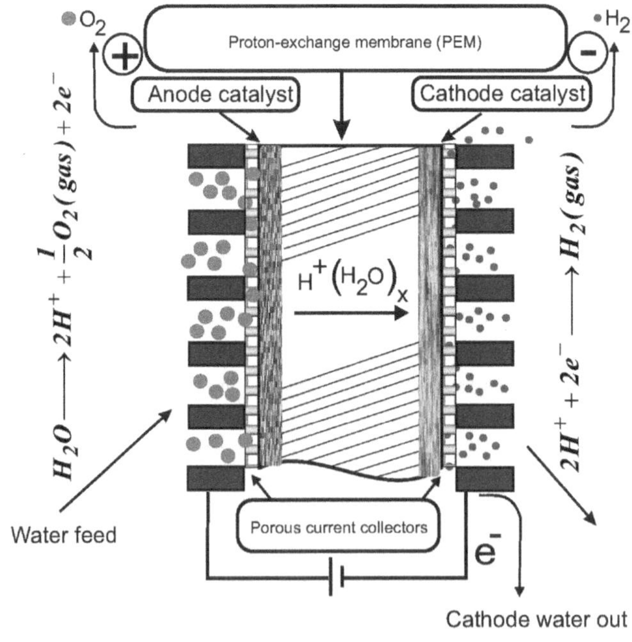

Figure 2.2 Schematic representation of the PEM cell design. *PEM*, proton exchange membrane. *Source*: Inspired by Bessarabov, D., and Millet, P. (2018). PEM water electrolysis. Academic Press. https://doi.org/10.1016/C2016-0-00574-0.

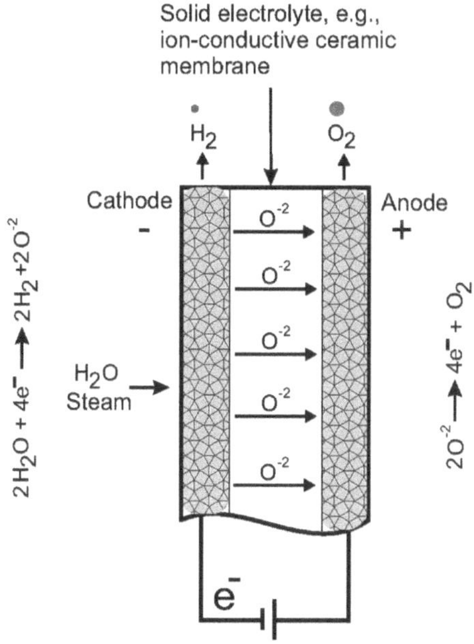

Figure 2.3 Basic working principle of a solid oxide electrolysis cell.
Source: Inspired by Bessarabov, D., and Millet, P. (2018). PEM water electrolysis. Academic Press. https://doi.org/10.1016/C2016-0-00574-0.

Alkaline, acidic and solid oxide WE can be performed, based on the type of electrolyte used. In Table 2.1, the advantages and disadvantages of the three different media used for electrolysis are presented (Carmo et al., 2013; Grigoriev et al., 2020). Electrolysis in acidic medium is also referred to as proton-exchange membrane water electrolysis (PEMWE), in alkaline medium as alkaline water electrolysis (AWE) and in solid oxide medium as solid oxide electrolysis (SOE). SOE is not addressed in this chapter.

Some features of PEMWE are now reported, including degradation mechanisms (Bessarabov and Millet, 2018; Feng et al., 2017). PEMWE is a mature technology at the 1–5 MW scale. It offers the following: high-purity gas delivery (99.999%); high compactness; commercial operation at up to 3.0 A cm^{-2}; demonstration of the performance on laboratory scale at 10 A cm^{-2} (Minnaar, De Beer, and Bessarabov, 2020); high-energy efficiency (70%–80% higher heating value at 1 A cm^{-2}), depending on the operating current density and temperature; and durability of continuous operation (> 60,000 h). Systems with increased capacities are becoming commercially available —systems of up to 7–10 MW are now available on the market and development of up to 100 MW is underway. Further features include the following: excellent flexibility and reactivity at system level for operation with transient power sources and operation under pressure (200 bar demonstrated, 350 bar prototypes; Bessarabov and Millet, 2018), or under a pressure difference (usually oxygen released at atmospheric pressure while compressed hydrogen is delivered).

Table 2.1 Three different types of electrolysis by the type of electrolyte medium (Carmo et al., 2013; Grigoriev et al., 2020).

Type of water electrolyser	**Alkaline**	**PEM**	**Solid oxide**
Technology status	Mature technology		Lab-scale, R&D
T range (°C)	Ambient-120		700–1000
Electrolyte	25–30 wt.% $(KOH)_{aq}$	Perfluorosulfonic acid	Y_2O_3-ZrO_2, Sc_2O_3-ZrO_2, MgO-ZrO_2, CaO-ZrO_2
Charge carrier	OH^-	H^+	O^{2-}
Overall reaction	$H_2O \rightarrow H_2 + 1/2O_2$		
Anode reaction	$2OH^- \rightarrow 1/2O_2 + H_2O + 2e^-$	$H_2O \rightarrow 1/2O_2 + 2e^- + 2H^+$	$O^{2-} - 2e^- \rightarrow 1/2O_2$
Cathode reaction	$2H_2O + 2e^- \rightarrow H_2 + 2OH^-$	$2H^+ + 2e^- \rightarrow H^2$	$H_2O + 2e^- \rightarrow H_2 + O^{2-}$
Anode catalyst	Ni_2CoCO_4, La-Sr-CoO_3, Co_3O_4	r/Ru oxide	(La, Sr) MnO_3, (La, Sr) (Co, Fe)O_3
Cathode catalyst	Nickel foam/Ni-stainless steel, Ni-Mo/ZrO_2-TiO_2	Platinum	Ni-YSZ or Ni-GDC Cermet
Separator	Asbestos, polysulfone-bonded polyantimonic acid, ZrO_2 on polyphenylsulfone, NiO, polysulfone impregnated with Sb_2O_5 polyoxide	Polymer membrane	Ceramic
Sealant	Metallic	Synthetic rubber/ fluoroelastomer	Glass and vitro ceramics
Current distributor	Ni	Ti	Ferritic stainless steel (Crofer APU)
Containment material	Nickel-plated steel	Stainless steel	Stainless steel
P range (bar)	1–200	1–350 (700)	1–5
Conventional current density (A cm^{-3})	0.2–0.5	0–3 (up to 20)	0–2

(*Continued*)

Table 2.1 (Continued)

Type of water electrolyser	Alkaline	PEM	Solid oxide
Efficiency (%) (at *i* A cm^{-3}/Ucell V/*T*°C)	60–80 (0.2–0.5/2.0/80)	80 (1.0/1.8/65)	100 (3.6/1.48/950)
Durability (h)	100,000	10,000–50,000	500–2000
H_2O specification	Liquid	`> 10 M Ω cm	Steam
Load cycling	Medium	Good	Good
Stop/go cycling	Weak	Good	Weak
T cycling	Weak	Good	Weak

AWE is the oldest and most mature, electrolysis technology used for hydrogen production. AWE is >100 years old and one of the simplest and advanced methods used for hydrogen production. It is also the most economical of the above-mentioned electrolysis types. An AWE system operates at low temperatures (40°C–90°C), and the electrolyte used is a liquid alkaline solution (containing NaOH or KOH). Several multimegawatt industrial electrolysers are used in the industry for the large-scale production of hydrogen (Bessarabov and Millet, 2018). The lifetime of these systems (several tens of thousands of hours of operation) can be considered as meeting industrial expectations for continuous operation. The advantages of AWE include its low cost, mature stage of development and long durability; disadvantages include low hydrogen gas purity and low operational current densities—there remain the main concerns (Rashid et al., 2015).

For various reasons, the CapEx of PEM technology (1700–2500 € kW^{-1} for 300–500 kWe systems) is still higher than the CapEx of alkaline technology (900–1700 € kW^{-1} for 300 kWe to 5 MWe systems). The main reasons include the expensive cell components and nonautomated manufacturing and assembly procedures. In the long term, to meet target CapEx costs, we need to replace, or reduce, the requirement for expensive platinum group metals (PGMs). Regarding the operating expenditure, in kWh/kgH_2 (common to all WE technologies), the energy cost is mainly due to the cost of electricity, which dictates the cost of gas produced by electrolysis.

Safety issues (common to all WE technologies) include the following: management of hydrogen–oxygen stoichiometric gas mixtures, gas cross-contamination during operation and management of caustic products.

A schematic representation of the zero-gap cell design with a microporous separator is shown in Fig. 2.1.

A schematic representation of the PEM cell design is shown in Fig. 2.2.

The basic working principle of a SOE cell is shown in Fig. 2.3.

The overall reaction for WE is presented in Eq. (2.1).

$$2H_2O + \text{electric current} \Rightarrow 2H_2 + O_2 \tag{2.1}$$

The anode half reaction for WE is referred to as the oxygen evolution reaction (OER). It is referred to as one of the most important anodic processes in the electrolysis of various aqueous solutions (Matsumoto and Sato, 1986; Suen et al., 2017).

Reactions in acid media are given below:

Acidic medium

Cathode reaction:

$$4H^+ + 4e^- \Rightarrow 2H_2, \quad E_c^0 = 0\text{ V} \tag{2.2}$$

Anode reaction:

$$2H_2O_{(l)} \Rightarrow O_{2(g)} + 4H^+ + 4e^-, \quad E_a^0 = 1.23\text{ V} \tag{2.3}$$

Proposed mechanism in acidic medium

$$* + H_2O_{(l)} \Rightarrow OH* + H^+ + e^- \tag{2.4}$$

$$OH* + OH^- \Rightarrow O* + H_2O_{(l)} + e^- \tag{2.5}$$

$$2O* \Rightarrow 2* + O_{2(g)} \tag{2.6}$$

$$O* + H_2O_{(l)} \Rightarrow OOH* + H^+ + e^- \tag{2.7}$$

$$OOH* + H_2O_{(l)} \Rightarrow * + O_{2(g)} + H^+ + e^- \tag{2.8}$$

The * in all the above and following equations denotes the surface of the catalytic material under consideration.

Reactions in alkaline media are given below:

Alkaline medium

Cathode reaction:

$$4H_2O + 4e^- \Rightarrow 2H_{2(g)} + 4OH^-, \quad E_c^0 = -0.83\text{ V} \tag{2.9}$$

Anode reaction:

$$4OH^- \Rightarrow 2H_2 + 2H_2O_{(l)} + 4e^-, \quad E_a^0 = -0.40\text{ V} \tag{2.10}$$

Proposed mechanism in alkaline medium

$$* + OH^- \Rightarrow OH* \tag{2.11}$$

$$OH* + OH^- \Rightarrow O* + H_2O_{(l)} \tag{2.12}$$

$$2O* \Rightarrow 2* + O_{2(g)} \tag{2.13}$$

$$O* + OH^- \Rightarrow OOH* + e^- \tag{2.14}$$

$$OOH* + OH^- \Rightarrow * + O_{2(g)} + H_2O_{(l)} \tag{2.15}$$

Various research groups have proposed different reaction mechanisms for the OER. For example, recently, Dickens et al. proposed an expanded OER pathway in an acidic medium, useful for performing kinetic analysis (Dickens, Kirk, and Nørskov, 2019). It includes intermediate pathways, such as the water adsorption and deprotonation of adsorbed OOH on the catalyst surface. The proposed mechanism is as follows:

$$* + H_2O_{(l)} \Rightarrow H_2O^* \tag{2.16}$$

$$H_2O^* \Rightarrow OH^* + H^+ + e^- \tag{2.17}$$

$$OH^* \Rightarrow O^* + H^+ + e^- \tag{2.18}$$

$$O^* + H_2O_{(l)} \Rightarrow OOH^* + H^+ + e^- \tag{2.19}$$

$$OOH^* \Rightarrow O_2{}^* + H^+ + e^- \tag{2.20}$$

$$O_2{}^* \Rightarrow * + O_{2(g)} \tag{2.21}$$

Typically, the intermediate pathways are not included in the four-step OER pathway, as presented below (Govindarajan et al., 2019):

$$* + H_2O_{(l)} \Rightarrow OH^* + H^+ + e^- \tag{2.22}$$

$$OH^* \Rightarrow O^* + H^+ + e^- \tag{2.23}$$

$$O^* + H_2O_{(l)} \Rightarrow OOH^* + H^+ + e^- \tag{2.24}$$

$$OOH^* \Rightarrow * + O_{2(g)} + H^+ + e^- \tag{2.25}$$

The different pathways presented in Fig. 2.4, as well as the results of several studies, suggest that the OER is a heterogeneous process. The OER reaction proceeds via O, OH and OOH intermediates that adsorb on the catalyst surface. The mechanisms include similar intermediates, such as *OH and *O. The major difference pertains to the formation of oxygen, which entails the oxygen molecule forming either 2O* or *OOH intermediates, resulting in $O_{2(g)}$. A schematic of the OER is presented in Fig. 2.4 for both acidic and alkaline conditions. The catalyst surface is represented by M. The various pathways for $O_{2(g)}$ formation are depicted using green, red and blue arrows.

2.2 Electrocatalyst modelling: state of the art

The application of modelling techniques to understand materials within academia and industry is currently commonplace. Modelling techniques include electronic

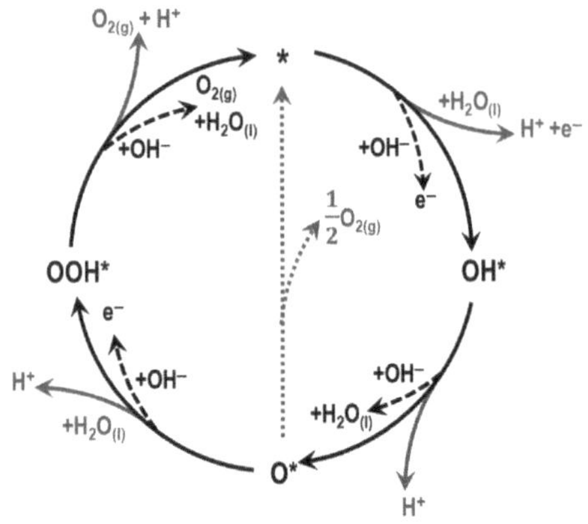

Figure 2.4 The OER mechanism for acidic (solid grey line) and alkaline (dotted black line) conditions. The solid black line and dotted grey line indicate that the OER entails the formation of a peroxide (OOH*) intermediate and two adjacent oxo (O*) intermediates, respectively to produce oxygen.
Source: Reprinted with permission from Suen, N.T. et al. (2017). Electrocatalysis for the oxygen evolution reaction: Recent development and future perspectives, Chemical Society Reviews. Royal Society of Chemistry, *46*(2), 337–365. https://doi.org/10.1039/c6cs00328a.

structure calculations [using density functional theory, (DFT), the tight binding method, and others], force field, molecular dynamics (MD) and quantum Monte Carlo simulation (Sarwar et al., 2015). These techniques are applicable in the study of electrocatalytic materials to relate the geometric structure and electronic properties to the activity and stability—thus acting as a predictive and investigative tool. Theoretically, an understanding and the development of materials takes place at different length scales: the macroscopic, mesoscopic and nanoscopic. Different process features of WE occur at these different length scales.

A major limiting factor towards the further development of WE technology is the slow rate of the OER. The OER requires the transfer of four electrons and four protons. This results in inefficiency, at the anode and cathode, with significant overpotentials. Several experimental and theoretical methods have been applied towards an understanding of the OER in different media (Bessarabov and Millet, 2018; Govindarajan et al., 2018; Mavrokefalos and Patzke, 2019; Pavlovic et al., 2016; Siracusano et al., 2017; Thomsen et al., 2015; Wang et al., 2017; Weber et al., 2019; Willinger et al., 2017).

Over the past two decades, rigorous attempts have been made towards the design of electrocatalysts using computational methods. The key drivers in the use of computational methods are improvements in algorithms, efficient generic codes, reduced cost of computing facilities, as well as increased computational power. By making use of computational modelling, significant discoveries have been made towards the improvement of electrocatalysts—which is (otherwise) rather challenging. Different computational strategies have been proposed towards determining an optimal OER catalyst under different conditions, such as ultra-high vacuum conditions and others.

Under ultra-high vacuum conditions, the use of DFT to achieve a better fundamental understanding of the reaction pathways, diffusion barriers, activation energies and reaction kinetics is an area of ongoing interest. DFT has been applied to the generation of scaling relations, which allows for the rapid screening of possible electrocatalytic materials (Govindarajan et al., 2019; Nørskov et al., 2004, 2009). Under more realistic conditions, Grajciar et al. showed that the following approaches provide significant theoretical understanding of catalytic materials: global optimisation techniques, *ab initio* constrained thermodynamics, biased MD, microkinetic models of reactions and machine learning (Grajciar et al., 2018). The use of empirical force field to model a large-scale catalytic performance of iridium oxide (IrO_2) surfaces and nanostructures is currently feasible, with prediction of the oxygen binding energy (a well-known descriptor for OER activity) (Sen et al., 2015).

Thus theoretical techniques have been applied to obtain a more detailed understanding of chemical reactions that occur on surfaces, predict material properties and offer significant insight into both technology and to industrial processes (Sarwar et al., 2015). Grajciar et al. have discussed potential computational strategies that offer potential, towards bridging the gap between experimental investigations and computational models (see above) (Grajciar et al., 2018). Other possible techniques include DFT, MD and quantum Monte Carlo. At different length scales, different techniques are important to the understanding of WE. This will now be addressed in more detail.

2.3 Overview of modelling techniques and modelling length scales

The type of data generated using both experimental and theoretical methods is increasing in terms of complexity and abundance. The storage and analysis of such information is a critical aspect, as it relates to the conclusions drawn. The use of recently described computational modelling techniques allows the capturing, interpretation, pattern identification and, ultimately, the drawing of correlation from these data (Schleder et al., 2019). Likewise, the application of various strategies, such as high-throughput data generation, data screening, data mining, machine learning for novel designs and more realistic models, is attracting increasing attention, also in the design of modern OER catalysts, with particular emphasis on IrO_2 (Ulissi et al., 2016).

The creation of models for materials prediction, understanding and exploration, as well as different length scales, is paramount. These computational strategies have the potential to bridge the gap between experimental investigations and computational models at various length scales. This section offers an elucidation on the different modelling techniques that are available, specifically, modelling techniques and modelling length scales. Here, the modelling of the OER and the IrO_2 catalyst is addressed, using a multiscale approach, which connects the atomistic process (electrochemistry) and the microscopic and mesoscopic processes (transport).

Applying a multiscale physical transient approach, Oliveira et al. have given a detailed account of the elementary OER kinetics at the anode, the nonequilibrium character of the nanoscale catalyst–electrolyte interface, microstructural resolution of the charge and oxygen transport at the microscale and mesoscale (Oliveira et al., 2012).

They also considered the use of IrO_2 and RuO_2 as catalytic materials in PEMWE and found that RuO_2 exhibited better catalytic performance.

There are several techniques relevant to adequately describing the multiscale nature of modelling, required to well describe catalytic materials. These techniques, applicable to the atomistic, molecular, coarse-grained and continuum scale, as presented in the review by Zhang and Bieberle-Hütter, are now described (Zhang and Bieberle-Hütter, 2016).

Fig. 2.5 presents a schematic of multiscale modelling methods at different length scales: the microscopic, mesoscopic and macroscopic scales. There are three groups/regions (see also Table 2.2). The microscopic scale entails the atomistic and molecular region, the mesoscopic scale entails the molecular and coarse-grained region, whereas the macroscopic scale entails the coarse-grained and continuum.

1. *Atomistic scale*: Atomistic scale (*ab initio*) techniques involve the modelling of materials at the atomic level (Å). DFT is currently one of the most widely used modelling technique to describe, predict and understand material properties at the atomistic level; kinetic and thermodynamic properties can be derived from DFT-based calculations. An *ab initio* method can be applied to study the dynamics of molecular systems by combining MD and DFT—this is referred to as *ab initio* MD. An example of the *ab initio* MD is Car–Parrinello MD, which can be applied to carry out realistic computational modelling of complex molecular systems, processes and chemical reactions.
2. *Molecular scale*: The molecular scale involves the modelling of materials at the nanoscale level. This is a level above the atomistic scale; it entails clusters of tens to hundreds of atoms, at the atomic scale. The dynamics of matter can be determined from the solution of a n-body problem. The structure and motion of its constituent building blocks, which determine the properties of matter and materials, can also be determined. In this scale, MD calculations can be used to obtain sound descriptions. Using this method, the trajectories of atoms/molecules can be tracked and evaluated. Reactive force-field molecular

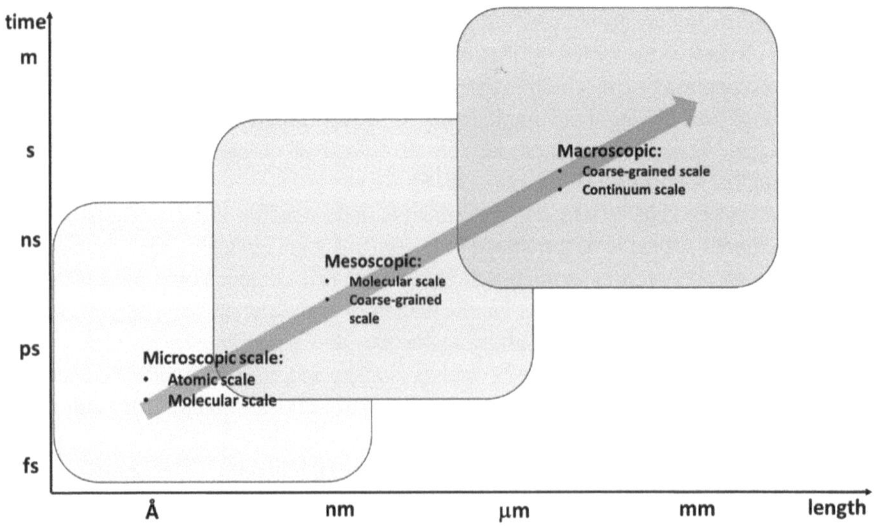

Figure 2.5 Schematic of multiscale modelling methods at different length scales.

Table 2.2 Different length scales and their discerning phenomena.

Scales	Tools	Phenomena
Microscopic	Density functional theory, Molecular dynamics, Tight binding, Force field, Quantum Monte Carlo	Electronic structure, Energetics, Charge transfer, Magnetisation, etc.
Mesoscopic	Molecular dynamics, Tight binding, Force field, Kinetic Monte Carlo	Mass transport (diffusion), Structure and properties (crystal growth and morphologic), etc.
Macroscopic	Bulter–Volmer equation, Ordinary and partial differential equations (ODE & PDE), Computational fluid dynamics	Heat flow and mass transfers

dynamics (ReaxFF MD) is one of the most popular technique applicable to MD calculations to determine chemical reactions using atomistic potentials. These potentials are based on the approach developed by Van Duin et al. (Van Duin et al., 2001). ReaxFF MD takes advantage of the relationship between the bond distance and bond order, as well as the bond order and bond energy, to determine the dissociation of bonds to separate atoms.

3. *Coarse-grained scale*: The coarse-grained scale deals with large system sizes and simulation time scales. Thus compared with the atomistic scale, this scale is referred to as coarse-grained. This region falls between the microscopic and the mesoscopic to macroscopic scales, where 'pseudo atoms' represents the groups of atoms. A very popular method applied in this scale is kinetic Monte Carlo (kMC), which describes a number of phenomena, such as transport (diffusion), structures and properties of materials (e.g. crystal growth), or equilibrium and nonequilibrium chemistry (catalysis). kMC is applicable to systems with potential energy surfaces with minima, and the evolution of these systems entails hopping between neighbouring minima. kMC accounts for these 'hops' as stochastic processes.
4. *Continuum scale*: In the continuum scale, state-space modelling and simulations (SSMS) can be applied. The SSMS method, in control theory, can be applied to simulate complex and interdepending systems. It is a mathematical model of a physical system as a set of input, output and state variables related by first-order differential equations.

The above multiscale modelling comprises three groups: the microscopic, mesoscopic and macroscopic scales. The microscopic scale entails the atomistic and molecular region, the mesoscopic scale entails the molecular and coarse-grained region, whereas the macroscopic scale entails the coarse-grained and continuum region (see Fig. 2.5 and Table 2.2).

2.3.1 Macroscopic length scale and macroscopic quantities

At the macroscopic length scale, different modelling techniques are applied, such as computational fluid dynamics (CFD), empirical models, analytical models and others (Atlam, 2009; Atlam and Kolhe, 2011; Marangio, Santarelli, and Calì, 2009; Toghyani et al., 2019).

An analytical model such as the Butler–Volmer equation offers an important relationship in the description of electrochemical kinetics. According to this equation, electrical current at an electrode depends on the electrode potential, considering that both a cathodic and an anodic reaction occur on the same electrode. CFD involves the application of a numerical method and data structures to simulate heat transfer and motion of fluid. This technique is of particular importance for modelling material properties at the macroscopic scale and modelling the flow and transport of mass as well as fluid (Grigoriev et al., 2011).

Choi et al. applied the Butler–Volmer kinetics to the electrodes and the transport resistance in a solid polymer electrolyte to determine the relationship between the terminal voltage applied to the electrolysis cell and the current density in terms of the Nernst potential, exchange current densities and conductivity of polymer electrolyte (Choi, Bessarabov, and Datta, 2004). Their model, consistent with findings of previous studies, showed that the limiting factor is the high anodic overpotential, which is responsible for most of the supplied energy. Rahim et al. reviewed the modelling and mass transport activities in the area of PEMWE and showed that the diffusion overpotential, which is important for determining the efficiency, was ignored in various models (Abdol Rahim et al., 2016). Thus more detailed studies on the gas cross-over and water mass transport are required.

Olivier et al. reviewed several articles on electrolysis modelling involving multiphysics phenomena and the electrolysis stack, as well as different domains: electrical, thermal, fluid and thermochemical (Olivier, Bourasseau, and Bouamama, 2017a). Various empirical and analytical models were evaluated, with the classification based on set criteria. The review also revealed that complex and nonlinear models attributed to different energies coupling, as well as nonstationarity and spatiotemporal dynamics, characterise electrolysis systems.

The significance of developing input–output models suited for future developments is a necessity in the field of modelling, especially in the understanding of phenomena, control design, system diagnostics and prognostics. A dynamic and multiphysics model of the PEM electrolysis system with a graphical modelling formalism, referred to as bond graphs modelling, has also been reported (Olivier, Bourasseau, and Bouamama, 2017b). The advantages include the following: four levels of modelling with one tool, the physical meaning of parameters, a multiphysics approach and the addition of new parameters without restarting the entire model. This form of modelling can be applied to control, sizing and diagnosis analysis. This model holds promise in the design of PEM electrolysis.

A review by Falcão and Pinto on PEM water electrolysers focuses on the use of models to predict macroscopic quantities such as cell voltage, including reversible voltage; activation losses; ohmic losses and mass transport losses (Falcão and Pinto, 2020). Hernández-Gómez et al. reviewed models applicable to investigating the electrical domain of PEM water electrolysers (Hernández-Gómez, Ramirez, and Guilbert, 2020). Several of the concepts pertaining to an understanding of the Faraday voltage and energy efficiencies are presented in Fig. 2.6.

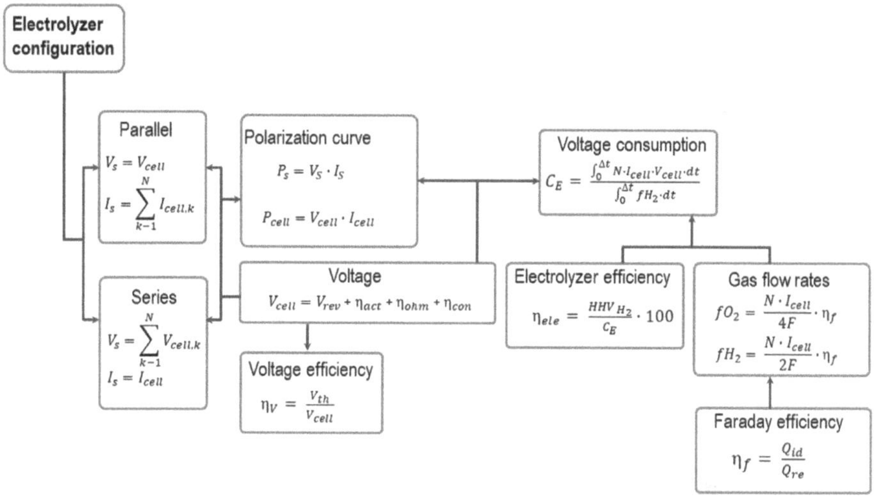

Figure 2.6 Connections between different models to give the electrolyser efficiency. *Source*: Reprinted with permission from Hernandez-Gomez et al., 2020.

At the macroscopic length scale, various quantities of interest can be modelled with the possibility of experimental validation. Some of these quantities are presented below.

2.3.1.1 Reversible cell potential

The Nernst overpotential depends on the variation of the applied pressure, the Gibbs free energy during the reaction, as well as the concentration of the reactants and products. The Gibbs free energy varies with the temperature. However, a constant reversible potential (E^0_{rev}) value of 1.23 V is taken when modelling the current−voltage characteristic of a PEM water electrolyser. Thus $E^0_{rev} = \frac{\Delta G}{nF} = 1.23\text{V}$, (Bessarabov et al., 2015), where ΔG, F and n are the Gibbs free energy of reaction, the Faraday constant and the number of electrons transferred during the reaction (in moles) respectively. Provided no external heat is supplied to the system, and the energy required for the reaction to proceed (ΔH^0_R) can be delivered via electric energy. Then, the current voltage, which is higher than the E^0_{rev}, is referred to as the thermoneutral voltage, and at standard state, it is given as $E^0_{thv} = \frac{\Delta H^0_R}{nF} = 1.48$ V (Choi et al., 2004). Faraday's law relates the amount of gas generated in a given electrochemical process to the electrical charge of the cell. The expression $Q = nzF$, where Q and z are the charge (in Coulomb) and charge number, summarises the above expression.

2.3.1.2 Voltage

The polarisation curve gives a common relationship between the current density and the voltage. The electrolyser voltage (V) is given as: $V = E + V_{Act} + V_{Ohm}$ (Görgün, 2006), where E, V_{Act} and V_{Ohm} are the open-circuit voltage, activation polarisation and ohmic polarisation respectively. This was expanded as $V = N(E + V_{Act,a} + V_{Act,c} + iR_{cell})$ (Nafeh, 2011), where N is the number of electrolyser cells, and $V_{Act,a}$ and $V_{Act,c}$ are the anode and cathode overpotentials respectively. Other studies report reformulation of the voltage and current relationship, based on experimental studies (Atlam and Kolhe, 2011; Sossan et al., 2014).

2.3.1.3 Open-circuit voltage

The open-circuit voltage can be determined from the Nernst equation as $E_{cell} = E^0_{rev} + \frac{RT}{2F}(\ln(\frac{P_{H_2}P^{1/2}_{O_2}}{P_{H_2O}}))$, where T, P, F and E^0_{rev} are the temperature, partial pressure, Faraday constant, and reversible cell potential at standard temperature and pressure respectively. The temperature-dependent empirical expressions for E^0_{rev}, as given below, were determined by various authors (as in the review article by Falcão and Pinto, 2020).

$$E^0_{rev} = 1.5241 - 0.0012261T + 1.1858 \times 10^{-5}T \cdot \ln(T) + 5.6692 \times 10^{-7}T^2 \tag{2.26}$$

$$E^0_{rev} = 1.229 - 0.9 \times 10^{-3}(T - 298) \tag{2.27}$$

$$E^0_{rev} = 1.229 - 0.9 \times 10^{-3}\left(T - 298 + 2.3\frac{RT}{4F}\log\left(P^2_{H_2}P_{O_2}\right)\right) \tag{2.28}$$

(Choi et al., 2004)

The E_{cell} is reformulated as shown in Eqs. (2.29) and (2.30):

$$E_{cell} = 1.5184 - 0.0015421T + 9.523 \times 10^{-5}T \cdot \ln(T) + 9.84 \times 10^{-8}T^2 \tag{2.29}$$

$$E_{cell} = -(1 - k)\frac{\Delta G}{2F} \tag{2.30}$$

where ΔG and k are the change in Gibbs energy and percentage of proton back-permeation relative to the total proton permeation respectively. Open-circuit voltage and reversible potential can be determined using simplified equations provided that the electrolyser is operated at standard temperature and pressure, with minimal errors. Variations in temperature and pressure would require additional approximations for accurate results.

2.3.1.4 Activation overpotential

Losses attributed to the activation of the electrochemical reaction occurring at the anode and cathode can be expressed using the Butler–Volmer equation: $V_{Act} = \frac{RT_a}{a_aF}\text{arcsin}h\left(\frac{i}{2i_{0,a}}\right) + \frac{RT_c}{a_cF}\text{arcsin}h\left(\frac{i}{2i_{0,c}}\right)$, where the a_a and a_c are the charge transfer coefficients at the anode and cathode respectively, and the $i_{0,a}$ and $i_{0,c}$ are the exchange current densities at the anode and cathode respectively, (Abdin, Webb, and Gray, 2015). An approximate equation with a factor of 2 for the charge transfer was used to describe the activation overpotential, $V_{Act} = \frac{RT_a}{2a_aF}\operatorname{arcsin}h\left(\frac{i}{2i_{0,a}}\right) + \frac{RT_c}{2a_cF}\operatorname{arcsin}h\left(\frac{i}{2i_{0,c}}\right)$ (Choi et al., 2004). A further simplified equation is $V_{Act} = \frac{RT}{naF}\ln\left(\frac{i}{i_0}\right)$ (Saeed and Warkozek, 2015). However, the exchange current density coefficients for the anode and cathode differ; thus different models prefer adjustable values. These coefficients follow an Arrhenius-type expression (García-Valverde, Espinosa, and Urbina, 2012) for the $i_0 = i_{0,ref}\exp\left[-\frac{E_{act}}{R}\left(\frac{1}{T} - \frac{1}{T_{ref}}\right)\right]$, where $i_{0,ref}$ is the exchange current density evaluated at the reference temperature T_{ref}, and E_{act} is the activation energy of the electrode. This latter expression takes into account changes in temperature.

2.3.1.5 Mass transport overpotential

The mass transport overpotential can also be referred to as the diffusion overpotential that prevails when high-current density impedes the access of reactants to the active sites through the overpopulation of the reacting molecules, thus reducing the rate of reaction. $V_{Diff} = \frac{RT}{nF}\ln\left(\frac{C}{C_0}\right)$, where C_0 and C are the working concentrations, taken as the reference and oxygen/hydrogen concentration at the membrane–electrode interface. This expression can be applied to the anode and the cathode (Marangio et al., 2009). Another approach (Saeed and Warkozek, 2015) takes into account the maximum current density ($i_{\lim}$) given to the electrolyser, which shows an electrolyser's permitted maximum production rate: $V_{Diff} = \frac{RT}{nF}\ln\left(1 - \frac{i}{i_{\lim}}\right)$. As mentioned, the mass losses are negligible with insignificant changes to the estimated voltage at moderate/low current densities <1.6 A cm^{-2}.

2.3.1.6 Ohmic losses

Ohmic losses (V_{Ohm}) **are related to a material's resistance to the flow of protons; therefore the magnitude of such losses is dependent on the material properties. $V_{Ohm} = RI = \frac{\delta}{\sigma}I$, where R, δ and σ are the membrane resistance, thickness and conductivity respectively, and $\sigma = (0.00514\lambda - 0.00326)\exp\left(128\left(\frac{1}{303} - \frac{1}{T}\right)\right)$, where λ and T are the water content of the membrane and the electrolyser temperature (Görgün, 2006). This expression can be applied to diffusion and catalytic layers; however, it is primarily applied to the membrane.

2.3.2 Mesoscopic length scale

The relevant quantities of interest at the mesoscopic length scale are sometimes related to the macroscopic and microscopic scales, with typical assumption of the length scale in the region of ~2–20 nm. In comparison with the other two length scales (macroscopic and microscopic), the mesoscopic length scale has not been much studied for the OER. At this scale, a typical area of interest includes understanding the porous structure of the catalyst layer in a PEM or the nature of the membrane. Quantitative descriptions of process-independent factors (such as porosity, tortuosity, pore size distribution and multipoint statistics) and process-dependent factors (such as effective transport properties and effectiveness) can be determined.

Different modelling techniques can be applied to describe relevant quantities at this length scale, such as MD (which considers particles and makes use of first-principles technique), kMC (which considers particles and makes use of non-first-principles technique), pore scale model (which considers a continuum and makes use of geometry resolved technique) and agglomerate model (which considers a continuum and makes use of volume averaged technique). Most of these methods have been discussed in the previous sections (they are included here simply to recap). Noteworthy, at the microscopic scale the systems are treated as particulate, whereas at the macroscopic scale the systems are largely treated as continuum. Thus the mesoscopic scale, which overlaps both the microscopic and macroscopic length scales, can be addressed as a particle and a continuum, which allows for a different resolution of the system. However, this presents its own challenges, such as the choice of method and the regime within the mesoscopic scale to evaluate a property of interest.

Mukherjee et al. and Weber et al. showed that mass transport loss limits the performance of polymer electrolyte fuel cells (PEFCs) (Mukherjee, Wang, and Kang, 2009; Weber et al., 2014). The origin of these losses is liquid water transport and flooding phenomena in the constituent components. Similar principles applicable to mass transport loss in a PEFC are applicable to a PEMWE. Overall, only a few modelling studies at the mesoscopic scale have been carried out (Bessarabov et al., 2015) in comparison with studies at the macroscopic and microscopic scales in PEM WE. We do, however, expect new work to be done soon regarding the high-current density operation of PEMWE.

2.3.3 Microscopic length scale

As mentioned earlier, the DFT method is the mainstay in determining the electrochemical catalytic activity. Although electrochemical barriers in aqueous conditions are difficult to determine with DFT, several approximations are made that make them applicable, as will now be described. Several descriptors for the OER activity have been proposed in efforts to determine an optimal OER catalyst at the microscopic scale. These include the following.

1. DFT calculated adsorption energy (E_{Ads}) of oxygen on a given surface configuration. This descriptor can be applied to rapidly screen for OER activity. The optimal E_{Ads} for OER has been found to be +2.0 eV (Nørskov et al., 2004; Wu et al., 2019).
2. Overpotential, $\eta^{OER} = \{\max[(\Delta G_1, \Delta G_2, \Delta G_3, \Delta G_4)/e^-] - 1.23\text{V}\}$. The overpotential is the extra potential required to make all elementary steps of the OER mechanism exothermic. The standard potential is 1.23 V (Bessarabov et al., 2015).
3. Relationship between $\Delta G_{O*} - \Delta G_{OH*}$ and η^{OER}, to develop a scaling relationship—generally referred to as a Volcano-type plot. (See Fig. 2.7.) This plot is constructed using the reaction free energies determined from Eqs. (2.31)–(2.32). The tip of the volcano-type activity plot is representative of the optimal catalyst, with an error bar ranging from −0.17 to −0.57 V, represented with a grey scale. The error bar is related to the dispersion in the scaling relations. The left and right sides of the volcano plot give a strongly/weakly binding catalyst.
4. Relationship between $(\Delta G_{OOH} - \Delta G_{OH} - 2.46eV)/2e^- \rightarrow 0$. The necessary approximations for determining the free energy relations are the following.

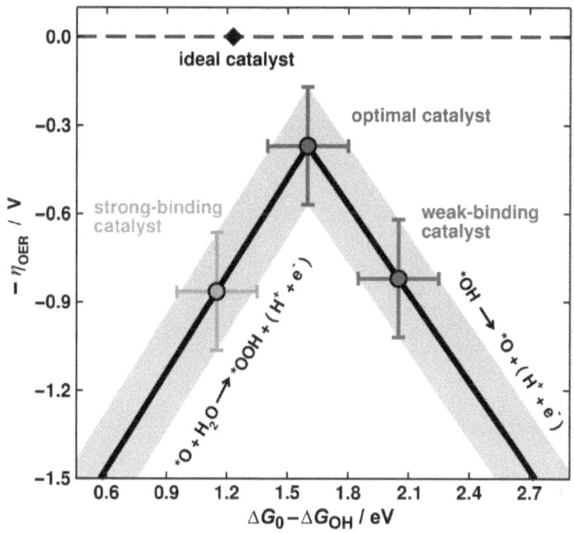

Figure 2.7 Volcano-type activity plot with error bars (grey-shaded region) due to adsorption–energy scaling relations. Examples of strong binding, weak binding and optimal catalysts are shown, together with the ideal catalyst. The top of the volcano is located at −0.37 ± 0.2 V.
Source: Reprinted with permission from Govindarajan, N. et al. (2019). Outlining the scaling-based and scaling-free optimization of electrocatalysts, ACS Catalysis. American Chemical Society, 9(5), 4218–4225. https://doi.org/10.1021/acscatal.9b00532.

1. Setting the reference electrode to that of the standard hydrogen electrode (SHE), the relationship between the chemical potential (the free energy per H) and that of the protons (H^+) and electrons (e^-) can be described taking the following into consideration. The assumption is that the solvated protons and electrons are in equilibrium with the hydrogen in the gas phase, $H^+ + e^- = \frac{1}{2}H_{2(g)}$, at standard condition [chemical potential $(\mu) = 0$, $pH = 0$, pressure $(p) = 1$ bar, $T = 298$ K]. Furthermore, calculation of the μ of an H atom in the gas phase is more straightforward than that of protons/electrons. The DFT standard hydrogen chemical potential is expressed as $\mu^0_{H_{2(g)}} = E_{DFT}^{H_{2(g)}} + ZPE^0_{H_{2(g)}} - TS^0_{H_{2(g)}}$. The standard entropy ($S^0$) is taken from the thermodynamics tables for gas-phase molecules.
2. The Gibbs free energy (ΔG_0) can be evaluated as $\Delta G_0 = \Delta E + \Delta ZPE - T\Delta S$, where ΔE is the reaction energy determined from DFT calculations. ΔE is used, rather than $G(T,p) = H(T,p) - TS(T,p)$, because at $T = 0$ the G and H are equivalent. The ΔZPE is the difference in zero point energies due to the reaction, and ΔS is the change in entropy, determined using DFT vibrational frequencies and the standard tables for gas-phase molecules.
3. The effect of bias on all the states involving an electron in the electrode can be accounted for by shifting the energy of this state, by $\Delta G_U = -eU$, where U is the electrode potential relative to the SHE.
4. $pH \neq 0$, the free energy of H^+ ions can be corrected for the concentration dependence of the entropy: $\Delta G_{pH}(pH) = -kT$. Thus taking the log will result in $\ln[H^+] = kT \cdot \ln 10 \times pH$.

5. The total free energy change of the reaction, $H_2O \Rightarrow 1/2O_2 + H_2$, is set to the experimental value of 2.46 eV. Based on assumption, calculation of the O_2 bond energy (which is difficult to determined accurately using DFT) is avoided. Thus $G^0_{H_2O_{(l)}} - \frac{1}{2}G^0_{O_{2(g)}} - \frac{1}{2}G^0_{H_{2(g)}} = 2.46\text{eV}$.
6. The adsorption energies of *O, *OH and *OOH are determined using $H_2O_{(g)}$ and $H_{2(g)}$ in the gaseous phase as reference states because they are appropriately determined using DFT calculations at 0 V, within the reversible hydrogen electrode scale. The following reactions apply:

$$* + H_2O_{(l)} \Rightarrow O^* + 2H^+ + e^- \tag{2.31}$$

$$* + H_2O_{(l)} \Rightarrow OH^* + H^+ + e^- \tag{2.32}$$

$$* + 2H_2O_{(l)} \Rightarrow OOH^* + 3H^+ + e^- \tag{2.33}$$

The free energies of reactions in Eqs. (2.22)–(2.25) are used to determine the elementary steps ΔG_1, ΔG_2, ΔG_3, ΔG_4 and the overpotentials are evaluated taking into consideration the various approximations given above and by applying Eqs. (2.31)–(2.33).

2.4 Ir-based compounds and their oxides: amorphous and crystalline phases

Ir-based compounds and their oxides are typical materials for use in the OER due to very high durability in acidic medium. Several studies have been carried out on the crystalline and amorphous phases of Ir-based compounds and their oxides, but these are not discussed further here. The synthesis of Sn-doped amorphous and crystalline IrO_2 compounds has been carried out. It has been shown that IrO_2 compound can exist in both phases (Arias-Egido et al., 2019; Thanawala et al., 2008). A schematic of the $IrO_2(110)$ rutile crystalline surface structure is presented in Fig. 2.8. Long-range order of the iridium and oxygen atoms can be observed, unlike in the amorphous phase with no long-range order.

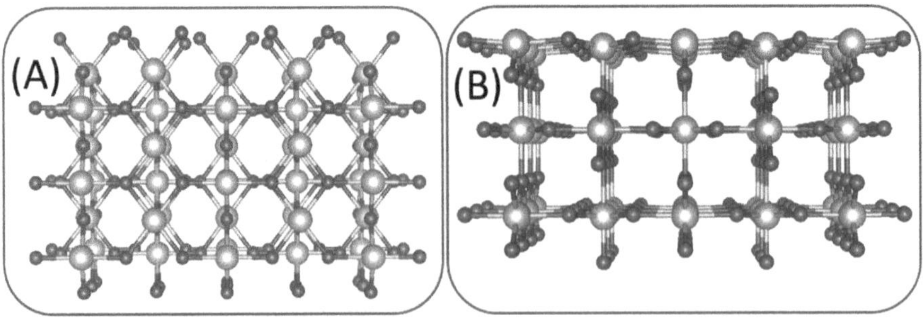

Figure 2.8 $IrO_2(110)$ rutile crystalline surface structure: (A) top view and (B) side view. (grey balls: oxygen atoms; dark grey balls: Ir atoms.)

2.4.1 Ir and its oxides: electrocatalysts

The OER occurs at the anode and is key in the development of WE. Currently, iridium (Ir) and ruthenium (Ru) are the main elements used for OER electrocatalysts in PEM water electrolysers. They offer tolerance in harsh PEMWE environments and good activity; however, Ir and Ru are costly rare-earth metals, with limited abundance. At present, the only known 5d transition metal (TM) with good OER activity is IrO_2. The state-of-the-art catalytic material is Ir and its oxides, but there are significant limitations towards its commercial use. These limitations pose a challenge to the wider application of Ir in PEMWE devices, particularly for large-scale conversions and the storage of renewable energy. One of the limitation is attributed to the scarcity of Ir metal, which is about 1/40 and 1/10 times the scarcity of gold and platinum respectively with a mass fraction of 0.001 ppm in crystal rock (Russell, Nuttall, and Fickett, 1973).

Iridium oxide (an oxide of iridium metal) is currently one of the most stable and active catalysts for the OER under acidic conditions. It is more stable than RuO_2 (although RuO_2 has shown higher activity for the OER). Thus several experimental and theoretical studies have been carried out on the use of IrO_2 as an electrocatalyst (Cherevko et al., 2016; Choe et al., 2018; Dickens et al., 2019; Geiger et al., 2016; Kahk et al., 2014; Kuo et al., 2017; Massué et al., 2017; Matz and Calatayud, 2017; Nong et al., 2018; Pavlovic et al., 2016; Pfeifer et al., 2016a; Ping, Galli, and Goddard, 2015; Siracusano et al., 2017; Spöri et al., 2019; Suen et al., 2017; Sun, Liu, et al., 2016; Tan et al., 2019; Yu et al., 2018; Zhang et al., 2008).

DFT-based calculations are currently the mainstay in the investigation of electronic structure properties as well as the calculation of OER reactions at the microscopic scale. As detailed above with standard Perdew−Burke−Ernzerhof (PBE)-generalised gradient approximation exchange−correlation functional results in consistent agreement with experimental X-ray photoelectron spectroscopy spectra compared to other higher level functionals (hybrid and PBE + U) (Ping et al., 2015). Furthermore, using computational methods, implicit solvation at constant potentials has been carried out to obtain a more detailed understanding of the atomistic reaction mechanism of the OER at the $IrO_2(110)$ surface, and reasonable comparison with experimental studies have been observed (Ping, Nielsen, and Goddard, 2017). Gauthier et al. carried out computational calculations with the explicit inclusion of water on rutile $IrO_2(110)$ and determined its influence on the OER (Gauthier et al., 2017). The geometry of the adsorbed intermediates is significantly altered as a result of the explicit solvent model, and the energies were converged with one to two bilayers of solvent. In the presence of the explicit solvent, the reaction pathway and energies are similar, with either O or *OH coadsorbates (Gauthier et al., 2017).

Structure−activity relationships have opened an avenue towards further understanding and improving IrO_x-based catalysts for the OER. The nature of surface species of IrO_x-based catalysts, starting with metallic Ir is a mixed-valent, conductive IrO_2 matrix during the OER that has O^{II-} and electrophilic O^{I-} species, as shown in various publications by Pfeifer et al. (2016a, 2016b, 2017). A positive interdependence between the electrophilic O^{I-} species and evolution of oxygen molecules has been established, implying that electrophilic O^{I-} sites are involved in the OER. The electrophilic O^{I-} species might assist in the formation of O–O bonds—likely to be the rate-determining step of the OER reaction (Pfeifer et al., 2017). Amorphous iridium oxyhydroxide $[Ir(OH)_3]$ surfaces readily host electrophilic O^{I-} reactive species, leading to greater activity—compared with Ir metal or crystalline rutile-type IrO_2 (Pfeifer et al., 2017). Furthermore, considering the energy feature of the amorphous iridium oxides, a preedge feature in the O K-edge has been identified as the O-2p-hole state, forming in conjunction with Ir^{III}. The O-2p-hole states give rise to O^{I-} species, shown to positively enhance the OER (Pfeifer et al., 2016a).

Operando X-ray absorption spectroscopy and X-ray photoelectron spectroscopy investigations of IrO_2 nanoparticles (NPs) (rod shape) indicate that the presence of iridium hydroxo (Ir-OH) surface species can be linked to strong OER activity. Larger IrO_2 particles (spherical shape) decrease the intrinsic OER activity. This decrease is associated with the predominant termination of the rod-shaped particles by highly ordered (110) facets, in addition to a limited diffusion within mesoporous features (Abbott et al., 2016). Thus smaller particles with rod-like shape showed increased activity compared with the larger particles due to the higher surface area and increased diffusion pathway; likewise, the conditions of synthesis can influence the stability and activity of the OER catalyst (Cherevko et al., 2014).

Key structural motifs associated with high OER activity of amorphous IrO_x hydroxides is related to the ratio of the corner- and edge-shared IrO_6 octahedra. In addition, the active and stable amorphous IrO_x hydroxide phases contain single-unit-cell-sized, hollandite-like, structural domains that are connected via undercoordinated oxygen and iridium atoms. The less-active phase contains a rutile-phase structural motif, which has been found to structurally collapse faster and become deactivated (Willinger et al., 2017). Pavlovic et al. used in situ Raman spectroscopy and theoretical computations to investigate the structure of anodic IrOx under water oxidation; they confirmed the formation of an $Ir=O$ species during the OER (Pavlovic et al., 2017). Decoupling of the structure sensitivity of Ir/IrO_x electrocatalysts towards the OER revealed that higher crystallinity and a large number of low-index crystal terminations would then exhibit increased resistance to dissolution. However, the higher crystallinity comes with a significant reduction in the intrinsic activity (Tan et al., 2019). This observation is concomitant with the understanding that higher intrinsic activity is observed for the amorphous IrO_2 lattice.

2.4.2 Ir and its oxides: defects and impurities

The presence of defects and impurities, which can be determined using modelling techniques, is known to alter the electronic and geometric properties of various materials (Obodo et al., 2017; Obodo et al., 2020; Obodo, Ouma, Gebreyesus et al., 2019; Ouma et al., 2019), as well as the IrO_2 lattice (Arias-Egido et al., 2019; Ghadge et al., 2017; Halck et al., 2014; Kim and Lee, 2019; Lee et al., 2018; Obodo et al., 2017, 2018; Obodo, Ouma, Obodo et al., 2019; Ouma et al., 2018,

2019; Reier et al., 2015; Sun et al., 2015; Sun, Liu, et al., 2016; Sun, Song, et al., 2016; ten Kortenaar et al., 1995).

As has been mentioned, the anodic OER is crucial for successful electrolysis—however, it remains somewhat of an obstacle because of the complex reactions involved at the electrochemical interface. Thus the introduction of defects and impurities is one method that can be used to overcome this obstacle in the IrO_2 lattice. Investigations into the introduction of different dopant atoms to improve the catalytic activity of Ir-based compounds have been carried out with different amounts of Ru (Ru is considered as the standard dopant choice) (Lyons and Floquet, 2011; Reksten et al., 2018). Ruthenium metal is rather expensive, but it has higher stability and low overpotential, compared with other dopant atoms, when used in combination with Ir metal (Mamaca et al., 2012). A mixture of iridium and ruthenium oxides has been shown to improve the stability and activity of an OER catalyst. The OER specific activity of RuO_2 with an IrO_2 core–shell nanocatalyst is threefold that of an IrO_2 catalyst at 1.48 V.

The catalytic resistance of IrO_2 can be reduced through the preparation method used, resulting in an enhanced Faradaic efficiency of the catalyst (Mamaca et al., 2012). Using DFT and experimental methods, it has been determined that the OER activity of an IrO_2 catalyst and RuO_2 with an IrO_2 core–shell nanocatalyst is limited by the formation of adsorbed O, whereas the adsorption of OOH limits the reaction rate on RuO_2. TiO_2 appears to be a better core for the enhancement of IrO_2 activity for oxygen evolution (Ma et al., 2018). Siracusano et al. synthesised IrO_x and $Ir_0Ru_{0.3}O_x$ crystallite ($\sim$5 nm) (Siracusano et al., 2015). The $Ir_0Ru_{0.3}O_x$ crystallite, compared with the IrO_x crystallite, at the same current density, showed an increased performance of 3.2 A cm^{-2} (at 1.85 V and 90°C), corresponding to a voltage gain of 0.1 V (Siracusano et al., 2015).

An experimental investigation into defective IrO_2 compounds containing Pb, Nb, Ln, Bi, La, Sr and Ca atoms (in varying proportions) revealed moderate to high OER activity, where the OER activity in these systems is attributed to the Ir^{4+}/Ir^{5+} couple (ten Kortenaar et al., 1995). Strong correlation of the Ir^{4+}/Ir^{5+} couple and increased OER activity was obtained as a result of increased distortion in the IrO_6 coordination geometry of the IrO_2 compound. This was attributed to the disappearance of the $J = 1/2$ subbands in the valence band, when Pb atoms are present in the IrO_2 lattice. The Pb atoms in the IrO_2 lattice result in increased distortion of the IrO_6 coordination geometry, accompanied by increased activity, whereas the Bi atoms in the IrO_2 lattice resulted in lower activity. The observed order of activity was Pb-Ir > IrO_2 > Bi-Ir.

A DFT study showed that distortion in the IrO_6 geometry optimised the binding strength between the Ir-5d and O-2p orbitals because of the broader d band width. Thus insight from the study of Sun et al. revealed that enhancement of the OER activity can be obtained as a result of doping or the structural manipulation of the IrO_6 octahedral geometry (Sun, Liu, et al., 2016; Sun, Song, et al., 2016). Furthermore, significant changes in the magnitude of the resistivity, as a function of the dopant content, have been observed, which can be related to the OER properties of IrO_2 compounds (Arias-Egido et al., 2019).

The catalytic activities of transition-metal-doped IrO_2 NPs (TM-IrO_2 NPs; TM = Cr, Mn, Fe, Co or Ni) showed high defect-related signals for Cr-IrO_2 and Mn-IrO_2, but lower for Fe-IrO and Co-IrO_2; they were not seen for Ni-IrO_2. Compared with bare IrO_2, the Cr-IrO_2 or Mn-IrO_2 had the highest catalytic activity for oxidation. The presence of these TM dopants in an IrO_2 lattice increases vacancy defects, which is proposed to enhance the oxidation activity (Lee et al., 2018).

In another study, nickel-leached IrNi@IrO_x metal oxide core–shell NPs were considered; the local geometric ligand environment and the electronic metal states of oxygen-coordinated Ir centres were investigated (Nong et al., 2018). Leaching of the Ni during catalyst activation was proposed, with the vacancy formation resulting in oxygen ligands in direct proximity to lattice vacancies. Improved activity was achieved due to a reduction in the kinetic barriers for the O–O bond formation. Thus the presence of targeted vacancies can be viewed as surface amorphization, with resultant high activity and stability.

2.5 Challenges and opportunities of using *ab initio* modelling of Ir and its oxides in the OER

As mentioned in our introduction, technology pertaining to sustainable hydrogen production, with the aim of creating clean energy, is currently an area that is attracting significant attention—here, the electrolysis of water is crucial. As it pertains to this, a better understanding and the rational design of material properties and characteristics, achieved through use of computational modelling methods, is of great importance. Good progress is being made in the use of machine learning to automate the search of IrO_2 compound surface structures, but further investigations are required in this area (Ulissi et al., 2016).

The modelling of crystalline and amorphous interfaces under different environmental conditions, such as different temperatures and pressures, in order to gain insight into the reaction kinetics of the OER, for which Ir and its oxides are used, presents an opportunity in realizing better electrocatalysts. Developing the adsorption kinetics relationship in the OER, with the inclusion of pH and different voltage configurations, as well as highlighting the significance of interfacial interaction towards electrocatalyst design should prove valuable.

Furthermore, further efforts regarding the modelling of NPs of different shapes and sizes are required, to assess structure–activity relationships. Optimisation of the different mesoporous structures for the diffusion of the various gaseous species should also more attention. To achieve a balance between high activity and stability of IrO_x-based compounds towards the OER electrochemical performance, the following aspects, among others, are still open to modelling: (i) a high ratio of the Ir(IV) to Ir (III) species and hydroxide content on the surface, allowing for more O–O bond formation; (ii) a high anodic charge and surface area, which would present more active sites for the electrochemical reactions active sites and (iii) the presence of homogeneous morphology of the anode catalyst layer, for ease of diffusion (Yu et al., 2018).

2.6 Future directions and open issues of low-temperature WE

2.6.1 Mergers, acquisitions and expansion

Renewable hydrogen can act as an enabler to link together many industries that require energy and chemical feedstock at the same time. Recently, many countries have announced ambitious plans to increase investments in the so-called 'hydrogen economy'.

Globally, there are many suppliers of WE technology (both PEM and alkaline), including, but not limited to, the following: Proton OnSite/Nel Hydrogen, 718th Research Institute of CSIC, Teledyne Energy Systems, Hydrogenics/Cummins, Suzhou Jingli, Beijing Zhongdian, McPhy, Siemens, TianJin Mainland Hydrogen Equipment, Areva H2gen, H-Tec Systems, Shandong Saksay Hydrogen Energy, Yangzhou Chungdean Hydrogen Equipment, Enapter, Giner/Plug Power, Idroenergy S.P.A., ErreDue S.p.A., ShaanXi HuaQin, Kobelco Eco-Solutions, ELB Elektrolysetechnik GmbH, ITM Power, Toshiba. Recent reviews of some state of the art (SoA) systems have been presented by Bessarabov and Millet and Grigoriev et al. (Bessarabov and Millet, 2018; Grigoriev et al., 2020).

Some recent business-related events and announcements in the area of the electrolytic hydrogen industry suggest that we are now observing an unprecedented high volume of activity in the sector, including the absorption of some relatively small companies that specialise in electrolysis by large industries, as well as large direct investments.

Examples include (but are not limited to) the following.

- Hydrogenics recently announced (2019): 'Hydrogenics to deliver world's largest PEM hydrogen electrolysis plant'. https://www.hydrogenics.com/2019/02/25/hydrogenics-to-deliver-worlds-largest-hydrogen-electrolysis-plant/
- In Switzerland, Nel Hydrogen Electrolyser, a division of Nel ASA, has received a purchase order for a 2 MW containerised PEM electrolyser, as part of a 30 MW framework agreement with Hydrospider AG, an affiliated company of H_2 Energy. https://www.greencarcongress.com/2019/02/20190226-nel.html
- The REFHYNE project is funded by the FCHJU/H2020 programme and will install and operate a 10 MW electrolyser from ITM Power at a large refinery in Rhineland, Germany, which is operated by Shell Deutschland Oils. A PEM WE system is aimed at operating at 3 A cm^{-2} current density. https://refhyne.eu/
- In the United States, Nikola orders enough electrolysis equipment from Nel to produce 40,000 kgH_2 day^{-1}. https://nikolamotor.com/press_releases/nikola-orders-enough-electrolysis-equipment-from-nel-to-produce-40000-kgs-of-hydrogen-per-day-79

Some recent acquisitions or investments in WE companies include the following:

- Man Energy Solutions recently announced (2019): 'Man Energy Solutions acquires 40% of hydrogen tech firm H-tec Systems'. March 26 (Renewables Now)—German Manufacturer Man Energy Solutions, a subsidiary of carmaker Volkswagen AG (ETR: VOW3), Tuesday announced it has agreed to acquire a 40% stake in domestic electrolysis technology company H-tec Systems. https://renewablesnow.com/news/man-energy-solutions-acquires-40-of-hydrogen-tech-firm-h-tec-systems-647972/

- 'Nel acquired Proton OnSite to create world's largest electrolyser company'. Nel ASA has acquired Proton Energy Systems Inc. (Proton OnSite) in a \$70 m deal that creates the world's largest electrolyser company. https://www.gasworld.com/nel-acquires-proton-onsite/2012338.article
- 'Cummins took over fuel cell and electrolyser maker Hydrogenics'. The US engine manufacturer Cummins has taken over the Canadian fuel cell manufacturer Hydrogenics Corporation. However, Air Liquide will remain on board as an investor in Hydrogenics. Cummins announced that the company would acquire all shares of the Canadian fuel cell and hydrogen technology developer and manufacturer, with the exception of the shares Air Liquide acquired in Hydrogenics earlier this year. https://www.electrive.com/2019/07/01/cummins-takes-over-fuel-cell-maker-hydrogenics/. {{{Fig. 2.9}}}

Yet another example of a SoA PEM hydrogen production installation is in Australia: the first power-to-gas and microgrid project (Jandakot, Australia; 2019). See Fig. 2.10. The system comprises a $30\ \mathrm{N\,m^3\,h^{-1}}$ PEM electrolyser, 300 kW power from PV and 400 kWh energy storage in batteries. Excess power is converted into hydrogen and blended in the natural gas microgrid pipeline.

- A very large alkaline electrolyser plant in operation at a polycrystalline silicon plant (Sarawak, Malaysia) is shown in Fig. 2.11. It produces approximately $5335\ \mathrm{N\,m^3\,h^{-1}}$ of hydrogen (about $11.5\ \mathrm{T\,day^{-1}}$) using 100% green electricity from hydropower at the Bakun Dam. Hydrogen is utilised for the chemical vapour deposition process, for highly purified silicon rods (Si 99.999999999%). The hydrogen supply is critical for the plant.

Other recent events include:

- Iwatani Corporation of America and ITM Power have signed a collaboration agreement for the deployment of multimegawatt electrolyser-based hydrogen energy systems in North America. Through the agreement, the Iwatani Corporation subsidiary and a UK-

Figure 2.9 Example of Hydrogenics product: 5 MW PEM HyLYZER system.
Source: Courtesy of Hydrogenics.

Figure 2.10 Power-to-gas and microgrid project in Jandakot (Australia), 2019.
Source: Courtesy of NEL.

Figure 2.11 Alkaline electrolytic hydrogen plant (Sarawak, Malaysia) used in the semiconductor industry.
Source: Courtesy of NEL.

based energy storage and clean fuel company hope to reduce pollutants and improve air quality by offering renewable hydrogen to the transportation, energy storage and renewable energy sectors in North America. The collaboration enables the two companies to share opportunities and, when/where a commercial opportunity exists, work together on an exclusive basis to deploy ITM Power's PEM electrolysers and Iwatani's gas handling and deployment solutions. https://www.gasworld.com/iwatani-and-itm-power-form-partnership/2018078.article

- Linde has confirmed that it has acquired a 20% stake in the UK-based clean energy firm ITM Power (October 2019). In addition to the investment, Linde will form a joint venture with ITM Power to implement projects based on ITM Power's technology. https://www.gasworld.com/linde-completes-minority-stake-in-itm-power/2017950.article
- Plug Power Inc., a provider of hydrogen engines and fuelling solutions enabling e-mobility, has completed the acquisitions of Giner ELX. https://ngtnews.com/plug-power-acquires-united-hydrogen-and-giner-elx

The key challenge in the area of electrolytic hydrogen production is still the cost of hydrogen, which depends on the initial capital expenditure (CapEx) and the cost of electricity. Existing hydrogen storage and transport technologies are also expensive and rather inefficient. Extremely high pressures or extremely low temperatures are required to achieve a high storage density. The Hydrogen Council Roadmap shows that, by 2050, the potential impact of hydrogen could be US$2,5 trillion in global revenues and 30 million jobs. https://hydrogencouncil.com/wp-content/uploads/2018/07/4.-PPT-McKinsey-FINAL.pdf

2.6.2 Fundamentals: modelling of the OER catalyst and operations

The modelling of iridium and its oxides for oxygen evolution offers several opportunities and challenges. It is proposed that the following will be some of the key drivers in the future directions of modelling OER catalysts.

- The development of nonprecious-metal-based OER catalysts is an area that is gaining traction and future research would expand on this (Anantharaj et al., 2016; Cao, Johnson, and Nath, 2019; Mavrokefalos and Patzke, 2019; Weng et al., 2018).
- The development of a different reaction mechanism for the newly developed OER catalyst, taking into account the chemistry of the surface and reactive species.
- Rational design of a catalytic surface close to experimental condition (albeit challenging), for which future research could make use of machine learning algorithms to elucidate this area and provide more robust results that are comparable with experimental observations.
- The development of a full multiscale approach using a state-of-the-art method for the three length scales, where data from the microscopic length scale would be fed into the mesoscopic length scale and then finally to the macroscopic length scale.

2.6.2.1 Safety

- Hydrogen diffuses through membranes and, when it comes into contact with oxygen, unwanted and potentially dangerous hydrogen concentrations can be established—leading to dangerous and potentially explosive situations.
- Gas cross-over mitigations strategy is receiving much attention in the research community.
- PEM technology offers very high hydrogen discharge pressure—for alkaline electrolysis, higher discharge pressure presents a challenge.

2.6.2.2 Cost reduction

- The matter of currently high costs is being addressed from various angles, including the materials used
- A few reviews relevant to this aspect have recently been published, for example, Ayers and Ayers et al. (Ayers, 2019; Ayers et al., 2019). One of the promising approaches includes increasing current density operation to reduce CapEx. (However for PEM technology, it could present a durability challenge.)

2.6.2.3 Durability

With the current trend in the reduction of PGMs in PEM technology, it is expected that the durability issues will be a great challenge.

Acknowledgements

This publication has emanated from research conducted with the financial support of the North-West University, Potchefstroom, South Africa, and the HySA-Infrastructure Centre of Competence, Faculty of Engineering, North-West University, Potchefstroom, South Africa.

Nomenclature

CapEx	capital expenditure
CFD	computational fluid dynamics
kMC	kinetic Monte Carlo
MD	molecular dynamics
NP/s	nanoparticle/s
OER	oxygen evolution reaction
PBE	Perdew–Burke–Ernzerhof
PEM	proton-exchange membrane
PEMWE	proton-exchange membrane water electrolyser
PGM	platinum group metal
ReaxFF MD	reactive force-field molecular dynamics
SHE	standard hydrogen electrode
SoA	state of the art
SSMS	state-space modelling and simulations
TM	transition metal
WE	water electrolysis

References

Abbott, D. F., et al. (2016). Iridium oxide for the oxygen evolution reaction: Correlation between particle size, morphology, and the surface hydroxo layer from operando XAS. *Chemistry of Materials*, *28*(18), 6591–6604. Available from https://doi.org/10.1021/acs.chemmater.6b02625.

Abdin, Z., Webb, C. J., and Gray, E. M. (2015). Modelling and simulation of a proton exchange membrane (PEM) electrolyser cell. *International Journal of Hydrogen Energy*, *40*(39), 13243–13257. Available from https://doi.org/10.1016/j.ijhydene.2015.07.129, Elsevier Ltd.

Abdol Rahim, A. H., et al. (2016). An overview of polymer electrolyte membrane electrolyzer for hydrogen production: Modeling and mass transport. *Journal of Power Sources*, *309*, 56–65. Available from https://doi.org/10.1016/j.jpowsour.2016.01.012, Elsevier B.V.

Anantharaj, S., et al. (2016). Recent trends and perspectives in electrochemical water splitting with an emphasis on sulfide, selenide, and phosphide catalysts of Fe, Co, and Ni: A review. *ACS Catalysis*, *6*(12), 8069–8097. Available from https://doi.org/10.1021/acscatal.6b02479.

Arias-Egido, E., Laguna-Marco, M. A., Piquer, C., Boada, R., and Diaz-Moreno, S. (2019). Toward the optimized spintronic response of Sn-doped IrO_2 thin films. *Advanced Functional Materials*, *29*(10), 10–15. Available from https://doi.org/10.1002/adfm.201806754.

Atlam, O. (2009). An experimental and modelling study of a photovoltaic/proton-exchange membrane electrolyser system. *International Journal of Hydrogen Energy*, *34*(16), 6589–6595. Available from https://doi.org/10.1016/j.ijhydene.2009.05.147, Elsevier Ltd.

Atlam, O., and Kolhe, M. (2011). Equivalent electrical model for a proton exchange membrane (PEM) electrolyser. *Energy Conversion and Management*, *52*(8–9), 2952–2957. Available from https://doi.org/10.1016/j.enconman.2011.04.007, Elsevier Ltd.

Ayers, K. (2019). The potential of proton exchange membrane–based electrolysis technology. *Current Opinion in Electrochemistry*, *18*, 9–15. Available from https://doi.org/10.1016/j.coelec.2019.08.008, Elsevier Ltd.

Ayers, K., et al. (2019). Perspectives on low-temperature electrolysis and potential for renewable hydrogen at scale. *Annual Review of Chemical and Biomolecular Engineering*, *10*, 219–239. Available from https://doi.org/10.1146/annurev-chembioeng-060718-030241.

Bessarabov, D., and Millet, P. (2018). *PEM water electrolysis*. Academic Press. Available from https://doi.org/10.1016/C2016-0-00574-0.

Bessarabov, D., et al. (2015). *PEM electrolysis for hydrogen production: Principles and applications*. CRC Press.

Cao, X., Johnson, E., and Nath, M. (2019). Identifying high-efficiency oxygen evolution electrocatalysts from Co-Ni-Cu based selenides through combinatorial electrodeposition. *Journal of Materials Chemistry A*, *7*(16), 9877–9889. Available from https://doi.org/10.1039/c9ta00863b.

Carmo, M., et al. (2013). A comprehensive review on PEM water electrolysis. *International Journal of Hydrogen Energy*, *38*(12), 4901–4934. Available from https://doi.org/10.1016/j.ijhydene.2013.01.151.

Cherevko, S., et al. (2014). Stability of nanostructured iridium oxide electrocatalysts during oxygen evolution reaction in acidic environment. *Electrochemistry Communications*, *48*, 81–85. Available from https://doi.org/10.1016/j.elecom.2014.08.027, Elsevier B.V.

Cherevko, S., et al. (2016). Oxygen evolution activity and stability of iridium in acidic media. Part 2. Electrochemically grown hydrous iridium oxide. *Journal of Electroanalytical Chemistry*, *774*, 102–110. Available from https://doi.org/10.1016/j.jelechem.2016.05.015, Elsevier B.V.

Choe, S., et al. (2018). Electrodeposited IrO_2/Ti electrodes as durable and cost-effective anodes in high-temperature polymer-membrane-electrolyte water electrolyzers. *Applied Catalysis B: Environmental*, *226*, 289–294. Available from https://doi.org/10.1016/j.apcatb.2017.12.037, Elsevier B.V.

Choi, P., Bessarabov, D. G., and Datta, R. (2004). A simple model for solid polymer electrolyte (SPE) water electrolysis. *Solid State Ionics*, *175*(1–4), 535–539. Available from https://doi.org/10.1016/j.ssi.2004.01.076.

Dickens, C. F., Kirk, C., and Nørskov, J. K. (2019). Insights into the electrochemical oxygen evolution reaction with ab initio calculations and microkinetic modeling: Beyond the limiting potential volcano'. *Journal of Physical Chemistry C*, *123*(31), 18960–18977. Available from https://doi.org/10.1021/acs.jpcc.9b03830, American Chemical Society.

Djinović, P., and Schüth, F. (2015). Energy carriers made from hydrogen. In P. T. Moseley, and J. Garche (Eds.), *Electrochemical energy Storage for renewable sources and grid balancing* (pp. 183–199). Amsterdam: Elsevier. Available from http://doi.org/10.1016/B978-0-444-62616-5.00012-7.

Falcão, D. S., and Pinto, A. M. F. R. (2020). 'A review on PEM electrolyzer modelling: Guidelines for beginners. *Journal of Cleaner Production*, *261*. Available from https://doi.org/10.1016/j.jclepro.2020.121184.

Feng, Q., et al. (2017). A review of proton exchange membrane water electrolysis on degradation mechanisms and mitigation strategies. *Journal of Power Sources*, *366*, 33–55. Available from https://doi.org/10.1016/j.jpowsour.2017.09.006, Elsevier B.V.

García-Valverde, R., Espinosa, N., and Urbina, A. (2012). Simple PEM water electrolyser model and experimental validation. *International Journal of Hydrogen Energy*, *37*(2), 1927–1938. Available from https://doi.org/10.1016/j.ijhydene.2011.09.027.

Gauthier, J. A., et al. (2017). Solvation effects for oxygen evolution reaction catalysis on IrO_2(110). *Journal of Physical Chemistry C*, *121*(21), 11455–11463. Available from https://doi.org/10.1021/acs.jpcc.7b02383.

Geiger, S., et al. (2016). Activity and stability of electrochemically and thermally treated iridium for the oxygen evolution reaction. *Journal of the Electrochemical Society*, *163*(11), 3132–3138. Available from https://doi.org/10.1149/2.0181611jes.

Ghadge, S. D., et al. (2017). Fluorine substituted (Mn, Ir)O_2:F high performance solid solution oxygen evolution reaction electro-catalysts for PEM water electrolysis. *RSC Advances*, *7*(28), 17311–17324. Available from https://doi.org/10.1039/c6ra27354h, Royal Society of Chemistry.

Görgün, H. (2006). Dynamic modelling of a proton exchange membrane (PEM) electrolyzer. *International Journal of Hydrogen Energy*, *31*(1), 29–38. Available from https://doi.org/10.1016/j.ijhydene.2005.04.001.

Govindarajan, N., et al. (2018). Does the breaking of adsorption-energy scaling relations guarantee enhanced electrocatalysis? *Current Opinion in Electrochemistry*, *8*, 110–117. Available from https://doi.org/10.1016/j.coelec.2018.03.025.

Govindarajan, N., et al. (2019). Outlining the scaling-based and scaling-free optimization of electrocatalysts. *ACS Catalysis*, *9*(5), 4218–4225. Available from https://doi.org/10.1021/acscatal.9b00532, American Chemical Society.

Grajciar, L., et al. (2018). Towards operando computational modeling in heterogeneous catalysis. *Chemical Society Reviews*, *47*(22), 8307–8348. Available from https://doi.org/10.1039/c8cs00398j.

Grigoriev, S. A., et al. (2011). 'High-pressure PEM water electrolysis and corresponding safety issues. *International Journal of Hydrogen Energy*, *36*(3), 2721–2728. Available from https://doi.org/10.1016/j.ijhydene.2010.03.058.

Grigoriev, S. A., et al. (2020). Current status, research trends, and challenges in water electrolysis science and technology. *International Journal of Hydrogen Energy*, *45*(49), 26036–26058. Available from https://doi.org/10.1016/j.ijhydene.2020.03.109, Elsevier Ltd.

Halck, N. B., et al. (2014). Beyond the volcano limitations in electrocatalysis-oxygen evolution reaction. *Physical Chemistry Chemical Physics*, *16*(27), 13682–13688. Available from https://doi.org/10.1039/c4cp00571f.

Hernández-Gómez, Á., Ramirez, V., and Guilbert, D. (2020). Investigation of PEM electrolyzer modeling: Electrical domain, efficiency, and specific energy consumption. *International Journal of Hydrogen Energy*, *45*(29), 14625–14639. Available from https://doi.org/10.1016/j.ijhydene.2020.03.195.

Kahk, J. M., et al. (2014). Understanding the electronic structure of IrO_2 using hard-X-ray photoelectron spectroscopy and density-functional theory. *Physical Review Letters*, *112* (11), 1–6. Available from https://doi.org/10.1103/PhysRevLett.112.117601.

Kim, J. Y., and Lee, H. (2019). Influence of pH modification on catalytic activities of metal-doped IrO_2 nanoparticles'. *Scientific Reports*, *9*(1), 1–9. Available from https://doi.org/10.1038/s41598-019-42358-9, Springer US.

Kuo, D. Y., et al. (2017). Influence of surface adsorption on the oxygen evolution reaction on $IrO_2(110)$. *Journal of the American Chemical Society*, *139*(9), 3473–3479. Available from https://doi.org/10.1021/jacs.6b11932.

Lee, H., et al. (2018). 'Comparative study of catalytic activities among transition metal-doped IrO_2 nanoparticles. *Scientific Reports*, *8*(1), 2–9. Available from https://doi.org/10.1038/s41598-018-35116-w, Springer US.

Lyons, M. E. G., and Floquet, S. (2011). Mechanism of oxygen reactions at porous oxide electrodes. Part 2—Oxygen evolution at RuO_2, IrO_2 and $Ir_xRu_{1-x}O_2$ electrodes in aqueous acid and alkaline solution'. *Physical Chemistry Chemical Physics*, *13*(12), 5314–5335. Available from https://doi.org/10.1039/c0cp02875d.

Ma, T. Y., Dai, S., and Qiao, S. Z. (2016). Self-supported electrocatalysts for advanced energy conversion processes. *Materials Today*, *19*(5), 265–273. Available from https://doi.org/10.1016/j.mattod.2015.10.012, Elsevier Ltd.

Ma, Z., et al. (2018). Reaction mechanism for oxygen evolution on RuO_2, IrO_2, and RuO_2@IrO_2 core-shell nanocatalysts. *Journal of Electroanalytical Chemistry*, *819*, 296–305. Available from https://doi.org/10.1016/j.jelechem.2017.10.062, Elsevier B.V.

Mamaca, N., et al. (2012). Electrochemical activity of ruthenium and iridium based catalysts for oxygen evolution reaction. *Applied Catalysis B: Environmental*, *111–112*, 376–380. Available from https://doi.org/10.1016/j.apcatb.2011.10.020, Elsevier B.V.

Marangio, F., Santarelli, M., and Calì, M. (2009). Theoretical model and experimental analysis of a high pressure PEM water electrolyser for hydrogen production. *International Journal of Hydrogen Energy*, *34*(3), 1143–1158. Available from https://doi.org/10.1016/j.ijhydene.2008.11.083.

Massué, C., et al. (2017). Reactive electrophilic O^{I-} species evidenced in high-performance iridium oxohydroxide water oxidation electrocatalysts. *ChemSusChem*, *10*(23), 4786–4798. Available from https://doi.org/10.1002/cssc.201701291.

Matsumoto, Y., and Sato, E. (1986). Electrocatalytic properties of transition metal oxides for oxygen evolution reaction. *Materials Chemistry and Physics*, *14*(5), 397–426. Available from https://doi.org/10.1016/0254-0584(86)90045-3.

Matz, O., and Calatayud, M. (2017). Periodic DFT study of rutile IrO_2: Surface reactivity and Catechol adsorption. *Journal of Physical Chemistry C*, *121*(24), 13135–13143. Available from https://doi.org/10.1021/acs.jpcc.7b01990.

Mavrokefalos, C. K., and Patzke, G. R. (2019). Water oxidation catalysts: The quest for new oxide-based materials. *Inorganics*, *7*(3), 29. Available from https://doi.org/10.3390/INORGANICS7030029.

Minnaar, C., De Beer, F., and Bessarabov, D. (2020). Current density distribution of electrolyzer flow fields: In situ current mapping and neutron radiography. *Energy and Fuels*, *34*(1), 1014–1023. Available from https://doi.org/10.1021/acs.energyfuels.9b03814.

Mukherjee, P. P., Wang, C. Y., and Kang, Q. (2009). 'Mesoscopic modeling of two-phase behavior and flooding phenomena in polymer electrolyte fuel cells. *Electrochimica Acta*, *54*(27), 6861–6875. Available from https://doi.org/10.1016/j.electacta.2009.06.066.

Nafeh, A. E.-S. A. (2011). Hydrogen production from a PV/PEM electrolyser system using a neural-network-based MPPT algorithm. *International Journal of Numerical Modelling*, *24*(3), 282–297. Available from https://doi.org/10.1002/jnm.778.

Nong, H. N., et al. (2018). A unique oxygen ligand environment facilitates water oxidation in hole-doped $IrNiO_x$ core–shell electrocatalysts. *Nature Catalysis*, *1*(11), 841–851. Available from https://doi.org/10.1038/s41929-018-0153-y, Springer US.

Nørskov, J. K., et al. (2004). Origin of the overpotential for oxygen reduction at a fuel-cell cathode. *Journal of Physical Chemistry B*, *108*(46), 17886–17892. Available from https://doi.org/10.1021/jp047349j.

Nørskov, J. K., et al. (2009). Towards the computational design of solid catalysts. *Nature Chemistry*, *1*(1), 37–46. Available from https://doi.org/10.1038/nchem.121.

Obodo, K. O., et al. (2017). Influence of transition metal doping on the electronic and optical properties of ReS_2 and $ReSe_2$ monolayers. *Physical Chemistry Chemical Physics*, *19*(29), 19050–19057. Available from https://doi.org/10.1039/c7cp03455e.

Obodo, K. O., et al. (2018). Influence of Tm, Ho and Er dopants on the properties of Yb activated $ZnTiO_3$ perovskite: A density functional theory insight. *Materials Research Express*, *5*(10). Available from https://doi.org/10.1088/2053-1591/aadaf2, IOP Publishing.

Obodo, K. O., et al. (2020). Controlling the electronic and optical properties of HfS_2 monolayers: Via lanthanide substitutional doping: A DFT + U study. *RSC Advances*, *10*(27), 15670–15676. Available from https://doi.org/10.1039/d0ra02464c, Royal Society of Chemistry.

Obodo, K. O., Ouma, C. N. M., Gebreyesus, G., et al. (2019). DFT + U studies of the electronic and optical properties of ReS_2 mono-layer doped with lanthanide atoms. *Materials Research Express*, *6*(10), 106307. (1–8). Available from https://doi.org/10.1088/2053-1591/ab3912.

Obodo, K. O., Ouma, C. N. M., Obodo, J. T., et al. (2019). First principles study of single and multi-site transition metal dopant ions in MoS_2 monolayer'. *Computational Condensed Matter*, *21*, e00419. Available from https://doi.org/10.1016/j.cocom.2019.e00419, Elsevier B.V.

Oliveira, L. F. L., et al. (2012). A multiscale physical model for the transient analysis of PEM water electrolyzer anodes. *Physical Chemistry Chemical Physics*, *14*(29), 10215–10224. Available from https://doi.org/10.1039/c2cp23300b.

Olivier, P., Bourasseau, C., and Bouamama, B. (2017a). Dynamic and multiphysic PEM electrolysis system modelling: A bond graph approach. *International Journal of Hydrogen Energy.*, *42*(22), 14872–14904. Available from https://doi.org/10.1016/j.ijhydene.2017.03.002, Elsevier Ltd.

Olivier, P., Bourasseau, C., and Bouamama, B. (2017b). Low-temperature electrolysis system modelling: A review. *Renewable and Sustainable Energy Reviews*, *78*, 280–300. Available from https://doi.org/10.1016/j.rser.2017.03.099.

Ouma, C. N. M., et al. (2018). *Ab initio* insights on the effect of embedding lanthanide atoms on nitrogenated holey doped graphene (g-C2N). *Journal of Materials Chemistry C*, *6*(15), 4015–4022. Available from https://doi.org/10.1039/c8tc00474a.

Ouma, C. N. M., et al. (2019). Insights on hydrogen evolution reaction in transition metal doped monolayer TcS_2 from density functional theory calculations. *Applied Surface Science*, *470*(November 2018), 107–113. Available from https://doi.org/10.1016/j.apsusc.2018.11.044, Elsevier.

Pavlovic, Z., et al. (2016). 'Probing the structure of a water-oxidizing anodic iridium oxide catalyst using raman spectroscopy. *ACS Catalysis*, *6*(12), 0–7. Available from https://doi.org/10.1021/acscatal.6b02343.

Pavlovic, Z., et al. (2017). The active site for the water oxidising anodic iridium oxide probed through: In situ Raman spectroscopy. *Chemical Communications*, *53*(92), 12414–12417. Available from https://doi.org/10.1039/c7cc05669a.

Pfeifer, V., et al. (2016a). The electronic structure of iridium oxide electrodes active in water splitting. *Physical Chemistry Chemical Physics*, *18*(4), 2292–2296. Available from https://doi.org/10.1039/c5cp06997a.

Pfeifer, V., et al. (2016b). The electronic structure of iridium and its oxides. *Surface and Interface Analysis*, *48*(5), 261–273. Available from https://doi.org/10.1002/sia.5895.

Pfeifer, V., et al. (2017). *In situ* observation of reactive oxygen species forming on oxygen-evolving iridium surfaces. *Chemical Science*, *8*(3), 2143–2149. Available from https://doi.org/10.1039/c6sc04622c.

Ping, Y., Galli, G., and Goddard, W. A. (2015). Electronic structure of IrO_2: The role of the metal d orbitals. *Journal of Physical Chemistry C*, *119*(21), 11570–11577. Available from https://doi.org/10.1021/acs.jpcc.5b00861.

Ping, Y., Nielsen, R. J., and Goddard, W. A. (2017). The reaction mechanism with free energy barriers at constant potentials for the oxygen evolution reaction at the $IrO_2(110)$ surface. *Journal of the American Chemical Society*, *139*(1), 149–155. Available from https://doi.org/10.1021/jacs.6b07557.

Rashid, M. M., et al. (2015). Hydrogen production by water electrolysis: A review of alkaline water electrolysis, PEM water electrolysis and high temperature water electrolysis. *International Journal of Engineering and Advanced Technology*, *4*(3), 2249–8958.

Reier, T., et al. (2015). Molecular insight in structure and activity of highly efficient, low-Ir Ir-Ni oxide catalysts for electrochemical water splitting (OER). *Journal of the American Chemical Society*, *137*(40), 13031–13040. Available from https://doi.org/10.1021/jacs.5b07788.

Reksten, A. H., et al. (2018). 'The oxygen evolution reaction mechanism at $Ir_xRu_{1-x}O_2$ powders produced by hydrolysis synthesis. *Journal of Electroanalytical Chemistry*, *819*, 547–561. Available from https://doi.org/10.1016/j.jelechem.2018.04.018, Elsevier B.V.

Russell, J. H., Nuttall, L. J., and Fickett, A. P. (1973). Hydrogen generation by solid polymer electrolyte water electrolysis. *American Chemical Society Division of Fuel Chemistry Preprints*, *18*(3), 24e40.

Saeed, E. W., and Warkozek, E. G. (2015). Modeling and analysis of renewable PEM fuel cell system. *Energy Procedia*, *74*, 87–101. Available from https://doi.org/10.1016/j.egypro.2015.07.527, Elsevier B.V.

Sarwar, M., et al. (2015). Atomic-scale modelling and its application to catalytic materials science. *Johnson Matthey Technology Review*, *59*(3), 257–283. Available from https://doi.org/10.1595/205651315X687975.

Schleder, G. R., et al. (2019). From DFT to machine learning: recent approaches to materials science—A review'. *Journal of Physics: Materials*, *2*(3), 032001. Available from https://doi.org/10.1088/2515-7639/ab084b, IOP Publishing.

Sen, F. G., et al. (2015). Towards accurate prediction of catalytic activity in IrO_2 nanoclusters via first principles-based variable charge force field. *Journal of Materials Chemistry A*, *3*(37), 18970–18982. Available from https://doi.org/10.1039/c5ta04678e.

Siracusano, S., et al. (2015). Nanosized IrOx and IrRuOx electrocatalysts for the O_2 evolution reaction in PEM water electrolysers'. *Applied Catalysis B: Environmental, 164*, 488–495. Available from https://doi.org/10.1016/j.apcatb.2014.09.005, Elsevier B.V.

Siracusano, S., et al. (2017). New insights into the stability of a high performance nanostructured catalyst for sustainable water electrolysis. *Nano Energy, 40*, 618–632. Available from https://doi.org/10.1016/j.nanoen.2017.09.014, Elsevier Ltd.

Sossan, F., et al. (2014). A model predictive control strategy for the space heating of a smart building including cogeneration of a fuel cell-electrolyzer system. *International Journal of Electrical Power and Energy Systems, 62*, 879–889. Available from https://doi.org/10.1016/j.ijepes.2014.05.040, Elsevier Ltd.

Spöri, C., et al. (2019). Experimental activity descriptors for iridium-based catalysts for the electrochemical oxygen evolution reaction (OER). *ACS Catalysis, 9*(8), 6653–6663. Available from https://doi.org/10.1021/acscatal.9b00648.

Suen, N. T., et al. (2017). Electrocatalysis for the oxygen evolution reaction: Recent development and future perspectives. *Chemical Society Reviews, 46*(2), 337–365. Available from https://doi.org/10.1039/c6cs00328a, Royal Society of Chemistry.

Sun, W., et al. (2015). An efficiently tuned d-orbital occupation of IrO_2 by doping with Cu for enhancing the oxygen evolution reaction activity. *Chemical Science, 6*(8), 4993–4999. Available from https://doi.org/10.1039/c5sc01251a.

Sun, W., Liu, J. Y., et al. (2016). OER activity manipulated by IrO_6 coordination geometry: An insight from pyrochlore iridates. *Scientific Reports, 6*(August), 1–10. Available from https://doi.org/10.1038/srep38429, Nature Publishing Group.

Sun, W., Song, Y., et al. (2016). Hollandite structure $K_{x\approx0.25}IrO_2$ catalyst with highly efficient oxygen evolution reaction. *ACS Applied Materials and Interfaces, 8*(1), 820–826. Available from https://doi.org/10.1021/acsami.5b10159.

Tan, X., et al. (2019). 'Decoupling structure-sensitive deactivation mechanisms of Ir/IrO_x electrocatalysts toward oxygen evolution reaction. *Journal of Catalysis, 371*, 57–70. Available from https://doi.org/10.1016/j.jcat.2019.01.018, The Author(s).

ten Kortenaar, M. V., et al. (1995). Oxygen evolution and reduction on iridium oxide compounds. *Journal of Power Sources, 56*(1), 51–60. Available from https://doi.org/10.1016/0378-7753(95)80008-5.

Thanawala, S. S., et al. (2008). Amorphous and crystalline IrO_2 thin films as potential stimulation electrode coatings'. *Applied Surface Science, 254*(16), 5164–5169. Available from https://doi.org/10.1016/j.apsusc.2008.02.054.

Thomsen, J. M., et al. (2015). Iridium-based complexes for water oxidation. *Dalton Transactions, 44*(28), 12452–12472. Available from https://doi.org/10.1039/c5dt00863h.

Toghyani, S., et al. (2019). Optimization of operating parameters of a polymer exchange membrane electrolyzer. *International Journal of Hydrogen Energy, 44*(13), 6403–6414. Available from https://doi.org/10.1016/j.ijhydene.2019.01.186, Elsevier Ltd.

Ulissi, Z. W., et al. (2016). Automated discovery and construction of surface phase diagrams using machine learning. *Journal of Physical Chemistry Letters, 7*(19), 3931–3935. Available from https://doi.org/10.1021/acs.jpclett.6b01254.

Van Duin, A. C. T., et al. (2001). ReaxFF: A reactive force field for hydrocarbons. *Journal of Physical Chemistry A, 105*(41), 9396–9409. Available from https://doi.org/10.1021/jp004368u.

Wang, P., et al. (2017). Oxygen evolution reaction dynamics monitored by an individual nanosheet-based electronic circuit. *Nature Communications, 8*(1), 1–7. Available from https://doi.org/10.1038/s41467-017-00778-z, Springer US.

Weber, A. Z., et al. (2014). 'A critical review of modeling transport phenomena in polymer-electrolyte fuel cells. *Journal of the Electrochemical Society*, *161*(12), F1254–F1299. Available from https://doi.org/10.1149/2.0751412jes.

Weber, T., et al. (2019). Potential-induced pitting corrosion of an IrO_2(110)-RuO_2(110)/Ru (0001) model electrode under oxygen evolution reaction conditions. *ACS Catalysis*, *9* (7), 6530–6539. Available from https://doi.org/10.1021/acscatal.9b01402, American Chemical Society.

Weng, C. C., et al. (2018). Nitrogen-doped defect-rich graphitic carbon nanorings with CoO_x nanoparticles as highly efficient electrocatalyst for oxygen electrochemistry. *ACS Sustainable Chemistry and Engineering*, *6*(11), 15811–15821. Available from https://doi.org/10.1021/acssuschemeng.8b04406.

Willinger, E., et al. (2017). Identifying key structural features of IrO_x water splitting catalysts. *Journal of the American Chemical Society*, *139*(34), 12093–12101. Available from https://doi.org/10.1021/jacs.7b07079.

Wu, X., et al. (2019). Metal-support interaction boosted electrocatalysis of ultrasmall iridium nanoparticles supported on nitrogen doped graphene for highly efficient water electrolysis in acidic and alkaline media'. *Nano Energy*, *62*, 117–126. Available from https://doi.org/10.1016/j.nanoen.2019.05.034, Elsevier Ltd.

Yu, H., et al. (2018). Nano-size IrO_x catalyst of high activity and stability in PEM water electrolyzer with ultra-low iridium loading. *Applied Catalysis B: Environmental*, *239*, 133–146. Available from https://doi.org/10.1016/j.apcatb.2018.07.064, Elsevier B.V.

Zhang, H., et al. (2008). Stability, structure, and electronic properties of chemisorbed oxygen and thin surface oxides on Ir(111). *Physical Review B—Condensed Matter and Materials Physics*, *78*(4), 1–12. Available from https://doi.org/10.1103/PhysRevB.78.045436.

Zhang, X., and Bieberle-Hütter, A. (2016). Modeling and simulations in photoelectrochemical water oxidation: From single level to multiscale modeling. *ChemSusChem*, *9*(11), 1223–1242. Available from https://doi.org/10.1002/cssc.201600214.

High-temperature electrolysis and co-electrolysis

3

Haoran Xu[1,2,3] *and Meng Ni*[3,4]
[1]State Key Laboratory of Clean Energy Utilization, Zhejiang University, Hangzhou, China, [2]Department of Chemical Engineering, Loughborough University, Loughborough, United Kingdom, [3]Building Energy Research Group, Department of Building and Real Estate, The Hong Kong Polytechnic University, Hong Kong, P.R. China, [4]Environmental Energy Research Group, Research Institute for Sustainable Urban Development (RISUD), The Hong Kong Polytechnic University, Hong Kong, P.R. China

3.1 Principle

A solid oxide electrolyser cell (SOEC) is an electrochemical device working at high temperatures (usually 600°C−1000°C) for electrolysis purposes (Chen, Chen, and Xia, 2014; Ebbesen and Mogensen, 2009; Laguna-Bercero, 2012). It has an all-solid-state structure with the electrolyte being sandwiched between the porous anode and cathode (Minh, 2004; Ormerod, 2003; Singhal, 2000). Driven by electricity, the oxidant (e.g. H_2O) is reduced in the cathode, where the fuel (e.g. H_2) is generated together with O^{2-} ions. The O^{2-} ions will transport through the dense electrolyte to the anode, where O_2 is formed through oxygen evolution reaction (OER) and electrons are released. As electrolyte is only ionic conductive, the released electrons will return to cathode through the external circuit to close the cycle.

Benefited from the simple working process, SOECs are easily scalable based on different demands, making them suitable for both centralized and distributed applications (Giddey et al., 2012; Kang and Ahn, 2017; Yano et al., 2007). Apart from the benefits brought by the simple structure, the high operating temperature facilitates the electrochemical reactions in SOECs and enables the use of non-noble metal catalyst, leading to a lower overall cost (Achenbach and Riensche, 1994; Wang et al., 2009; Xiao et al., 2014). In addition, the electricity requirement of SOEC is relatively low as part of the energy needed for electrolysis is provided in the form of heat (Luo, Shi, Li, and Cai, 2014; Ni, 2012; Stempien et al., 2015). With the non-noble catalysts, SOECs are capable of co-electrolyzing CO_2 and H_2O to generate syngas (H_2 and CO mixture), which broadens the application of SOECs and attracts their combination with other commercial technologies, such as Fischer−Tropsch (F−T) synthesis (Becker et al., 2012; Chen, Xu, and Ni, 2017; Wang et al., 2017).

The high electricity consumption of electrolysis has limited the development of electrolysis, as most of today's electricity is generated by conventional thermal

Power to Fuel. DOI: https://doi.org/10.1016/B978-0-12-822813-5.00008-4

power plants. These power plants produce electricity by burning fossil fuels with a complex working process, making the fossil fuel-to-hydrogen efficiency much lower than expected. To reduce the electricity consumption of electrolysis, the concept of fuel-assisted SOEC (FA-SOEC) is proposed. In the FA-SOEC, part of the driving force is shifted from electric potential to chemical potential by introducing reductants into anode to reduce the required potential for OER. In this way, the electricity consumption for electrolysis is significantly reduced (Giddey, Kulkarni, and Badwal, 2015; Luo, Shi, Li, and Ni, 2014; Tao, Butler, and Virkar, 2011).

This chapter will firstly introduce the working principle and the physical and chemical processes in the SOECs. Then, the processes and key parameters in SOECs electrolysis and co-electrolysis will be discussed. Meanwhile, a novel concept called fuel-assisted electrolysis will be introduced. Finally, the application of SOECs in the hybrid system for value-added fuel/chemical generation will be discussed.

3.1.1 Physical and chemical fields inside the solid oxide electrolyser cell

The operation of SOEC for high-temperature electrolysis combines various physical/chemical processes, including electrochemical reactions, chemical reactions, mass transport, momentum transport and heat transfer. These processes affect each other and determine the efficiency and safety in SOEC operation.

3.1.1.1 Electrochemical process

H_2O is usually used to generate H_2 in the SOEC. The electrochemical reduction reaction of H_2O in the SOEC cathode can be written as Eq. (3.1).

$$H_2O + 2e^- \Rightarrow H_2 + O^{2-} \tag{3.1}$$

CO_2 is another oxidant that attracts growing interest for electrolysis in the SOEC. Its electrochemical reduction reaction can be written as Eq. (3.2).

$$CO_2 + 2e^- \Rightarrow CO + O^{2-} \tag{3.2}$$

In above reactions, fuels (H_2 and CO) are generated in the cathode, whereas the oxygen ions (O^{2-}) will transport through the dense electrolyte to the anode, where gaseous O_2 is formed and the electrons are released as shown in Eq. (3.3).

$$2O^{2-} \Rightarrow O_2 + 4e^- \tag{3.3}$$

The overall reactions of electrolysing H_2O and CO_2 to fuels can be written as Eqs. (3.4) and (3.5).

$$H_2O \Rightarrow H_2 + 0.5\ O_2 \quad (3.4)$$

$$CO_2 \Rightarrow CO + 0.5\ O_2 \quad (3.5)$$

As these reactions are driven by electricity, the operating voltage can be calculated by reducing the activation overpotential (η_{act}) and ohmic overpotential (η_{ohmic}) from the equilibrium potential (E_{eq}), as shown in Eq. (3.6) (Xu, Chen, and Liu, 2016). It should be noted that in electrolysis, the value of applied voltage (V) is usually negative, which means the cell is consuming electricity instead of providing power.

$$V = E_{eq} - \eta_{act} - \eta_{ohmic} \quad (3.6)$$

Equilibrium potential

The equilibrium potentials (Nernst potential) are related to the thermodynamics of the reactions. For reactions in Eqs. (3.4) and (3.5), their equilibrium potentials $(E_{H_2O}$ and $E_{CO_2})$ can be calculated by Eqs. (3.7) and (3.8), respectively (Xu, Chen, and Tan, 2018):

$$E_{H_2O} = E^0_{H_2O} + \frac{RT}{2F} \ln \left[\frac{P^L_{H_2O}}{P^L_{H_2} \left(P^L_{O_2} \right)^{\frac{1}{2}}} \right] \quad (3.7)$$

$$E_{CO_2} = E^0_{CO_2} + \frac{RT}{2F} \ln \left[\frac{P^L_{CO_2}}{P^L_{CO} \left(P^L_{O_2} \right)^{\frac{1}{2}}} \right] \quad (3.8)$$

where $E^0_{H_2O}$ and $E^0_{CO_2}$ are standard potentials for the electrochemical reduction of H_2O and CO_2. Their values can be calculated by Eqs. (3.9) and (3.10) at high operating temperature, respectively (Xu et al., 2017). P^L is the local partial pressure of gas specie; R is the universal gas constant (8.3145 J mol^{-1} K^{-1}); T is the local temperature (K) and F is the Faraday constant (96,485 C mol^{-1}).

$$E^0_{H_2O} = -1.253 + 0.00024516\ T \quad (V) \quad (3.9)$$

$$E^0_{CO_2} = -1.46713 + 0.0004527\ T \quad (V) \quad (3.10)$$

For more general situation (e.g. in the FA-SOEC), the equilibrium potential can be calculated based on oxygen partial pressure (Stempien et al., 2014) as shown in Eq. (3.11).

$$E_{eq} = \frac{RT}{nF} \ln \left(\frac{\sum P^L_{O_2,ca}}{\sum P^L_{O_2,an}} \right) \quad (3.11)$$

For the pairs of H_2O/H_2 and CO_2/CO, their oxygen partial pressures can be expressed by Eqs. (3.12) and (3.13):

$$P^L_{O_2,(H_2O/H_2)} = \left(\frac{P^L_{H_2O}}{P^L_{H_2}} \cdot e^{\frac{\Delta G_{H_2O/H_2}}{RT}} \right)^2 \tag{3.12}$$

$$P^L_{O_2,(CO_2/CO)} = \left(\frac{P^L_{CO_2}}{P^L_{CO}} \cdot e^{\frac{\Delta G_{CO_2/CO}}{RT}} \right)^2 \tag{3.13}$$

where $\Delta G_{H_2O/H_2}$ and $\Delta G_{CO_2/CO}$ are the Gibbs free energy changes in reactions (3.4) and (3.5), respectively.

Activation overpotential

The activation overpotential stems from the activation barrier in the electrochemical reactions. It is related to the reaction type and the catalyst/electrode properties. The widely used Butler–Volmer equation is adopted to express the relationship of current density (i) and η_{act}, as shown in Eq. (3.14) (Xu, Chen et al., 2019):

$$i = i_0 \left\{ e^{\left(\frac{\alpha n F \eta_{act}}{RT}\right)} - e^{\left(-\frac{(1-\alpha) n F \eta_{act}}{RT}\right)} \right\} \tag{3.14}$$

where α is the electron transfer coefficient; n is the number of electrons transferred in each electrochemical reaction and i_0 is the exchange current density that can be further expressed by Eq. (3.15).

$$i_0 = \gamma e^{\left(-\frac{E_{act}}{RT}\right)} \tag{3.15}$$

Ohmic overpotential

Ohmic overpotential is caused by ionic and electronic resistances in the cell. They can be calculated by Ohm's law, as given in Eqs. (3.16) and (3.17):

$$i_l = -\sigma_l^{eff} \nabla(\varnothing_l) \tag{3.16}$$

$$i_s = -\sigma_s^{eff} \nabla(\varnothing_s) \tag{3.17}$$

where $\varnothing_l$ and $\varnothing_s$ are the electric potentials of ionic and electronic conductions, respectively; σ_l^{eff} and σ_s^{eff} are the effective conductivities of the porous electrodes. At given solid properties, σ_l^{eff} and σ_s^{eff} can be further calculated by Eqs. (3.18) and (3.19), respectively:

$$\sigma_l^{eff} = \sigma_l \cdot \frac{V_l}{\tau_l} \tag{3.18}$$

$$\sigma_s^{eff} = \sigma_s \cdot \frac{V_s}{\tau_s} \tag{3.19}$$

where σ_l and σ_s are the intrinsic conductivity of ionic and electronic phases. V_l and V_s are the volume fractions of ionic and electronic phases. τ_l and τ_s are the tortuosity of ionic and electronic phases. The ionic/electronic conductivities of typical materials used in SOECs are given in Table 3.1.

3.1.1.2 *Chemical process*

Chemical reactions in the solid oxide electrolyser cell

Chemical reactions are key processes in the SOEC and other reactors in the possible hybrid system. In the SOEC, the existence of catalyst (usually nickel) and high operating temperature favours many reactions.

In co-electrolysis process, where H_2O and CO_2 are both introduced to the SOEC cathode, there simultaneously exist H_2O, CO_2, H_2 and CO. In this situation, water

Table 3.1 Properties of typical solid oxide electrolyser cell materials (Eguchi, 1992, 1997; Fan, Yan, and Yan, 2011; Guo, 2013; Luo, Shi, Li, and Cai, 2014; Luo, Shi, Li, and Ni, 2014; Tai, 1995).

Parameters	Expressions ($S\ m^{-1}$)
Ionic conductivity	
CGO (dense)	$1.22 \times e^{\frac{16054}{RT}}$
GDC	$\frac{100}{T} \times 10^{\left(6.66071-\frac{5322.92}{T}\right)}$
LSCF	$\frac{100}{T} \times 10^{\left(2.51289-\frac{3036.75}{T}\right)}$
ScSZ	$6.92 \times 10^4 e^{\left(-\frac{9681}{T}\right)}$
SDC	$\frac{100}{T} \times 10^{\left(5.48077-\frac{3792.53}{T}\right)}$
YSZ	$3.34 \times 10^4 e^{\left(-\frac{10300}{T}\right)}$
Electronic conductivity	
Ag	$\frac{1.59\ e^8}{(0.0038\ T - 0.1134)}$
BSCF	$-24455 + 82.321\ T - \frac{8.28\ T^2}{1000} + \frac{2.7778\ T^3}{10000}$
LSCF	$\frac{100}{T} \times 10^{\left(4.32576+\frac{1204.26}{T}\right)}$
LSM	$4.2 \times 10^7 e^{\left(-\frac{1150}{T}\right)}$
Ni	$3.27 \times 10^6 - 1065.3\ T$

Ag, Silver; *BSCF*, barium strontium cobalt ferrite; *CGO (dense)*, gadolinium-doped ceria (dense); *GDC*, gadolinium-doped ceria; *LSCF*, lanthanum strontium cobalt ferrite; *LSM*, lanthanum strontium manganite; *Ni*, nickel; *ScSZ*, scandium stabilized zirconium; *SDC*, samaria-doped ceria; *YSZ*, yttrium stabilized zirconium.

gas shift gas reaction is unneglectable under nickel catalyst as shown in Eq. (3.20) (Smith, Loganathan, and Shantha, 2010).

$$CO + H_2O \Leftrightarrow CO_2 + H_2 \tag{3.20}$$

This forward reaction direction is favoured at low temperature, whereas at high temperature the reverse direction is favoured.

In the FA-SOEC, there may exist various chemical reactions at the anode, including steam methane reforming [SMR, Eq. (3.21)], dry methane reforming [Eq. (3.22)], Boudouard reaction [Eq. (3.23)] and steam gasification [Eq. (3.24)]. The kinetics of these reactions depends on the reactants and catalysts adopted in the SOEC anode (Chen et al., 2019; Xu, Chen, and Zhang, 2018).

$$CH_4 + H_2O \Rightarrow CO + 3H_2 \tag{3.21}$$

$$CH_4 + CO_2 \Rightarrow 2CO + 2H_2 \tag{3.22}$$

$$C + CO_2 \Rightarrow 2CO \tag{3.23}$$

$$C + H_2O \Rightarrow CO + H_2 \tag{3.24}$$

The widely used reaction kinetics and related parameters are listed in Table 3.2.

3.1.1.3 Mass transport process

The rate of mass transport is generally affected by the operating conditions and material properties. In large spaces, it is dominated by free diffusion of molecules. In the porous area, where the pore size is smaller than the molecular mean-free path, Knudsen diffusion becomes significant. The dusty-gas model is used to calculate mass transport of gas species ($N_{i/j}$), as shown in Eq. (3.25) (Suwanwarangkul et al., 2003):

$$\frac{N_i}{D_{ik}^{eff}} + \sum_{j=1,j\neq i}^{n} \frac{y_j N_i - y_i N_j}{D_{ij}^{eff}} = -\frac{1}{RT}\left(\nabla\left(y_j P\right) + \frac{B_0 y_j P}{\mu D_{ik}^{eff}} \nabla P\right) \tag{3.25}$$

where D_{ik}^{eff} is the Knudsen diffusion coefficient, D_{ij}^{eff} is molecular diffusion coefficient, y is the mole fraction of species, B_0 is the permeability, μ is the gas viscosity.

In steady state, the mass conservation can be described by Eq. (3.26),

$$\nabla \cdot N_j = R_j \tag{3.26}$$

where R_j is the mass source term.

Table 3.2 Kinetics of typical solid oxide electrolyser cell reactions.

Reactions (mol m^{-3} s^{-1})
Methane oxidization reaction rate (Jin et al., 2000; Xu, Chen, Tan et al., 2018)
$R = 5 \times 10^4 \times e^{\frac{-166000}{RT}} \times p_{CH_4} \times p_{O_2}$ (anode)
$R = 1.1 \times e^{\frac{-166000}{RT}} \times p_{CH_4} \times p_{O_2}$ (other area)
Hydrogen oxidization reaction rate (Xu, Chen, Tan et al., 2018)
$R = 5 \times 10^4 \times e^{\frac{-48484}{RT}} \times p_{H_2} \times p_{O_2}^{0.5}$
Carbon monoxide oxidization reaction rate (Xu, Chen, Tan et al., 2018)
$R = 5 \times 10^4 \times e^{\frac{-47773}{RT}} \times p_{CO} \times p_{O_2}^{0.5}$
Methane steam reforming reaction rate (Ni, 2012; Ni, Leung, and Leung, 2008)
$R = 2395 \times e^{\frac{-231266}{RT}} \times \left(p_{CH_4} \times p_{H_2O} - \frac{p_{H_2}^3 \times p_{CO}}{K_{pr}}\right)$
Water gas shift reaction rate (Ni, 2012)
$R_{WGS} = 0.0171 \times e^{\frac{-103191}{RT}} \times \left(p_{CO} \times p_{H_2O} - \frac{p_{H_2} \times p_{CO_2}}{K_{ps}}\right)$
Carbon gasification reactions (Huang et al., 2010)
$R_{C_{CO_2}} = \frac{K_1 p_{CO_2}}{1 + K_2 p_{CO} + K_3 p_{CO_2}}$
$R_{C_{H_2O}} = \frac{K_4 p_{H_2O}}{1 + K_5 p_{H_2} + K_6 p_{H_2O}}$
Parameters
$K_{pr} = 1.0267 \times 10^{10} \times e^{-0.2513 \times Z^4 + 0.3665 \times Z^3 + 0.5810 \times Z^2 - 27.134 \times Z + 3.277}$
$K_{ps} = e^{-0.2935 \times Z^3 + 0.6351 \times Z^2 + 4.1788 \times Z + 0.3169}$
$Z = \frac{1000}{T(K)} - 1$
$K_1 = 9.32 \times 10^{-4}$
$K_2 = 1.25 \times 10^{-3}$
$K_3 = 3.82 \times 10^{-5}$
$K_4 = 2.19 \times 10^{-3}$
$K_5 = 9.88 \times 10^{-4}$
$K_6 = 8.13 \times 10^{-5}$

3.1.1.4 Momentum transport process

The general Navier–Stokes equation with Darcy's term is used to calculate the fluid flow, as shown in Eq. (3.27),

$$\rho\frac{\partial u}{\partial t} + \rho u \nabla u = -\nabla p + \nabla\left[\mu\left(\nabla u + (\nabla u)^T\right) - \frac{2}{3}\mu\nabla u\right] - \frac{\varepsilon\mu u}{k} \tag{3.27}$$

where ρ is the gas density and u is the velocity vector.

3.1.1.5 Heat transfer process

General heat balance equation is used to calculate the heat transfer, as shown in Eq. (3.28),

$$\rho C_p u \cdot \nabla T + \nabla \cdot (-\lambda_{eff}\nabla T) = Q \tag{3.28}$$

where C_p is the fluid heat capacity at constant pressure, λ_{eff} is the effective thermal conductivity and Q is the general heat source term. In the porous structure of electrodes, λ_{eff} can be corrected by the porosity as shown in Eq. (3.29):

$$\lambda_{eff} = (1-\varepsilon)\lambda_s + \varepsilon\lambda_g \tag{3.29}$$

where λ_s and λ_g are the heat conductivities of solid and gaseous phases, respectively.

3.2 High-temperature electrolysis for syngas generation

Syngas (the mixture of H_2 and CO) is the feedstock to produce other gaseous/liquid fuels in the modern industry (e.g. F–T synthesis). High-temperature co-electrolysis of H_2O and CO_2 is a promising method to generate syngas. This method directly converts H_2O and CO_2 into H_2 and CO with high efficiency. With the supplement of renewable power, it provides a route to achieve a carbon-neutral society through CO_2 electrolysis. However, the unstable supplement of renewable energy makes electrolysis rely on the electricity generated by conventional thermal power plant that uses fossil fuels as energy source, which declines the overall energy efficiency. To overcome this, fuel-assisted electrolysis (e.g. FA-SOEC) becomes a more practical and eco-friendly choice at the current stage.

Compared with electrolysis process working at high voltage (usually < -1.5 V, negative value means power supply is needed), the fuel-assisted electrolysis can work at a much lower input voltage (< -0.5 V). Because the assistant fuel at anode helps decrease the potential required for OER process as shown in Fig. 3.1.

In some cases, the FA-SOEC can electrolyze oxidants without any electricity consumption ($V = 0$) and even provide power ($V > 0$). To reduce the overall cost in the FA-SOEC, cheap fuels are usually selected as the assistant fuel. Methane-assisted SOEC for H_2O and CO_2 co-electrolysis is an example as given below (Xu, Chen, Irvine et al., 2016).

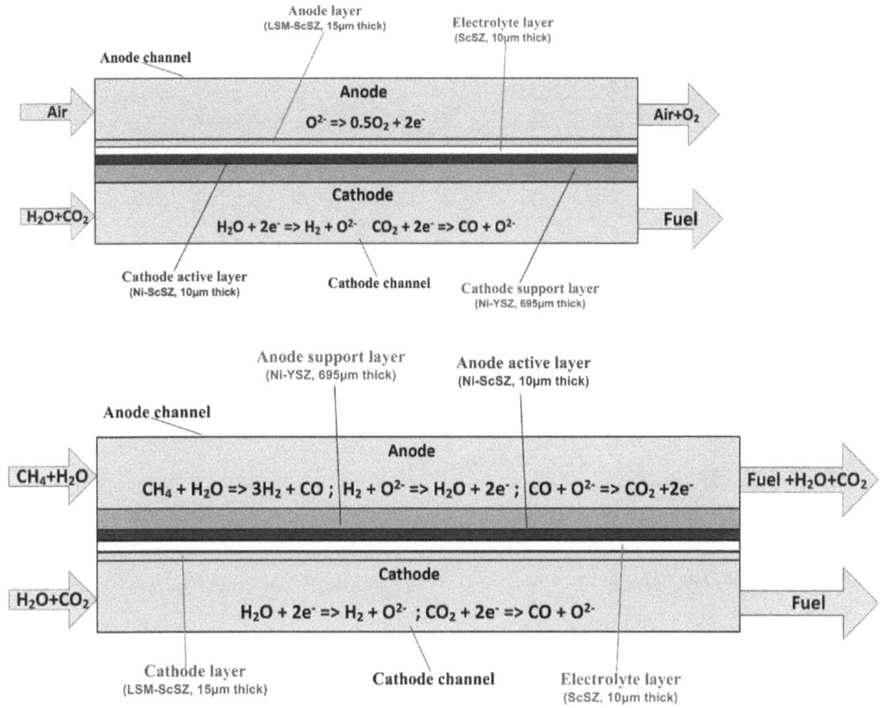

Figure 3.1 Comparison of the working process of SOEC and FA-SOEC, where methane is used as an assistance fuel in the FA-SOEC (Xu, Chen, and Irvine, 2016). *FA-SOEC*, Fuel-assisted solid oxide electrolyser cell; *SOEC*, solid oxide electrolyser cell.

3.2.1 Methane-assisted solid oxide electrolyser cell for H_2O and CO_2 co-electrolysis

Methane is cheap, easily available with high energy density per mass, thus very suitable to be used as the assistant fuel in the SOEC. Fig. 3.2 shows the comparison of I−V characteristics between a conventional SOEC and a CH_4-assisted SOEC for H_2O and CO_2 co-electrolysis at different operating temperatures.

The current density of both conventional SOEC and CH_4-assisted SOEC for CO_2/H_2O co-electrolysis increases with the increasing input voltage. Compared with conventional SOEC, CH_4-assisted SOEC operates at a much lower input voltage (e.g. −0.4 vs −1.2 V) to reach the same current density (e.g. 650 mA cm^{-2}), which means the CH_4-assisted SOEC needs a much smaller overpotential to drive the co-electrolysis process.

Moreover, the current densities of both conventional SOEC and CH_4-assisted SOEC for CO_2/H_2O co-electrolysis increase with the increasing operating temperature. Compared with conventional SOEC co-electrolysis, the performance of CH_4-assisted SOEC in H_2O and CO_2 co-electrolysis has a much higher improvement

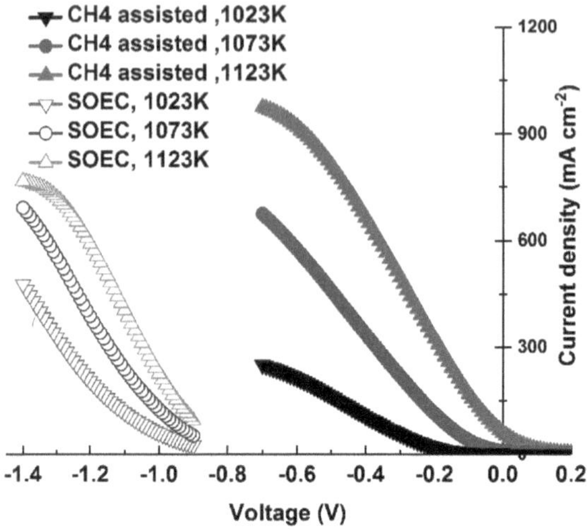

Figure 3.2 Effect of temperature on CH_4-SOEC and SOEC co-electrolysis: (A) at 0.7 and 1.4 V for CH_4-SOEC and SOEC, respectively; (B) at operating temperature of 1023K, 1073K and 1123K for CH_4-SOEC and SOEC (Xu, Chen, Irvine et al., 2016). *SOEC*, Solid oxide electrolyser cell.

with the increase of temperature. This advantage is mainly benefited from the quick increase of SMR rate at higher temperatures, where more CH_4 is converted into H_2 and CO for electrochemical reactions. In addition, a high operating temperature is found to improve the performance of CH_4-assisted SOEC in the whole range of applied voltages (from -0.7 to 0.2 V). It should also be noted that the CH_4-assisted SOEC can simultaneously co-electrolyze H_2O/CO_2 and produce electricity when it works at a high temperature.

3.2.2 Carbon-assisted SOEC for H_2O electrolysis and syngas generation

Solid carbons such as coal and biomass are cheap fuels with high energy density in volume. They are widely distributed all over the world. They are currently the largest energy sources for electricity production in conventional thermal power plants. However, the utilization of solid carbons is criticized for causing global warming and many other environmental problems, such as acid rain and air pollution. Therefore a clean and sustainable method of using solid carbons is urgently needed, such as solid oxide cells.

SOECs are eco-friendly due to their high efficiency and easy treatment of their off-gases. The utilization of solid carbon as the assistant fuel in the SOEC [the so-called carbon-assisted SOEC (CA-SOEC)] has raised great interest. In a CA-SOEC for H_2O electrolysis, solid carbon is introduced into the anode chamber for in situ gasification. With the increase of operating temperature (usually above 700°C), the carbon gasification reaction occurs and gaseous fuel (e.g. CO) is generated. With the gaseous fuel diffusing from carbon layer to the anode, it will be electrochemically oxidized, and the H_2O at cathode will be electrochemically reduced to generate H_2. After electrochemical reactions, the emitted gases from anode (e.g. CO_2) and cathode can be easily collected for the post-processes.

As shown in Fig. 3.3, the CA-SOEC for H_2O electrolysis can work both in fuel cell mode (voltage > 0) and electrolyser cell mode (voltage < 0). When the CA-SOEC is working in the fuel cell mode, it generates fuel and electricity simultaneously and its open-circuit voltage is about 0.7 V. The high current density of CA-SOEC (300 mA cm^{-2}) also indicates that it can generate hydrogen at a high rate even without external electrical power supply.

The concentration of H_2 at cathode outlet is found to be closely related with the applied voltage and operating temperature as shown in Fig. 3.4. At small output voltages (e.g. 0 and 0.2 V), the molar fraction of the H_2 at the cathode outlet increases significantly with increasing temperature, whereas at large output voltages (e.g. 0.4 and 0.6 V), the H_2 molar fraction at cathode outlet only increases slightly with the temperature increasing from 973K to 1273K. This is because the high temperature boosts the kinetics of electrochemical and chemical reactions inside the CA-SOEC, including the electrochemical reduction of H_2O, the electrochemical oxidation of CO and the gasification of solid carbon (Boudouard reaction). Among these kinetic improvements, the electrochemical reaction rates are directly boosted. The high Boudouard reaction rate helps produce more CO, which reduces oxygen partial potential at anode and reduces electricity consumption at required electrochemical reaction rate. Moreover, the high operating temperature increases the

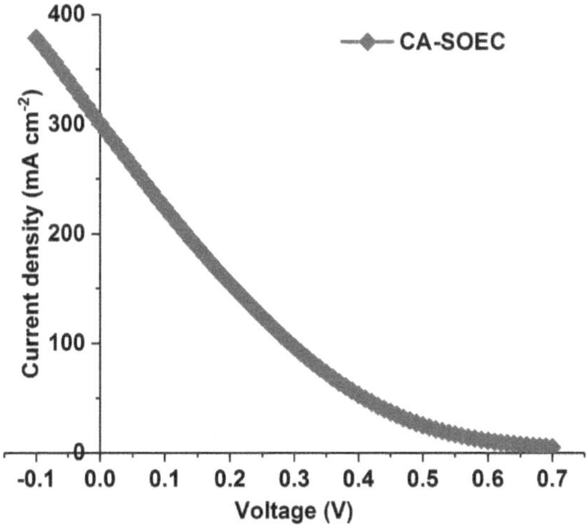

Figure 3.3 The I–V characteristics of the carbon-assisted SOEC (Xu, Chen, and Ni, 2016). *SOEC*, Solid oxide electrolyser cell.

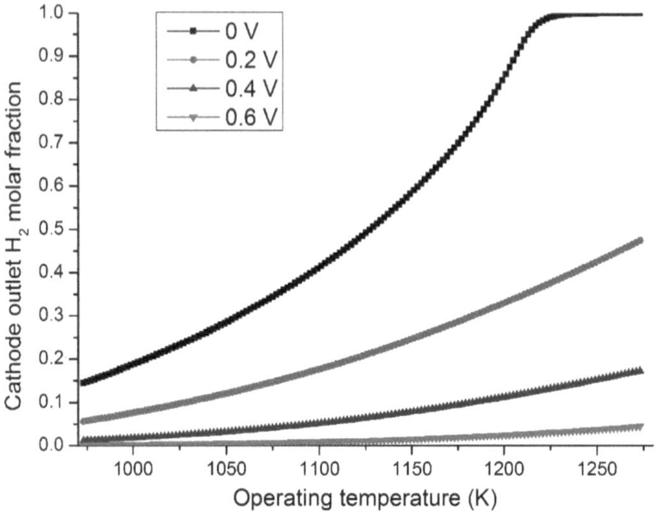

Figure 3.4 Effects of operating temperature on the H_2 mole fraction at cathode outlet at different applied voltages (Xu et al., 2016).

cell's ionic/electronic conductivities, thus reduces the ohmic overpotential in the ion/electron transportation.

Compared with adjusting the operating temperature, the adjustment of applied voltage is more effective in controlling the electrochemical kinetics, because the voltage (potential) is the driving force in the electrolysis process. As can be found in Fig. 3.4, the molar fraction of H_2 at cathode outlet in the 0 V case is much higher than that in other voltage cases. Therefore a negative (or low output) voltage and a high operating temperature are favoured to obtain a high molar concentration of the target fuel at the SOEC cathode.

It is also found that the molar fraction of CO at the anode outlet is very high (> 0.95) at small applied voltages as shown in Fig. 3.5. It indicates that the CO_2, which was electrochemically oxidized from CO at anode, is reduced to CO through carbon gasification reaction. In this way, the CA-SOEC not only generates H_2 in cathode through H_2O electrolysis but also generates CO in anode through Boudouard reaction. And in proper operating conditions, a high concentration of H_2 and CO can be achieved. In a special way, the CA-SOEC realized the generation of syngas while produces H_2 and CO at two different electrodes. It also means we can generate syngas with the desired H_2/CO ratio and overcome the problem in controlling the H_2O/CO_2 composition in conventional SOEC co-electrolysis process. For comparison, in the conventional chemical reactors based on reforming reaction to convert carbon or hydrocarbon fuels for syngas production, both H_2 and CO are produced in the same chamber and it is usually difficult to control the gas composition. To this end, it indicates that CA-SOEC has a great potential for electricity, H_2 and CO tri-generation at high temperatures. This

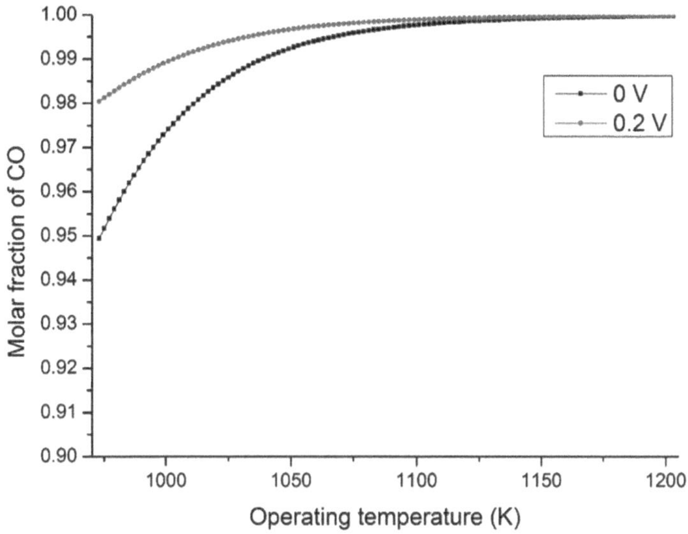

Figure 3.5 Effects of operating temperature on the CO mole fraction at anode outlet at 0 and 0.2 V applied voltages (Xu et al., 2016).

method will greatly facilitate subsequent F–T processes for value-added fuel/chemical generation.

3.3 Combined SOEC and F–T system for low-carbon fuel generation

To reverse global warming and achieve a carbon-neutral society, effective CO_2 capture storage and utilization methods are needed (Krekel et al., 2018). SOECs are promising technologies for CO_2 reduction to generate CO. Moreover, SOECs can co-electrolyse CO_2 and H_2O to generate syngas as have been introduced in above sections, which is the feedstock to generate more valuable hydrocarbons, for example through F–T process (Chen et al., 2016). The generated hydrocarbons vary from C_1 (methane) to C_{5+} in distribution. Usually methane is less wanted for industry (Cao et al., 2018) and it can be used to assist the SOEC to reduce the electricity cost (Xu, Chen, Irvine et al., 2016). In this section, a hybrid system combining the CH_4-assisted SOEC and the F–T reactor is proposed to generate hydrocarbons from CO_2.

3.3.1 Chemical reactions in F–T reactor

F–T synthesis process is the key process to convert the fuels generated from SOECs into different chemicals and products. The product from the F–T synthesis process is affected by the catalyst and operating conditions such as the operating temperature and pressure. For model simplification, the reactions in the F–T process are expressed as given in Eqs. (3.30)–(3.37) (Moazami et al., 2015).

$$CO + 3H_2 \Rightarrow CH_4 + H_2O \tag{3.30}$$

$$2CO + 4H_2 \Rightarrow C_2H_4 + 2H_2O \tag{3.31}$$

$$2CO + 5H_2 \Rightarrow C_2H_6 + 2H_2O \tag{3.32}$$

$$3CO + 7H_2 \Rightarrow C_3H_8 + 3H_2O \tag{3.33}$$

$$4CO + 9H_2 \Rightarrow n - C_4H_{10} + 4H_2O \tag{3.34}$$

$$4CO + 9H_2 \Rightarrow i - C_4H_{10} + 4H_2O \tag{3.35}$$

$$6.05\ CO + 12.23\ H_2 \Rightarrow C_{6.05}H_{12.36} + 6.05\ H_2O \tag{3.36}$$

$$CO + H_2O \Rightarrow CO_2 + H_2 \tag{3.37}$$

The reaction kinetics for these reactions can be calculated in the form of Eq. (3.38) (Rahimpour and Elekaei, 2009).

$$R_i = 0.278 k_i \exp\left(-\frac{E_i}{RT}\right) P_{CO}^m \cdot P_{H_2}^n \tag{3.38}$$

Related parameters can be found in Table 3.3.

Table 3.3 Reaction kinetic parameters for F–T reactor (Rahimpour and Elekaei, 2009).

Reaction number	m	n	k	E
30	−1.0889	1.5662	1.43×10^5	83,423.9
31	0.7622	0.0728	5.16×10^1	65,018.0
32	−0.5645	1.3155	2.47×10^1	49,782.0
33	0.4051	0.6635	4.63×10^{-1}	34855.5
34	0.4728	1.1389	4.74×10^{-3}	27,728.9
35	0.8204	0.5026	8.32×10^{-3}	25,730.1
36	0.5850	0.5982	2.32×10^{-2}	23,564.3
37	0.5742	0.7100	4.11×10^2	58,826.3

3.3.2 Description of the hybrid system

The hybrid system combines an SOEC and an F–T reactor, as shown in Fig. 3.6. In the SOEC, CO_2 and H_2O are introduced to the cathode and CH_4 is introduced to the anode together with H_2O to prevent methane coking. The produced syngas is then introduced to the F–T reactor for hydrocarbons generation.

3.3.3 Results and discussion

For the SOEC sub-model, experimental data conducted by Luo et al. (2014) were adopted, and the same cell structure and material properties are applied in the numerical model. The kinetic parameters of the sub-model are validated by comparing the I–V characteristics of the experimental data with simulation results. A good agreement is achieved, as shown in Fig. 3.7A. In the F–T sub-model, the kinetic model proposed by Rahimpour and Elekaei (2009) was adopted. The comparison between experimental data and numerical results is given in Fig. 3.7B for model validation.

As shown in Figs. 3.8 and 3.9, the proportion of hydrocarbons generated from F–T reactor is very sensitive to the mole fraction of CO_2 at the inlet of the SOEC cathode, because the CO proportion in the syngas generated from the SOEC is affected by the CO_2 concentration. The CO proportion further affects the hydrocarbon proportion as determined by the kinetics of reactions in Eqs. (3.30)–(3.37). Generally, the CO/H_2 ratio from the SOEC is improved with the increase of inlet CO_2 mole fraction. In addition, a high CO_2 mole fraction increases its utilization rate but results in less proportion of hydrocarbons in the

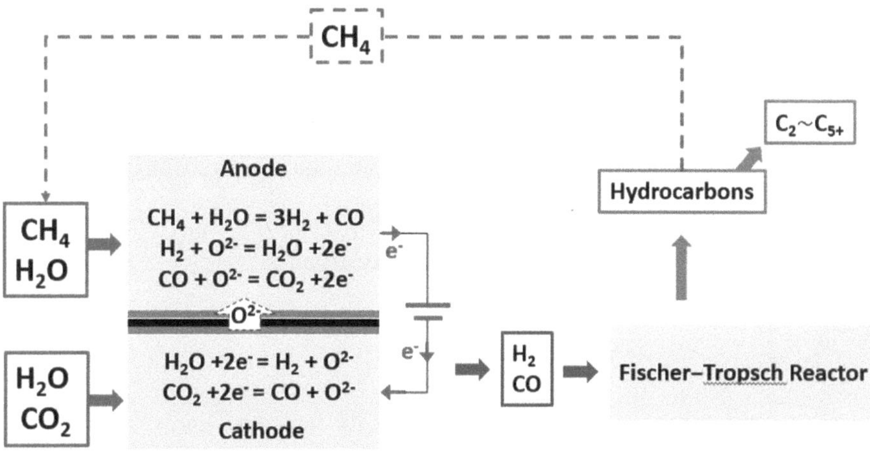

Figure 3.6 Schematic of combined CH_4-assisted SOEC and F–T reactor system (Xu, Maroto-Valer et al., 2019). *F–T*, Fischer–Tropsch; *SOEC*, solid oxide electrolyser cell.

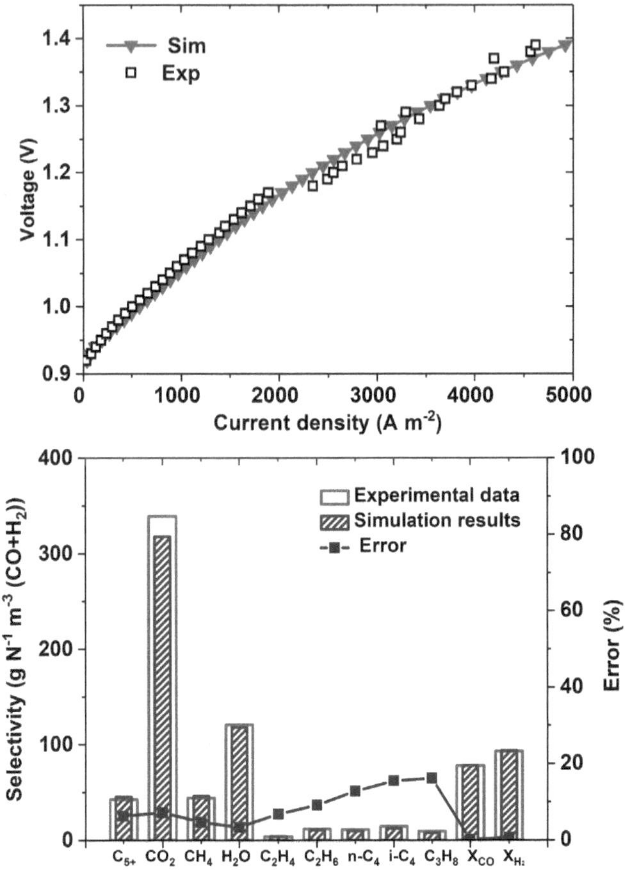

Figure 3.7 Model validation of the kinetics used in the SOEC model and the F–T reactor model (Xu, Maroto-Valer et al., 2019). *F–T*, Fischer–Tropsch; *SOEC*, solid oxide electrolyser cell.

F–T products, because the CO/H_2 ratio has great effects on the methane selectivity and chain-growth probability (Malek Abbaslou, Soltan Mohammadzadeh, and Dalai, 2009).

The distribution of C atoms in carbon fuels (CO plus C_1 to C_{5+}) is given in Fig. 3.10. The proportions of carbon fuels are calculated as:

$$P_c = \frac{C_i}{C_{total}} \tag{3.39}$$

where C_i is the number of C atoms contained in the fuel and C_{total} is the total number of C atoms in all fuels.

Generally, these proportions are significantly affected by the CO_2 mole fraction at the SOEC inlet. Most the C atoms (30%) are in the form of CH_4 at 0.5 CO_2

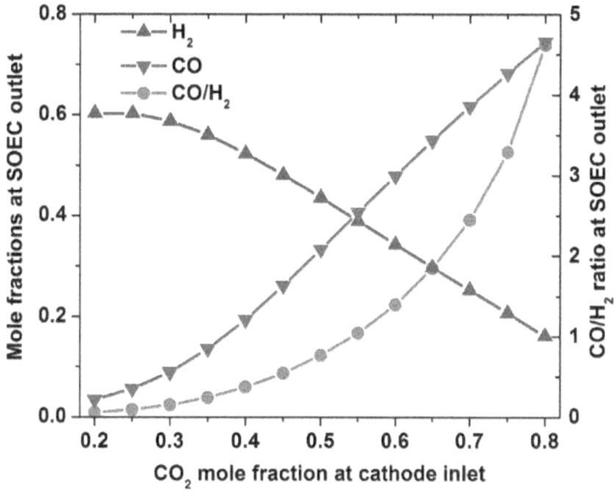

Figure 3.8 Effects of CO_2 mole fraction at cathode inlet on H_2 and CO mole fractions at SOEC outlet (Xu, Maroto-Valer et al., 2019). *SOEC*, Solid oxide electrolyser cell.

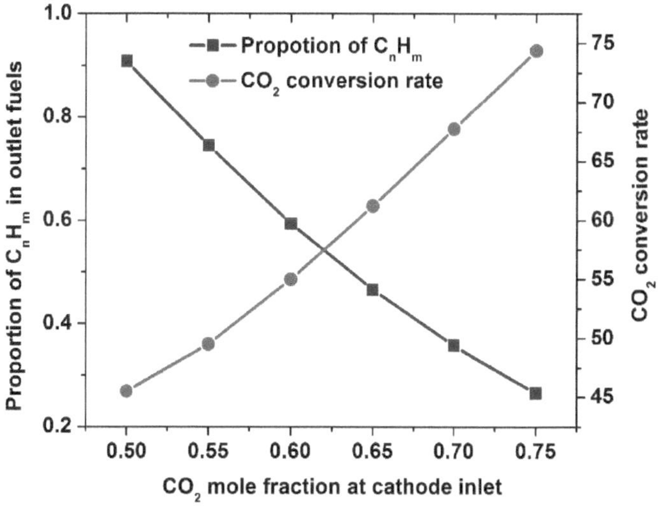

Figure 3.9 Effects of CO_2 mole fractions at cathode inlet on proportion of hydrocarbons in the outlet fuels and CO_2 conversion rate (Xu, Maroto-Valer et al., 2019).

mole fraction case. At high inlet CO_2 case (0.65), the C atoms contained in CH_4 decrease dramatically to only 4%. On the other hand, the proportion of CO increases from 9% to 53% with the inlet CO_2 mole fraction increasing from 0.5 to 0.65. As to the proportion the C_{5+}, it keeps relatively stable at different inlet CO_2 mole fractions.

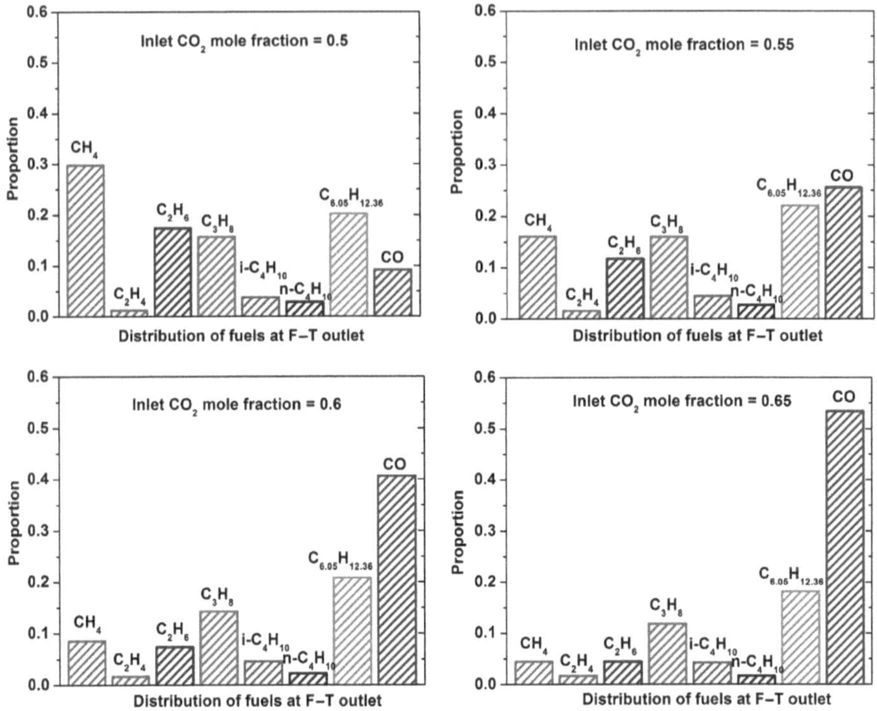

Figure 3.10 Effects of CO_2 mole fractions at cathode inlet on the distribution of fuels at F−T outlet (Xu, Maroto-Valer et al., 2019). *F−T*, Fischer−Tropsch.

3.4 Conclusion

In this chapter, the multi-physics processes of high-temperature electrolysis and co-electrolysis through the SOEC are introduced, including the chemical reactions, electrochemical reaction, mass transport, momentum transport and heat transfer. In addition, the application of SOEC in the hybrid system, where the SOEC is combined with the F−T process, is introduced. Considering the optimization in the SOEC and the hybrid system, the key parameters that affect the performance in their operation are fully discussed, including the temperature, applied voltage, inlet gas flow rate and inlet gas composition. Especially, the fuel-assisted electrolysis process (FA-SOEC) is introduced. The FA-SOEC combines the characteristics of a fuel cell and an electrolyser cell and shows its unique advantage in reducing the electricity consumption for H_2O/CO_2 electrolysis. SOEC provides an alternative solution to solving our energy and environmental problems, including the energy crisis, global warming and air pollution. SOECs and their hybrid systems are promising technologies in achieving a sustainable human civilization.

Nomenclature

Abbreviations	
CA-SOEC	carbon-assisted SOEC
F–T	Fischer-Tropsch
SMR	steam methane reforming
SOEC	solid oxide electrolysis cell
Roman	
$\mathbf{B_0}$	permeability coefficient, m^2
$\mathbf{C_i}$	the number of C atoms contained in the fuel
$\mathbf{C_p}$	fluid heat capacity at constant pressure
$\mathbf{C_{total}}$	total number of C atoms in all fuels
D_{ik}^{eff}	Knudsen diffusion coefficient of i, $m^2\ s^{-1}$
D_{ij}^{eff}	molecular diffusion coefficient of i, $m^2\ s^{-1}$
$\mathbf{E_{act}}$	activation energy, $J\ mol^{-1}$
$\mathbf{E_{co}}$	equilibrium potential for carbon monoxide oxidization, V
E_{CO}^0	standard equilibrium potential for carbon monoxide oxidization, V
E_{H_2}	equilibrium potential for hydrogen oxidization, V
$E_{H_2}^0$	standard equilibrium potential for hydrogen oxidization, V
$\mathbf{E_{eq}}$	equilibrium potential, V
F	Faraday constant, 96,485 $C\ mol^{-1}$
ΔG	Gibbs free energy change, $J\ mol^{-1}$
i	current density, $A\ m^{-2}$
$\mathbf{i_0}$	exchange current density, $A\ m^{-2}$
n	number of electrons transferred per electrochemical reaction
$\mathbf{N_i}$	flux of mass transport, $kg\ m^{-3}\ s^{-1}$
P	pressure, Pa
$\mathbf{P_c}$	proportions of carbon fuels
P_{CO}^L	local partial pressure of carbon monoxide, Pa
$P_{CO_2}^L$	Local partial pressure of carbon dioxide, Pa
$P_{H_2}^L$	local partial pressure of hydrogen, Pa
$P_{H_2O}^L$	local partial pressure of steam, Pa
$P_{O_2}^L$	local partial pressure of oxygen, Pa
Q	heat, $J\ s^{-1}$
R	gas constant, 8.314 $J\ mol^{-1}\ K^{-1}$
$\mathbf{R_i}$	reaction rate of equation i, $mol\ m^{-3}\ s^{-1}$
$\mathbf{R_j}$	mass source term, $mol\ m^{-3}\ s^{-1}$
t	time, s
T	temperature, K
u	velocity field, $m^3\ s^{-1}$
V	voltage, V
$\mathbf{V_l}$	volume fractions of ionic phase
$\mathbf{V_s}$	volume fractions of electronic phase
$\mathbf{y_i}$	molar fraction of component i
Greek letters	
α	charge transfer coefficient
γ	pre-exponential factor
ε	porosity

η	efficiency
η_{act}	activation overpotential, V
η_{ohmic}	ohmic overpotential, V
κ	permeability, m^2
λ	thermal conductivity, W m^{-1} K^{-1}
μ	dynamic viscosity of fluid, Pa s
ρ	fluid density, kg m^{-3}
σ	conductivity, S m^{-1}
$\varnothing$	potential, V
Subscripts	
an	anode
ca	cathode
H_2O/H_2	reaction pairs of H_2O/H_2
CO_2/CO	reaction pairs of CO_2/CO
g	gas phase
l	ionic phase
s	electronic phase
Superscripts	
0	parameter at equilibrium conditions
eff	effective
L	local

References

Achenbach, E., and Riensche, E. (1994). Methane/steam reforming kinetics for solid oxide fuel cells. *Journal of Power Sources*, *52*(2), 283–288. Available from https://doi.org/10.1016/0378-7753(94)02146-5.

Becker, W. L., et al. (2012). Production of Fischer–Tropsch liquid fuels from high temperature solid oxide co-electrolysis units. *Energy*, *47*(1), 99–115. Available from https://doi.org/10.1016/j.energy.2012.08.047.

Cao, P., et al. (2018). Tuning dry reforming of methane for F-T syntheses: A thermodynamic approach. *Applied Energy*, *227*, 190–197. Available from https://doi.org/10.1016/j.apenergy.2017.08.007.

Chen, B., et al. (2016). Modelling of one-step methanation process combining SOECs and Fischer-Tropsch-like reactor. *Journal of the Electrochemical Society*, *163*(11), F3001–F3008. Available from https://doi.org/10.1149/2.0011611jes.

Chen, B., et al. (2019). Combined methane reforming by carbon dioxide and steam in proton conducting solid oxide fuel cells for syngas/power co-generation. *International Journal of Hydrogen Energy*, *44*(29), 15313–15321. Available from https://doi.org/10.1016/j.ijhydene.2019.02.244.

Chen, B., Xu, H., and Ni, M. (2017). *Modelling of SOEC-FT reactor: Pressure effects on methanation process*, . *Applied Energy* (185, pp. 814–824). Available from http://doi.org/10.1016/j.apenergy.2016.10.095.

Chen, L., Chen, F., and Xia, C. (2014). *Direct synthesis of methane from* CO_2-H_2O *co-electrolysis in tubular solid oxide electrolysis cells. Energy and Environmental Science* (7, pp. 4018–4022). Available from http://doi.org/10.1039/C4EE02786H.

Ebbesen, S. D., and Mogensen, M. (2009). Electrolysis of carbon dioxide in solid oxide electrolysis cells. *Journal of Power Sources*, *193*(1), 349–358. Available from http://doi.org/10.1016/j.jpowsour.2009.02.093.

Eguchi, K. (1997). Ceramic materials containing rare earth oxides for solid oxide fuel cell. *Journal of Alloys and Compounds*, *250*(1–2), 486–491. Available from https://doi.org/10.1016/S0925-8388(96)02629-1.

Eguchi, K., et al. (1992). Electrical properties of ceria-based oxides and their application to solid oxide fuel cells. *Solid State Ionics*, *52*(1–3), 165–172. Available from https://doi.org/10.1016/0167-2738(92)90102-U.

Fan, B., Yan, J., and Yan, X. (2011). *The ionic conductivity, thermal expansion behavior, and chemical compatibility of La0.54Sr0.44Co0.2Fe 0.8O3-δ as SOFC cathode material, . Solid State Sciences* (13, pp. 1835–1839). Available from https://doi.org/10.1016/j.solidstatesciences.2011.07.007.

Giddey, S., et al. (2012). A comprehensive review of direct carbon fuel cell technology. *Progress in Energy and Combustion Science*, *38*(3), 360–399. Available from https://doi.org/10.1016/j.pecs.2012.01.003.

Giddey, S., Kulkarni, A., and Badwal, S. P. S. (2015). Low emission hydrogen generation through carbon assisted electrolysis. *International Journal of Hydrogen Energy*, *40*(1), 70–74. Available from https://doi.org/10.1016/j.ijhydene.2014.11.033.

Guo, Y., et al. (2013). An all porous solid oxide fuel cell (SOFC): A bridging technology between dual and single chamber SOFCs. *Energy and Environmental Science*, *6*(7), 2119–2123. Available from https://doi.org/10.1039/c3ee40131f.

Huang, Z., et al. (2010). Kinetic studies of char gasification by steam and CO2 in the presence of H2 and CO. *Fuel Processing Technology*, *91*(8), 843–847. Available from https://doi.org/10.1016/j.fuproc.2009.12.020.

Jin, W., et al. (2000). Experimental and simulation study on a catalyst packed tubular dense membrane reactor for partial oxidation of methane to syngas. *Chemical Engineering Science*, *55*(14), 2617–2625. Available from https://doi.org/10.1016/S0009-2509(99)00542-4.

Kang, S., and Ahn, K.-Y. (2017). Dynamic modeling of solid oxide fuel cell and engine hybrid system for distributed power generation. *Applied Energy*, *195*, 1086–1099. Available from https://doi.org/10.1016/j.apenergy.2017.03.077.

Krekel, D., et al. (2018). *The separation of CO2 from ambient air – A techno-economic assessment. Applied Energy* (218, pp. 361–381). Available from http://doi.org/10.1016/J.APENERGY.2018.02.144.

Laguna-Bercero, M. A. (2012). Recent advances in high temperature electrolysis using solid oxide fuel cells: A review. *Journal of Power Sources*, *203*, 4–16. Available from https://doi.org/10.1016/j.jpowsour.2011.12.019.

Luo, Y., Shi, Y., Li, W., and Cai, N. (2014). *Comprehensive modeling of tubular solid oxide electrolysis cell for co-electrolysis of steam and carbon dioxide. Energy* (70, pp. 420–434). Available from http://doi.org/10.1016/j.energy.2014.04.019.

Luo, Y., Shi, Y., Li, W., Ni, M., et al. (2014). Elementary reaction modeling and experimental characterization of solid oxide fuel-assisted steam electrolysis cells. *International Journal of Hydrogen Energy*, *39*(20), 10359–10373. Available from https://doi.org/10.1016/j.ijhydene.2014.05.018.

Malek Abbaslou, R. M., Soltan Mohammadzadeh, J. S., and Dalai, A. K. (2009). Review on Fischer–Tropsch synthesis in supercritical media. *Fuel Processing Technology*, *90*(7), 849–856. Available from https://doi.org/10.1016/j.fuproc.2009.03.018.

Minh, N. Q. (2004). Solid oxide fuel cell technology—Features and applications. *Solid State Ionics*, *174*(1), 271–277. Available from https://doi.org/10.1016/j.ssi.2004.07.042.

Moazami, N., et al. (2015). Catalytic performance of cobalt–silica catalyst for Fischer–Tropsch synthesis: Effects of reaction rates on efficiency of liquid synthesis. *Chemical Engineering Science*, *134*, 374–384. Available from https://doi.org/10.1016/j.ces.2015.05.025.

Ni, M. (2012). 2D thermal modeling of a solid oxide electrolyzer cell (SOEC) for syngas production by H2O/CO2 co-electrolysis. *International Journal of Hydrogen Energy*, *37*(8), 6389–6399. Available from https://doi.org/10.1016/j.ijhydene.2012.01.072.

Ni, M., Leung, D. Y. C., and Leung, M. K. H. (2008). Modeling of methane fed solid oxide fuel cells: Comparison between proton conducting electrolyte and oxygen ion conducting electrolyte. *Journal of Power Sources*, *183*(1), 133–142. Available from https://doi.org/10.1016/j.jpowsour.2008.04.073.

Ormerod, R. M. (2003). *Solid oxide fuel cells. Chemical Society Reviews* (32, pp. 17–28). Available from http://doi.org/10.1039/B105764M.

Rahimpour, M. R., and Elekaei, H. (2009). A comparative study of combination of Fischer–Tropsch synthesis reactors with hydrogen-permselective membrane in GTL technology. *Fuel Processing Technology*, *90*(6), 747–761. Available from https://doi.org/10.1016/j.fuproc.2009.02.011.

Singhal, S. C. (2000). Advances in solid oxide fuel cell technology. *Solid State Ionics*, *135* (1), 305–313. Available from https://doi.org/10.1016/S0167-2738(00)00452-5.

Smith, R. J. B., Loganathan, M., and Shantha, M. S. (2010). A review of the water gas shift reaction kinetics. *International Journal of Chemical Reactor Engineering*, 8.

Stempien, J. P., et al. (2014). Physical principles for the calculation of equilibrium potential for co-electrolysis of steam and carbon dioxide in a Solid Oxide Electrolyzer Cell (SOEC). *Electrochimica acta*, *147*, 490–497. Available from https://doi.org/10.1016/j.electacta.2014.09.144.

Stempien, J. P., et al. (2015). Production of sustainable methane from renewable energy and captured carbon dioxide with the use of Solid Oxide Electrolyzer: A thermodynamic assessment. *Energy*, *82*, 714–721. Available from https://doi.org/10.1016/j.energy.2015.01.081.

Suwanwarangkul, R., et al. (2003). Performance comparison of Fick's, dusty-gas and Stefan–Maxwell models to predict the concentration overpotential of a SOFC anode. *Journal of Power Sources*, *122*(1), 9–18. Available from https://doi.org/10.1016/S0378-7753(02)00724-3.

Tai, L. W., et al. (1995). Structure and electrical properties of La1-xSrxCo1-yFeyO3. Part 2. The system La1-xSrxCo0.2Fe0.8O3. *Solid State Ionics*, *76*(3–4), 273–283. Available from https://doi.org/10.1016/0167-2738(94)00245-N.

Tao, G., Butler, B., and Virkar, A. (2011). Hydrogen and power by fuel-assisted electrolysis using solid oxide fuel cells. *ECS Transactions*, *35*(1), 2929–2939. Available from https://doi.org/10.1149/1.3570292.

Wang, Y., et al. (2009). Performance and effective kinetic models of methane steam reforming over Ni/YSZ anode of planar SOFC. *International Journal of Hydrogen Energy*, *34* (9), 3885–3893. Available from https://doi.org/10.1016/j.ijhydene.2009.02.073.

Wang, Y., et al. (2017). *High temperature solid oxide* H_2O/CO_2 *co-electrolysis for syngas production*, . *Fuel Processing Technology* (161, pp. 248–258). Available from http://doi.org/10.1016/j.fuproc.2016.08.009.

Xiao, J., et al. (2014). Deactivation of nickel-based anode in solid oxide fuel cells operated on carbon-containing fuels. *Journal of Power Sources*, *268*, 508–516. Available from https://doi.org/10.1016/j.jpowsour.2014.06.082.

Xu, H., et al. (2017). Modeling of direct carbon solid oxide fuel cells with H2O and CO2 as gasification agents. *International Journal of Hydrogen Energy*, *42*(23), 15641–15651. Available from https://doi.org/10.1016/j.ijhydene.2017.05.075.

Xu, H., Chen, B., et al. (2019). The thermal effects of all porous solid oxide fuel cells. *Journal of Power Sources*, *440*227102. Available from https://doi.org/10.1016/j.jpowsour.2019.227102.

Xu, H., Chen, B., Irvine, J., et al. (2016). Modeling of CH4-assisted SOEC for H2O/CO2 co-electrolysis. *International Journal of Hydrogen Energy*, *41*(47), 21839–21849. Available from https://doi.org/10.1016/j.ijhydene.2016.10.026.

Xu, H., Chen, B., Liu, J., et al. (2016). Modeling of direct carbon solid oxide fuel cell for CO and electricity cogeneration. *Applied Energy*, *178*, 353–362. Available from https://doi.org/10.1016/j.apenergy.2016.06.064.

Xu, H., Chen, B., and Ni, M. (2016). Modeling of direct carbon-assisted solid oxide electrolysis cell (SOEC) for syngas production at two different electrodes. *Journal of the Electrochemical Society*, *163*(11), F3029–F3035. Available from https://doi.org/10.1149/2.0041611jes.

Xu, H., Chen, B., Tan, P., et al. (2018). Modeling of all porous solid oxide fuel cells. *Applied Energy*, *219*, 105–113. Available from https://doi.org/10.1016/j.apenergy.2018.03.037.

Xu, H., Chen, B., Zhang, H., et al. (2018). Experimental and modeling study of high performance direct carbon solid oxide fuel cell with in situ catalytic steam-carbon gasification reaction. *Journal of Power Sources*, *382*, 135–143. Available from https://doi.org/10.1016/j.jpowsour.2018.02.033.

Xu, H., Maroto-Valer, M. M., et al. (2019). Low carbon fuel production from combined solid oxide CO2 co-electrolysis and Fischer-Tropsch synthesis system: A modelling study. *Applied Energy*, *242*, 911–918. Available from https://doi.org/10.1016/j.apenergy.2019.03.145.

Yano, M., et al. (2007). Recent advances in single-chamber solid oxide fuel cells: A review. *Solid State Ionics*, *177*(39–40), 3351–3359. Available from https://doi.org/10.1016/j.ssi.2006.10.014.

Power to methane

4

Eleonora Bargiacchi[1], *Daniele Candelaresi*[2] *and Giuseppe Spazzafumo*[2]
[1]Department of Energy, Systems, Territory and Constructions Engineering, University of Pisa, Pisa, Italy, [2]Department of Civil and Mechanical Engineering, University of Cassino and Southern Lazio, Cassino, Italy

Injecting synthetic methane into existing natural gas pipelines is probably the closest solution to obtain a vast decarbonisation of the energy system. As shown in Fig. 4.1, this solution is suitable to store large quantities of energy and for a long time (hourly to seasonal storage). Therefore it allows to solve the problem of storing a large amount of energy, which is the main obstacle to a greater diffusion of systems based on renewable sources.

The production of synthetic methane involves the methanation of a carbon source using hydrogen. Methanation is an exothermic process that combines hydrogen and carbon oxides generating methane and water, according to reactions (4.1) and (4.2):

$$CO + 3H_2 \Rightarrow CH_4 + H_2O \quad \Delta H_{298K} = -206.1\ \text{kJ mol}^{-1} \tag{4.1}$$

$$CO_2 + 4H_2 \Rightarrow CH_4 + 2H_2O \quad \Delta H_{298K} = -165.1\ \text{kJ mol}^{-1} \tag{4.2}$$

This process has already been used downstream of coal gasification to obtain a gas with a calorific value higher than that of the syngas produced. In fact, hydrogen and carbon monoxide, which are the main components of the syngas produced by coal gasification, have a net combustion enthalpy of 241.8 and 283 kJ mol^{-1} respectively, whereas for methane it is 802.6 kJ mol^{-1} (Perry and Green, 1998).

Actually, reaction (4.2), called Sabatier reaction and discovered in 1902 (Sabatier and Senderens, 1902), proceeds indirectly through the reverse water–gas shift reaction which reduces CO_2 (Stangeland et al., 2017; Ullmann, 1989):

$$CO_2 + H_2 \Rightarrow CO + H_2O \quad \Delta H_{298K} = +41.2\ \text{kJ mol}^{-1} \tag{4.3}$$

With a view to a profound decarbonisation of the energy and industrial systems, power plants, refineries, steel industry and cement factories are enormous sources of carbon, even too large for current power to methane technologies. Furthermore, the gases emitted by the aforementioned sectors, although characterised by a high concentration of carbon dioxide, also contain substances that could harm the methanation process, for example sulphur, which can deactivate the catalyst.

Power to Fuel. DOI: https://doi.org/10.1016/B978-0-12-822813-5.00001-1

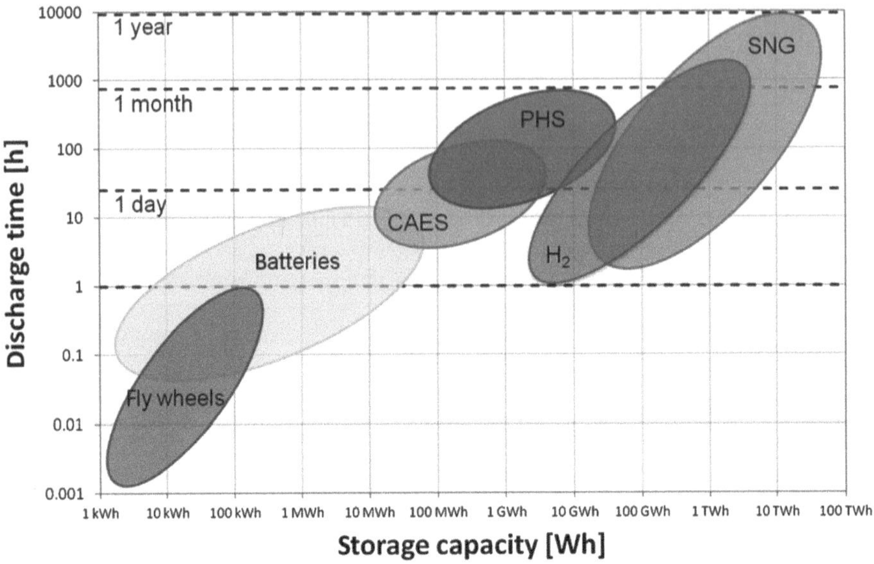

Figure 4.1 Comparison among different energy storage systems. CAES, compressed air energy storage; PHS, pumped hydro storage; SNG, substitute natural gas (Schaaf, Grünig et al., 2014).

Other carbon dioxide sources are represented by biomass combustion and syngas and biogas from various organic sources (e.g. from biomass gasification, waste treatment or anaerobic digestion plants), although potentially available in smaller amount if compared to industrial point sources (see Table 4.1).

The syngas produced by gasification of lignocellulosic biomass mainly contains carbon monoxide and hydrogen, usually without particular harmful substances. The presence of monoxide instead of dioxide can be very advantageous: reaction (4.1) is more direct and production of a mole of methane requires only 3 moles of hydrogen. Furthermore, it is possible to gasify biomass using directly hydrogen (hydro-gasification).

Unlike syngas, biogas is mainly made up of methane (50%–70%) and carbon dioxide (30%–50%) and may contain small amounts of sulphur and silicon compounds (Köppel, Götz, and Graf, 2009; Yentekakis and Goula, 2017). Such a biogas can be used directly, possibly after purification, for methanation, or CO_2 can be used downstream of the upgrading process from biogas to biomethane.

Whatever the carbon source is, after water vapour condensation, the resulting gas from the process could be methane containing only a small amount of carbon oxides and hydrogen. Such a gas is usually called substitute of natural gas (SNG). Often it is also referred to as synthetic natural gas, but we personally do not like this second option because synthetic and natural have opposite meanings. From a technical point of view, SNG could be distributed and used to replace natural gas using the same infrastructures and the same energy conversion devices without particular adjustments. This is because its composition easily falls within the composition limits of a natural gas (see Table 4.2). However, in general, the technical specifications required

Table 4.1 Typical volumetric flows and required power for electrolysis of different carbon sources (Götz et al., 2016).

CO_2 source		Biogas plant	Biomass gasification	Industrial processes	Power plant
Typical feed gas ($m^3\ h^{-1}$) (STP), dry	CO_2	500	2100	30,000	300,000
	CO	-	2500	-	-
	H_2	-	4000	-	-
	CH_4	500	1100	-	-
Additional H_2 from electrolysis ($m^3\ h^{-1}$)		2000	11,900	120,000	1,200,000
Power demand for electrolysis (MW) (5 kWh m^{-3} H_2)		10	59.5	600	6000
Produced CH_4 (total) ($m^3\ h^{-1}$)		1000	5700	30,000	300,000
Produced CH_4 (total) (MW)		11	63	332	3320

STP, Standard temperature and pressure.

Table 4.2 Typical composition of natural gas (Uniongas, 2020).

Component	Range (mole %)
Methane	87.0–98.0
Ethane	1.5–9.0
Propane	0.1–1.5
Iso-butane	Trace–0.3
Normal-butane	Trace–0.3
Iso-pentane	Trace–0.04
Normal-pentane	Trace–0.04
Hexanes plus	Trace–0.06
Nitrogen	0.2–5.5
Carbon dioxide	0.05–1.0
Oxygen	Trace–0.1
Hydrogen	Trace–0.05
Sulphur	Trace–0.002
Properties	Range (MJ m^{-3})
Gross heating value, dry basis	36.0–40.2
Wobbe number	47.5–51.5

for the injection of SNG are determined by the legislation of each country (Thema et al., 2019).

Two different methanation routes are possible: chemical or biological.

4.1 Chemical route

Although from a thermodynamic point of view the methanation process is favoured by low temperatures, these hinder the process from a kinetic point of view. In fact, methanation reactions are characterised by a high activation energy and therefore at low temperatures, the number of reacting particles that have an adequate kinetic energy value is very low and the reaction proceeds very slowly. It is therefore essential to use catalysts that are capable of lowering the reaction activation energy to keep the temperature low and favour the carbon conversion to methane rather than the opposite reaction (methane reforming).

4.1.1 Catalysts

The catalysts used in the methanation process are solid (heterogeneous catalysis) and the catalyst is generally deposited on a porous support which has a triple purpose:

- increase the active surface;
- guarantee mechanical resistance;
- prevent the dispersion of the catalyst.

The catalyst choice must take into account two fundamental properties: selectivity and activity, which together determine how effective a catalyst is in promoting a certain reaction without favouring competing reactions. The catalysts that can be used for hydrogenation reactions are, of decreasing activity, ruthenium, nickel, rhodium, palladium and platinum (Gao et al., 2015). Obviously, the choice must also take into account the economic factor and Table 4.3 shows the current (18 November 2020) price of the mentioned metals.

Nickel is the most used catalyst because, in addition to being the least expensive and showing a good activity, it is also endowed with a remarkable selectivity towards the methanation reaction (Martínez et al., 2019).

It is also important to define the shape, size and degree of porosity, characteristics that are related to the type of reactor.

The typical unsupported catalyst is Raney nickel, which is produced by treating a nickel–aluminium alloy with concentrated sodium hydroxide. The hydroxide

Table 4.3 Cost of metals used as catalysts (Infomine, 2020).

	Pt	Rh	Pd	Ru	Ni
€ per kg	28 €/kg	441,500	70,519	8047	13.27

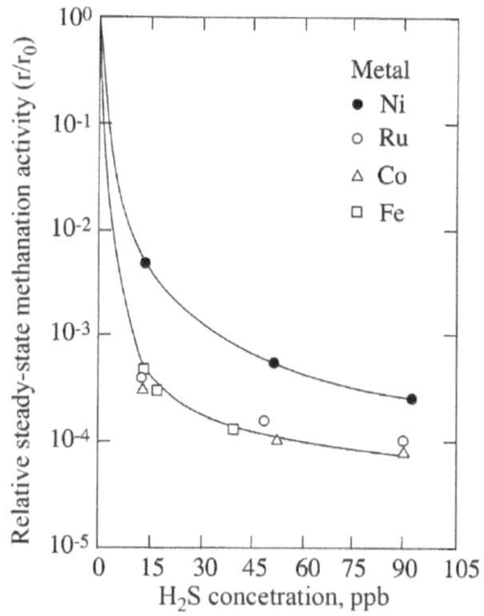

Figure 4.2 Effect of H_2S concentration on the catalytic activity of Ni and other metals. Reaction conditions: 100 kPa, 400°C; 1% CO + 99% H_2 for Co, Fe and Ru; 4% CO + 96% H_2 for Ni (Bartholomew, 2001).

causes the dissolution of aluminium so that the residual copper takes on the appearance of a metal sponge with an extremely fine porous structure and, therefore, with a high specific surface.

In the case of supported catalyst, however, it is deposited, often in the form of nanoparticles, on porous substrates of alumina (Al_2O_3), silica (SiO_2) or zeolites, minerals consisting of aluminium and silicon oxides.

Although the catalyst does not participate in the overall reaction, its activity tends to decrease over time with a consequent reduction in the reaction rate. This fact involves the need for catalyst periodic replacement and, consequently, an additional expense that further strengthens the choice of nickel.

However, nickel is particularly subject to poisoning by sulphur compounds. These are obviously not present if pure carbon dioxide is used, but they can be present in case of methanation of a biogas or a syngas. Fig. 4.2 shows how the reaction rate decreases as the presence of sulphur increases. Therefore it is necessary that the feed gas has an overall sulphur content of less than 1 ppm (Bartholomew, 2001; Bartholomew, Agrawal, and Katzer, 1982). Fixed-bed reactors, due to the high temperatures reached, also suffer from the presence of hydrocarbons heavier than methane which can decompose forming coke deposits which deactivate the catalyst (Bartholomew, 1982; Gao et al., 2012).

4.1.2 Operational parameters

Besides the low tolerance against the presence of those impurities, a problem could consist in temperature fluctuations when the process is subject to variability in feed flows, since the latter can cause a drift or a cooling down of the reaction

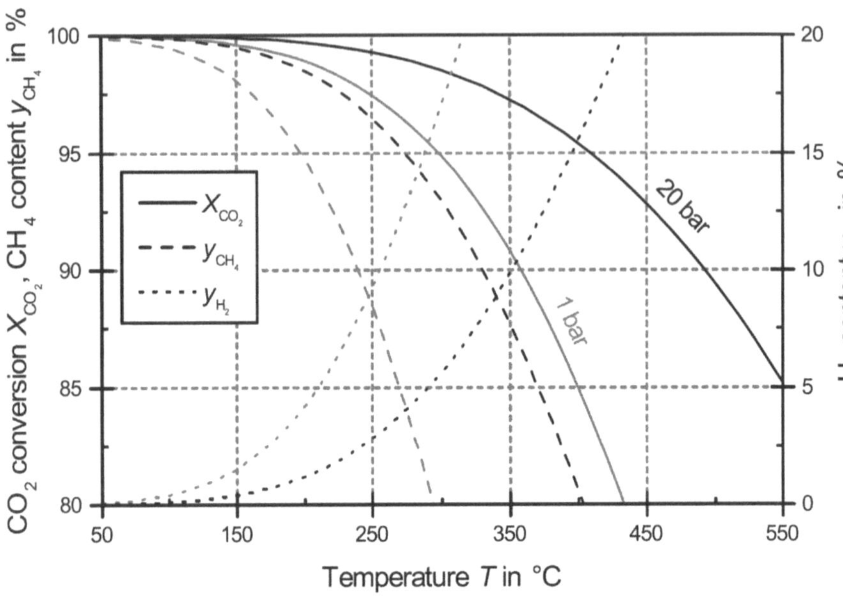

Figure 4.3 CO_2 methanation: equilibrium conversion and H_2 and CH_4 content for stoichiometric feeding (Götz et al., 2016).

(Bailera et al., 2017; Götz et al., 2016). However, Lefebvre et al. (2015) have shown that the load changes limiting factor is more related to the process control system than to the process itself.

Another parameter that affects the reaction progress is pressure. Since both reactions (4.1) and (4.2) take place with a reduction of molecules, a high pressure is usually adopted. Fig. 4.3 shows the influence of temperature and pressure on the equilibrium conversion of a stoichiometric mixture of hydrogen and carbon dioxide (Götz et al., 2016).

Optimal values of temperature and pressure are respectively in the range 250°C – 400°C and around 30 bar (Foscolo, Gallucci, and Micheli, 2011; Gao et al., 2012).

Finally, the composition of the supplied gas has, within certain limits, considerable importance. Fig. 4.4 shows the methanation yield versus reaction temperature, when the ratio between carbon dioxide and hydrogen in the mixture fed to the reactor is equal to 1:3, 1:4 and 1:5 (Assettati et al., 2010). It is interesting to observe that a temperature around 200°C is needed to start the reaction and that above 300°C the yield starts to drop due to competition from the methane reforming reaction. The stoichiometric ratio between carbon dioxide and hydrogen is 1:4 [see reaction (4.2)] so that for a supply ratio of 1:3 the maximum yield is lower than 80%. At around 300°C there is no significant difference between the yield obtained with a ratio of 1:4 and a ratio of 1:5, whereas for higher temperatures the presence of over-stoichiometric hydrogen slows down the methane reforming reaction and the yield is kept higher.

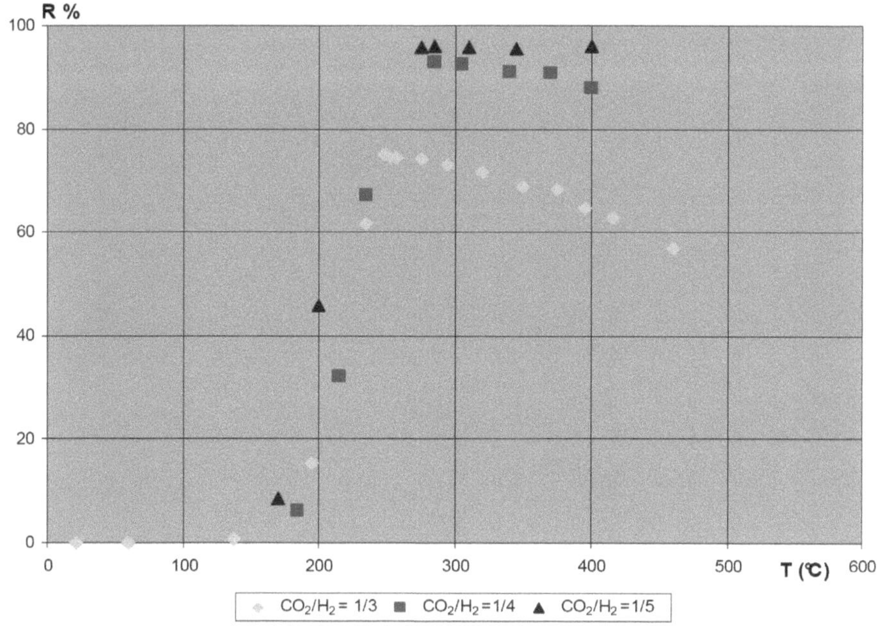

Figure 4.4 Methanation yield versus temperature and CO_2/H_2 supply ratio.
Source: Adapted from Assettati, A. et al. (2010) Sperimentazione di sistemi di trattamento e conversione del syngas. Rome: ENEA.

Being the process exothermic, a very important aspect is heat removal during the process to keep the temperature within its optimal range. With reference to a stoichiometric mixture of carbon dioxide and hydrogen, the heat released by a cubic metre of mixture in standard condition is about 1473 kJ and therefore the heat flux to be removed depends on the gas flow fed to the reactor. A parameter used to normalise the gas flow to the reactor size is the gas hourly space velocity (GHSV):

$$GHSV = \frac{F_{V,g,in}}{V_R} \tag{4.4}$$

where $F_{V,g,in}$ is the volumetric flow in standard condition of the CO_2 and H_2 stoichiometric mixture, and V_R is the reactor volume.

4.1.3 Reactor structures

The existing methanation reactors can be grouped into the following types:

- fixed bed
- fluidised bed
- three-phase
- structured

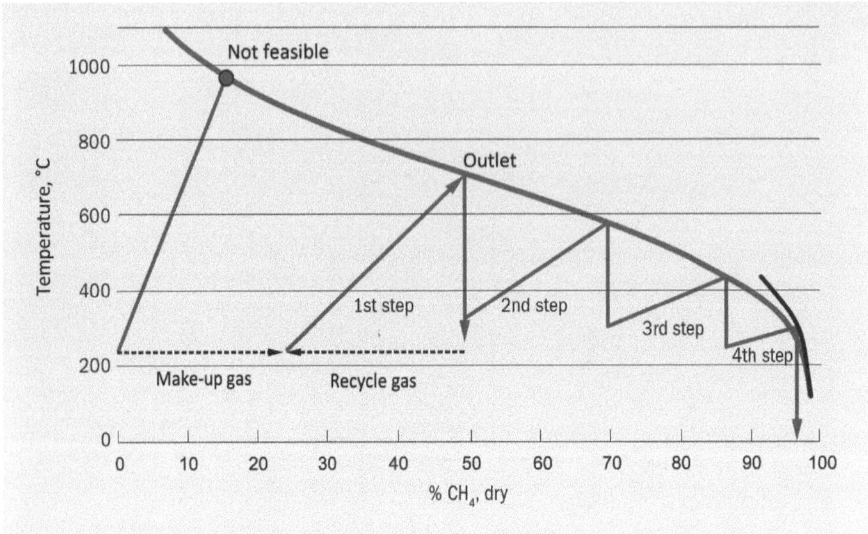

Figure 4.5 Equilibrium curve for methanation process for a specific temperature and pressure (Jensen et al. 2011).

When an adiabatic fixed-bed reactor is used, the temperature increases; this is favourable to the process kinetic since it reduces the need for catalyst but is unfavourable to reaction equilibrium. Moreover, the catalyst is subjected to high temperatures, higher than 700°C, with the risk of cracking or sintering. To obtain a high degree of conversion, a series of reactors with intermediate cooling is required. Due to the larger amount of converted carbon oxides, the first reactor is the one where the most critical operating conditions are reached. For this reason, a partial recycle of its products may be appropriate. Fig. 4.5 shows that, by feeding an adiabatic reactor with a methane-free mixture, the reactor temperature reaches very high values that may be harmful to the catalyst and unfavourable to the chemical equilibrium (Jensen, Poulsen, and Andersen, 2011). By a partial recycle of the reactor products at the inlet, a positive effect is obtained, despite the initial presence of a greater quantity of products which could inhibit the direct reaction. Indeed, the addition of the recycle lowers the reaction temperature, which causes a more favourable effect than the unfavourable effect caused by a higher percentage of methane in the feed. In this way, an almost complete carbon conversion can be achieved with three to four adiabatic reactors.

Such a concept was applied, for example, in the Topsøe recycle energy-efficient methanation process.

To avoid the complexity of a multistage process, a single cooled fixed-bed reactor can be used, from which heat is extracted continuously so that the temperature is maintained at the optimal value. The typical layout of this kind of reactor is a vessel containing several tubes filled with catalyst. The gas flows inside the tubes, whereas the cooling fluid (water, oil or air) flows outside of the tubes. In this case, a single reactor allows to obtain a high rate of conversion. On the other hand, the heat is extracted at a lower temperature and this reduces the heat recovery efficiency, unless there is the need for low-temperature heat (Bailera et al., 2020; Eigenberger, 1992; Kopyscinski, Schildhauer, and Biollaz, 2010).

A technologically mature alternative to fixed-bed reactors is the fluidised-bed reactor. In this type of reactor, the speed of gases and the size of catalyst particle are such to give the whole the appearance of a fluid. In these conditions, the heat transfer is favoured and the bed temperature is almost uniform, which makes the heat removal simpler than in the fixed-bed reactor (Kopyscinski et al. 2010; Kopyscinski, Schildhauer, and Biollaz, 2011). The disadvantages consist in the need of controlling the gas speed (which must be high enough to maintain fluidisation, but not too high to avoid elutriation of the catalyst), and in the risk of catalyst deactivation due to friction generated by fluidisation (Bartholomew, 2001).

The other two types of reactor are the result of more recent concepts which are still under development.

The three-phase reactor provides for the suspension of fine catalyst particles in a heat-carrying liquid, generally an oil, which gives the mass a high thermal capacity (Held et al., 2020). Therefore as in the case of the fluidised-bed reactor, the heat developed by the reaction can be removed more easily, making the temperature control easier. The main drawbacks of this reactor architecture are the liquid–gas mass transfer resistance and the possibility of liquid decomposition and evaporation (Lefebvre et al., 2015).

Structured reactors are monolithic reactors, that is consisting of a metal mass crossed by channels on whose internal surface the catalyst is deposited (Eigenberger, 1992). Thanks to this metal structure, heat is transferred more rapidly in the radial direction up to two to three orders of magnitude compared to conventional fixed-bed reactors, reducing the possibility of the formation of hot spots (Götz et al., 2016; Held et al., 2020). In case of micro-structured reactors, the pressure drop is also reduced. However, the catalyst deposition inside channels and the replacement of deactivated catalyst is very complex (Liu et al., 2012).

A significant element in terms of cost is the hydrogen storage. Even assuming that there is a constant flow of the carbon donor gas, using the electricity surplus for the production of hydrogen, the availability of the latter varies over time. Therefore to operate the methanation reactor in stationary conditions, it is necessary to have a sufficiently large hydrogen storage, with a consequent increase in both the system complexity and costs.

A methanation reactor that is capable of operating dynamically can eliminate or at least reduce the need for hydrogen storage. On the other hand, this involves critical aspects due to temperature variability when the gas flow varies. In fact, strong temperature changes increase the risk of catalyst cracking or sintering (Bartholomew, 2001). However, several researchers asserted that a change in the gas flow rate is rapidly absorbed also by a catalytic system (González and Schaub, 2015; Götz et al., 2016; Lefebvre et al., 2015). Furthermore, the liquid mass present in the three-phase reactor is able to moderate the temperature variations resulting from load variations. Contrarily, fixed-bed reactors are slower in responding to dynamic fluctuations compared to other reactor types due to their lower temperature control capability, therefore they are less suitable for dynamic operation. However, some authors remark that the catalyst is not significantly affected by load variations (Götz et al., 2016).

While for a fixed-bed reactor a minimum load of 40% is suggested, gas flow rate variations ranging between 25% and 100% were proved to be tolerated by a three-phase reactor (Lefebvre et al., 2015). If the hydrogen flow is insufficient to achieve the minimum acceptable conditions for the reactor operation, it is necessary to switch to standby mode. However, the catalyst tends to oxidise rapidly if exposed to a CO_2 atmosphere (Mutz et al., 2015), which is why it would be preferable for the reactor to contain hydrogen during the standby phase. This implies the availability of a storage system for the carbon donor gas.

4.2 Biological route

During a biomass anaerobic digestion process, there are four subsequent conversion stages performed by different microorganisms (Fig. 4.6): hydrolysis, acidogenesis, acetogenesis and methanogenesis. During hydrolysis, the biomass, consisting of

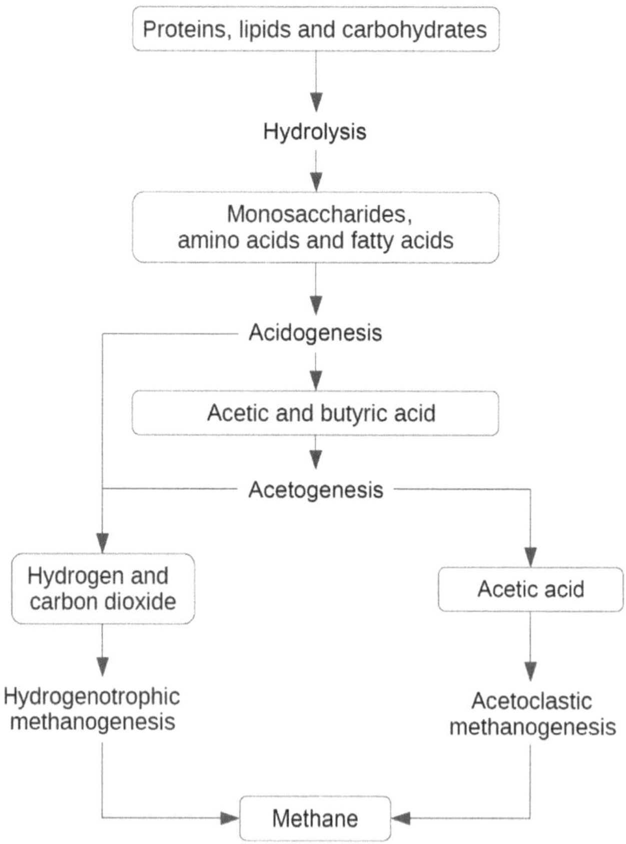

Figure 4.6 Anaerobic digestion process.

complex organic substrates (proteins, lipids and carbohydrates), is attacked by fermentative hydrolytic bacteria (e.g. Clostridium) which transforms it into simpler and more soluble compounds such as monosaccharides, amino acids and fatty acids. These compounds are then subjected by fermenting bacteria to an acidogenic fermentation process that transforms them into volatile fatty acids such as acetic acid and butyric acid and starts also to produce carbon dioxide and hydrogen. The generated acids are then further transformed by acetogenic bacteria into acetic acid. Finally, the methanogenic microorganisms generate methane by decomposing acetic acid and combining carbon dioxide and hydrogen (Burkhardt and Busch, 2013; Pavlostathis and Giraldo-Gomez, 1991). An environment that is hostile to methanogenic microorganisms could prevent this last phase and could make it possible to obtain hydrogen.

For example, the chain of reactions leading from sucrose to methane can be the following:

$$\text{Hydrolisis:} \quad C_{12}H_{22}O_{11} + H_2O \Rightarrow 2C_6H_{12}O_6 \tag{4.5}$$

$$\text{Acidogenesis:} \quad \begin{array}{c} C_6H_{12}O_6 + 2H_2O \Rightarrow 2C_2H_4O_2 + 2CO_2 + 4H_2 \\ C_6H_{12}O_6 \Rightarrow C_4H_8O_2 + 2CO_2 + 2H_2 \end{array} \tag{4.6}$$

$$\text{Acetogenesis:} \quad C_4H_8O_2 + 2H_2O \Rightarrow 2C_2H_4O_2 + 2H_2 \tag{4.7}$$

$$\text{Methanogenesis:} \quad \begin{array}{ll} C_2H_4O_2 + H_2O \Rightarrow CH_4 + H_2CO_3 & \text{(acetoclastic methanogenesis)} \\ H_2CO_3 \Rightarrow H_2O + CO_2 & \\ CO_2 + 4H_2 \Rightarrow CH_4 + 2H_2O & \text{(hydrogenotrophic methanogenesis)} \end{array} \tag{4.8}$$

Methanogenesis usually occurs as a final step of a process of decaying organic matter in an anaerobic environment like landfills, swamps, intestines of animals, aquatic sediments, oil fields and sea depths.

In particular, hydrogenotrophic methanogenesis is a process performed by a group of microorganisms (Archaea) in mesophilic or thermophilic conditions (20°C–70°C). These microorganisms obtain energy for their growth from a metabolic route which consumes hydrogen. Since oxygen is toxic to the microorganisms, some oxidised compounds are used to oxidise hydrogen. The simplest compounds utilised by these bacteria are the carbon oxides, which are converted into methane and water.

Since methanogenesis is the final step in the process of biomass decomposition, methanogenic Archaea can tolerate impurities containing nitrogen and sulphur and are not significantly affected by the presence of traces of oxygen although the process is anaerobic (Götz et al., 2016; Seifert et al., 2013). As a consequence, hydrogenotrophic methanogenesis can be applied also as a further step after a process like anaerobic digestion or biomass gasification after providing the adequate amount of hydrogen. However, if the product gas does not fall within the composition specifications suitable for the gas grid, it is necessary to purify it after the methanation.

The conversion efficiency depends on parameters related to the used microorganism (type and concentration) and the reactor (type, operating temperature and pressure). The substrate acidity is also an important parameter. The conversion efficiency is evaluated in relation to the GHSV (4.4), to the methane concentration in the gas obtained and to the methane formation rate (MFR):

$$MFR = \frac{\Delta F_{V,CH_4}}{V_R} \tag{4.9}$$

As with three-phase reactors, it is necessary to take into account the gas–liquid mass transfer resistance and this aspect has a significant influence on the type of reactor. To increase the effective reaction speed, and consequently the MFR, two parameters can be improved: the mass transfer coefficient, which can be increased for example by agitation; and the gas solubility, which can be increased by regulating the pressure (Götz et al., 2016).

Usually, reactors used for biological methanation are continuous stirred tank reactors (CSTRs), in which the reaction takes place in an aqueous solution that is mechanically agitated to favour the gas–liquid mass transfer. In particular, when stirring rates increases, the transfer of hydrogen (much less soluble than carbon dioxide) to the liquid increases (Peillex et al., 1988). A drawback is the power consumption for the agitation (Stone et al., 2017).

The gases are supplied to the vessel from the bottom and rise in the liquid mass coming into contact with microorganisms. Besides increasing gases solubility, working at a higher pressure than atmospheric also increases the contact surface between gases and the Archaea by reducing the size of gas bubbles. On the other hand, a pressure increase causes an increase in CO_2 concentration in the substrate and consequently a reduction in pH: it is therefore necessary to carry out an adequate control of acidity to ensure adequate conditions for the microorganisms (Martin et al., 2013; Nishimura et al., 1992; Wang et al., 2013).

Due to the low reaction rate (GHSV $< 10\ h^{-1}$), CSTRs are characterised by a large volume (Bailera et al., 2017). Although this is an advantage from the point of view of system thermal inertia, which allows an easier temperature control, it increases the overall system costs.

As part of a power-to-methane process, it is possible to consider two distinct cases: a simple biomass anaerobic digester or an integrated system comprising the anaerobic digester.

Regarding the last one, the product of anaerobic digestion of biomass is a gas mainly formed by methane and carbon dioxide. Usually, this gas is purified by separating the carbon dioxide to obtain pure methane. This process is called biogas upgrading and the resulting gas is called biomethane. However, the separated carbon dioxide is a waste product, and it could also be converted into methane. Since a reactor in which the methanation process takes place is already available, it is advisable to save money and to avoid carbon dioxide separation by directly supplying the hydrogen flow rate required for the conversion of the excess CO_2 to the reactor. Of course, it is also possible to use the same reactor to convert additional carbon dioxide made available from another source.

About the increase of CO_2 conversion to methane, Götz et al. (2016) reported data from MicrobEnergy GmbH which observed a final methane content up to 75%. Other data were reported by Luo et al. (Luo and Angelidaki, 2012; Luo et al., 2012) who experimented carbon dioxide conversion inside a small CSTR at atmospheric pressure and 55°C: 80% of the additional supplied hydrogen allowed a reduction of carbon dioxide from 38% to 15%. Although the advancement of the methanation reaction could be increased by improving the gas–liquid mass transfer of hydrogen, these results prove that converting all the available carbon dioxide is impossible.

By operating biological methanation in a reactor that is specifically dedicated to this process, it is possible to directly use carbon dioxide and hydrogen and to adjust the process parameters in an optimised way for the hydrogenotrophic methanogenic microorganisms. It is also possible to adopt reactors of a different type from CSTR which, although still at an experimental level, could prove to be more efficient.

Using a CSTR with GHSV of 28.7 h^{-1}, Nishimura et al. (1992) obtained an MFR of 21.3 h^{-1}, with a methane content in the product gas of only 13.4%. Still using a CSTR, in this case with GHSV of 120 h^{-1}, a better result was obtained by Seifert, Rittmann, and Herwig (2014): 21.3 h^{-1} with a methane content in the produced gas equal to 60%. By reducing the GHSV to 30 h^{-1}, the same researchers increased the methane content in the produced gas by up to 85%, while significantly reducing the MFR (Seifert et al., 2014). However, these values are too low for an SNG.

Even changing the reactor type, it seems that high concentration values of methane in the produced gas cannot be reached, except at the expense of production speed.

In fixed-bed reactors, the Archaea cells can be immobilised on a support as a biofilm (Jee, Nishio, and Nagai, 1988), naturally present inside an organic matter bio-filter (Nikiema et al., 2005), or artificially inoculated into an inorganic porous filter (Ganendra et al., 2015).

Also for these reactors, the gases are fed from below, and the liquid, which derives from the production of water from the reaction and from the residue of the nutrient solution that must be periodically supplied to the reactor, is drained from below (Limbri et al., 2014; Nikiema et al., 2005).

Regarding operating parameters, fixed-bed reactors can operate at ambient pressure and temperature (Ganendra et al., 2015; Götz et al., 2016).

Using a fixed-bed reactor, Jee et al. (1988) obtained a methane content of 34% with an MFR of 4.6 h^{-1}.

Trickled-bed reactors also contain solid material or filters. In this case, however, there is a continuous circulation of liquid that from the top of the reactor drips down to the discharge. These reactors can operate with both free and immobilised Archaea and produce a gas with a higher methane content, with the same GHSV, compared to fixed-bed reactors (Stone et al., 2017).

In the absence of the agitator action, the mixing between microorganisms and gases is less uniform with the consequence that the GHSV is much lower (Stone et al., 2017). With a trickle-bed reactor it was possible to obtain a methane content of 98%, but with a GHSV of even 0.3 h^{-1} (Burkhardt, Koschack, and Busch, 2014).

On the other hand, the energy requirements of both fixed-bed reactors and dropped-bed reactors are very low because they are limited to the feed pump consumption (Thema et al., 2019).

4.3 Comparison among available technologies

The first consideration to be made concerns the available technologies level of development: only fixed-bed catalytic methanation is a mature and already commercial technology. Both the catalytic alternatives and biological methanation as a dedicated process are still at a laboratory or pilot-scale experimentation level (Götz et al., 2016).

As can be seen from Table 4.4, due to different operating temperatures, the reaction rate for catalytic methanation is much higher than that found for biological methanation. This has a great influence on the reactor volume and, consequently, on the plant capital cost.

Therefore it can be said that the technology to be adopted depends on the plant size and the carbon source. If the latter is a pure or purified carbon dioxide source, catalytic methanation is to be preferred and the choice may fall on isothermal fluidised-bed or three-phase reactors for smaller sizes, whereas for the large sizes fixed-bed reactors are more suitable. The latter is also preferable in the case of lignocellulosic biomass as a source of carbon. In particular, the adiabatic reactor is convenient when the biomass is subjected to hydro-gasification as the syngas obtained already has a high methane content and no more than two reactors in series are required. Finally, if the biomass is treated by anaerobic digestion, it is possible to inject hydrogen directly into the digester thus operating a biological methanation. As the size increases, it is instead appropriate to separate the digestion phase from that of methanation and the latter can be carried out in a catalytic reactor.

The choice of steady state or dynamic operation depends on cost analysis since it involves the presence and the size of a hydrogen storage. Although a methanation reactor does not work at its optimum during transient operation, its performance depends more on the auxiliary equipment than on the reactor itself. This also entails a certain difficulty in extrapolating the behaviour of a commercial-sized system from the experiments carried out in the laboratory, especially to define the sustainable minimum load below which the reactor must be stopped. Even when the reactor is able to operate with very low flow rates, as in the case of the biological process, it is not economically convenient to go lower a certain minimum load. Götz et al. (2016) report minimum load values ranging from 10% for biological reactors up to 40% for fixed-bed adiabatic reactors.

In the case of a catalytic reactor, the shutdown requires the expulsion of residual carbon oxides, a result that can be obtained by flushing the reactor with hydrogen or inert gas. Furthermore, it must be taken into consideration that at the new start-up energy and time are required to bring the reactor again at the operating temperature. For this reason, if the shutdown duration is limited, it will be preferable to keep the reactor in temperature rather than cool it down.

Table 4.4 Key parameters of the different reactor configurations of biological and catalytic methanation (Bailera et al., 2020).

	Biological methanation			Catalytic methanation				
	CSTR	**FBR**	**TBR**	**AFB**	**CFB**	**FR**	**3PR**	**SR**
Temperature (°C)	20–70	20–65	37–55	250–700	250–500	300–400	300–350	250–300
Pressure (bar)	1–3	1	1	5–100	5–100	1–12	20	1–20
Load range (%)	10–100	0–100	0–100	40–100	30–100	50–100	20–100	30–100
GHSV (h^{-1})	10–300	1–30	<1	<5000	<6000	$<60,000$	<1000	<3000
CH_4 content (%)	5–85	15–90	94–98	>90	>90	>90	>90	>90
No. of stages	1	1	1	2–7	1–2	1–2	1–2	1–2
Heat use	Low	Low	Low	High	High	Regular	Regular	Regular
Tolerance to impurities	High	High	High	Low	Low	Low	Regular	Low

CSTR, Continuously stirred tank reactor; *FBR*, fixed-bed reactor; *TBR*, trickle-bed reactor; *AFB*, adiabatic fixed-bed; *CFB*, cooled fixed-bed; *FR*, fluidised reactor; *3PR*, three-phase reactor; *SR*, structured reactor.

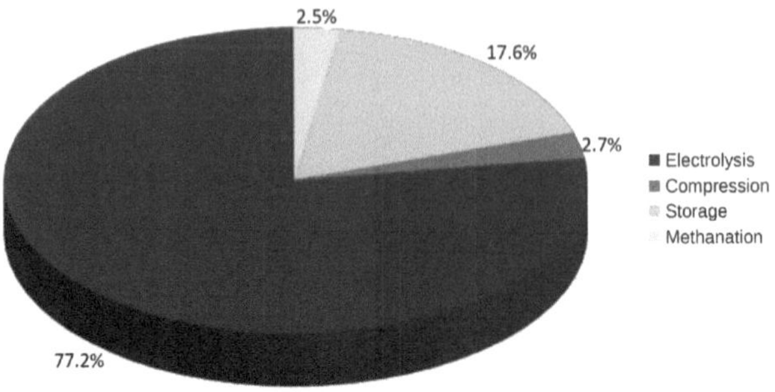

Figure 4.7 Cost-sharing of a methanation plant.

From an energy balance point of view, no catalytic reactor needs a stirring system, not even the three-phase reactor, although it contains liquid. In fact, thanks to the operating temperature, hydrogen solubility in the liquid is sufficient even in the absence of agitation. On the contrary, any type of biological reactor needs some stirring device that can be a real stirrer, a feed or a recirculation pump.

Another aspect of the energy comparison concerns the possibility of recovering the heat produced by methanation. In fact, the methanation reactions (4.1) and (4.2) are strongly exothermic, and it is therefore necessary to identify an efficient recovery system to improve the overall process efficiency. In the case of catalytic process, the available heat temperature is above 300°C and this allows the heat to be recovered to produce electricity or steam for industrial or residential use. Conversely, the heat produced in a biological methanation reactor is usually available at a very low temperature, usually around 60°C. Therefore it can only be used for direct heat recovery, for example to heat the anaerobic digestion section in case the methanation phase is separated, or for direct use on-site (i.e. space heating).

Although costs related to electrolysis are rather consolidated, there is limited information on costs related to methanation and misaligned with each other. However, the cost attributable to the methanation unit is very low compared to the other costs of a power to methane plant, as shown in Fig. 4.7, obtained by considering the average of two cases analysed by Götz et al. (2016). The cost by far dominant is the electrolyser one and therefore its correct sizing based on the energy surplus to be accumulated is essential. The cost of the hydrogen storage is also significant, for which reason sometimes it can be convenient to let the methanation reactor work in dynamic behaviour since the higher cost of the latter can be easily compensated by the savings connected to the volume reduction of the hydrogen storage.

Regarding the plant operation, on the one hand the costs of electricity and the carbon dioxide supply have to be considered, whereas on the other hand any revenues from the sale of oxygen and heat recovery should be assessed. It is practically impossible to make a comprehensive assessment without going into the individual specific cases. In fact, in addition to the differences in costs that may exist between one country and another, it must be considered that

- when electricity corresponds to a surplus on the network its cost should be very low, even zero;

- carbon dioxide would certainly have a cost if purchased pure, but its cost could be very low (i.e. as residuals from a biomass treatment plant or industrial carbon sources);
- oxygen has different prices depending on the buyer and required purity;
- heat recovery offers different economic advantages in relation to the product of the recovery itself.

As a consequence, the cost of produced SNG is extremely variable depending mainly on the carbon source, full load hours and electricity cost (Table 4.5). In any case, the SNG cost is higher than the current cost of natural gas and often higher than the cost of biomethane.

Despite this, some fundamental aspects must be taken into account. First of all, the natural gas cost could rise significantly if reserves start to decline or if carbon taxes are applied to fossil fuels. Furthermore, SNG production might reduce the environmental footprint if compared to its fossil alternative (Bargiacchi et al., 2021). Finally, the use of SNG on a large scale can accelerate the transition to a system based on hydrogen as an energy vector, because it allows to build a large

Table 4.5 Overview of SNG generation costs produced via power to methane.

Full load hours	Assumptions	SNG costs in Eurocent/ kWh SNG	Source
1200	80 MW electrical power input; Energy integration with lignite power plant	60	Buchholz et al. (2014)
	0–9 ct/kWh electricity	19–50	Müller-Syring et al. (2013)
	5 ct/kWh electricity; 110 MW SNG output	27–30	Schaaf, Graf et al. (2014)
	0–5 ct/kWh electricity; 10 €/t O_2	10–16	Vandewalle, Bruninx, and D'Haeseleer (2015)
3000	5 ct/kWh electricity; 110 MW SNG output	13.5–17	Schaaf, Graf et al. (2014)
	0–3.3 ct/kWh electricity	6.3–21	Brunner and Thomas (2014)
	0–5 ct/kWh electricity; 10 €/t O_2	4–8	Vandewalle et al. (2015)
-	Heat and O^2 utilisation	16.5–39.2	E&E Consultant, HESPUL, and Solagro (2014)

SNG, Substitute of natural gas.
Source: Adapted from Götz, M. et al. (2016) Renewable power-to-gas: A technological and economic review, Renewable Energy, 85, pp. 1371–1390. https://doi.org/10.1016/j.renene.2015.07.066.

hydrogen production infrastructure and to overcome current bottlenecks for the use of hydrogen on a large scale.

4.4 System integration

Methanation process can be applied for direct conversion of carbon dioxide that is captured from other processes or can be integrated in a system including fossil fuels or biomass treatment. Fig. 4.8 shows a generic integrated layout for SNG production from biomass or carbon dioxide. In the latter case, the plant comprises hydrogen production by electrolysis, to which methanation, condensation and separation of water and heat recovery sections are added. Starting from biomass, it is necessary to obtain a gas and three options are schematised: anaerobic digestion, gasification and hydro-gasification.

Actually, the system integration could be even more complex, because the surplus electricity deriving from a renewable energy source could be due to a lack of demand or to a problem of grid instability. Unless there is the possibility of using heat directly, in the first case heat recovery should be used to produce electricity to increase the hydrogen production from the electrolyser. In the second case, the plant also has the task of generating high-quality power for the grid: cogeneration of power and fuel is more efficient than producing fuel alone to be partially used to generate the required power.

Some analyses on the energy balance and life cycle assessment (LCA) have been carried out for a plant based on gasification of hard-wood biomass and devoted to produce a SNG. However for such a SNG, the final hydrogen content is actually much higher than that commercially allowed, but technically it could be distributed

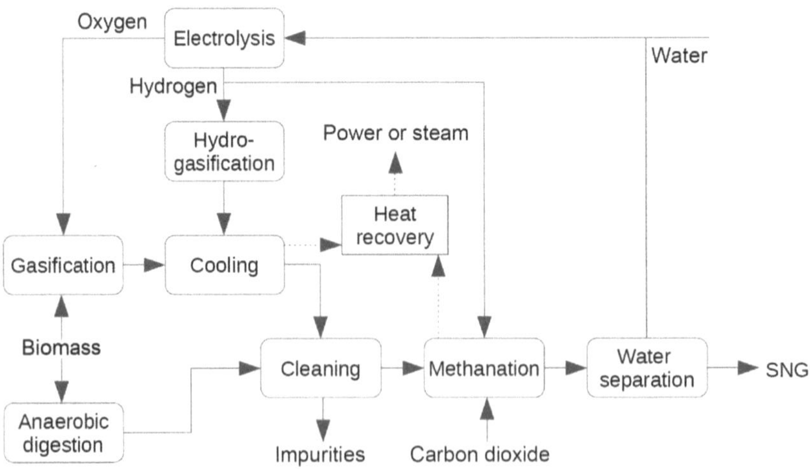

Figure 4.8 Generic integrated plant layout for SNG production starting from biomass or captured carbon dioxide. *SNG*, Substitute of natural gas.

using the existing pipeline and used in natural gas-fuelled vehicles (Bargiacchi et al., 2021; Bargiacchi, Frigo, and Spazzafumo, 2018, 2019; Frigo and Spazzafumo, 2020).

In case there is no need for power, hydro-gasification has been considered as the preferable treatment, whereas to cogenerate power the possibility to use oxygen as a gasifying agent has been selected and some different power units have been considered.

Fig. 4.9 shows the layout with a hydro-gasifier. An over-stoichiometric flow of hydrogen is fed to the hydro-gasifier in such a way that the produced syngas is mainly composed by methane, hydrogen and water, with a small although not negligible amount of carbon oxides. Therefore a subsequent methanation process is required, but only one adiabatic reactor is needed to convert almost completely the residual carbon oxides into methane.

When cogeneration of power is needed, biomass is gasified using part of the produced electrolytic oxygen, and the obtained syngas can be burnt into a power unit, for example a gas turbine or an internal combustion engine. Being further electrolytic oxygen available, oxy-combustion in the power unit offers the advantage to obtain an exhaust gas composed almost exclusively of carbon dioxide and steam, so that almost pure CO_2 can be easily obtained by steam condensation and separation. On the other hand, a problem arises from the high combustion temperature achieved and a temperature moderator is required to replace the role of nitrogen in the air; this issue could be easily solved by a partial recycling of carbon dioxide. Another solution proposed for gas turbine is the water injection. Finally, it is also possible to supply the syngas to a high-temperature fuel cell stack.

In all these cases, all the carbon present in the biomass is converted to carbon dioxide, and the methanation section in the case of adiabatic reactors is shown in Fig. 4.10, requiring only a partial recycle at the entrance of the first methanator.

Due to the reaction exothermicity, a certain amount of heat is available and must be extracted from the different reactors. To optimise the heat recovery, heat exchangers can be divided into two groups, namely at high and at low temperatures. A bottoming Brayton cycle working with air and two pressure levels can be designed and its power can be increased by producing steam to be injected into the high-pressure turbine (Fig. 4.11).

Table 4.6 reports the result obtained from the energy analysis.

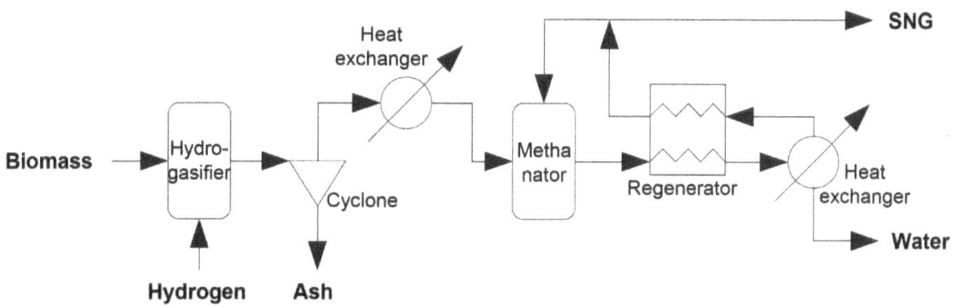

Figure 4.9 Layout of biomass hydro-gasification followed by methanation (Frigo and Spazzafumo, 2020).

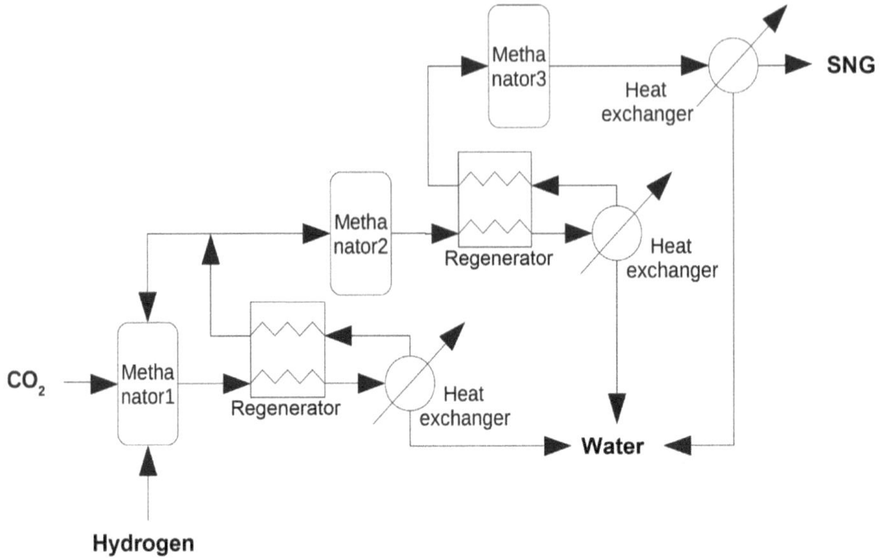

Figure 4.10 Methanation section (Frigo and Spazzafumo, 2020).

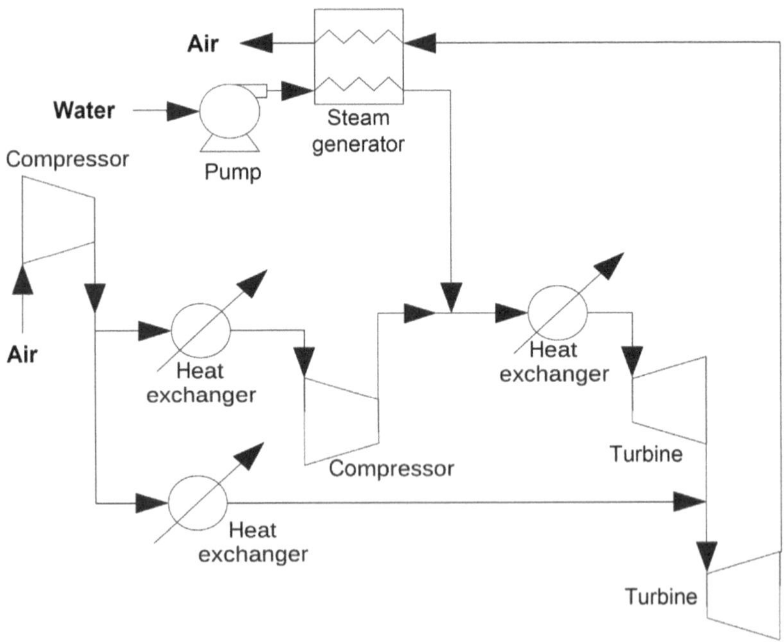

Figure 4.11 Bottoming plant for heat recovery (Frigo and Spazzafumo, 2020).

Table 4.6 Energy balance (based on HHV).

	Hydro-gasification	Cogeneration
Biomass (kW)	3911	
Electrolysis (kW)	8184	16044
SNG (kW)	8931	8904–8935
Electrical power (kW)	0	1523–2914
Efficiency	73.8%	52.4%–59.4%
Chemical input (biomass)	32%	20%
Chemical output (SNG)	100%	75%–85%

HHV, Higher heating value; *SNG*, substitute of natural gas.

4.5 Environmental impacts of substitute natural gas

As many other synthetic fuels, the environmental impacts of SNG production were investigated by several authors in recent years, by means of LCA methodology. This section does not mean to provide a full, comprehensive state of the art of the environmental assessment of power-to-methane systems, rather to provide relevant references and point out the challenges to be addressed.

Regarding chemical methanation, Zhang et al. (2017) investigated the environmental performance of SNG and power to methane systems using LCA and considering different carbon sources (CO_2 from atmosphere, from cement plant, from coal power plant) as well as different technologies for electricity supply and electrolysis, while using thermochemical methanation. The main result is that the energy mix for electricity production and the CO_2 source dominates the life cycle emissions and therefore the reduction potential of this new technology in comparison with natural gas. On the same topic, Hoppe, Bringezu, and Thonemann (2016) concluded that SNG production from an industrial CO_2 source (in this case from raw biogas separation) in the German scenario saves greenhouse gas emissions if compared to natural gas. On the other, Sternberg and Bardow (2016) argued that the power-to-gas pathways imply higher global warming impact (GWI) and fossil depletion than conventional natural gas even in a 2050 electricity mix while proving promising GWI reduction with the power-to-syngas routes. Collet et al. (2017) compared different scenarios for biogas production with SNG production via power to gas, proving that impacts of power to gas are higher than with biogas upgrading for continuous operation but lower than natural gas impacts.

Regarding SNG production from biomass, Felder and Dones (2007) evaluated the ecological impacts of SNG production from wood with its utilisation in current heating systems and cars. They found that SNG outperforms the oil derivatives and the fossil natural gas when the consumption of fossil resources is strongly weighted in the assessment. Tasca et al. (2019) performed an LCA of biomethane production from gasification of poplar short rotation forestry wood chips and anaerobic digestion of microalgae biomass. Skorek-Osikowska, Martín-Gamboa, and Dufour (2020) conducted a combined technical, economic and environmental assessment of the production of biomethane from three different pathways: manure fermentation and upgrading, manure fermentation and biogas methanation using renewable hydrogen from electrolysis and biomass gasification with syngas methanation. Both Tasca and Skorek-Osikowska concluded that no solution outperforms the others on all impact categories and that in any case the proposed solutions still need improvements to be economically competitive with natural gas. Gerber, Gassner, and Maréchal (2011) proposed a methodology to integrate LCA in techno-economic models, and they applied it to a thermo-economic model developed for the optimisation of SNG and electricity production from lignocellulosic biomass. Recently, Singlitico, Goggins, and Monaghan (2019) conducted a comprehensive review of LCA papers dealing with the production of biomethane, produced by anaerobic digestion, and SNG, produced by methanation and gasification.

Finally, the layouts mentioned in the previous section were analysed under environmental criteria, selecting global warming potential (GWP), non-renewable cumulative energy demand (CEDnr) and acidification potential (AP) as the main impact indicators, due to their relevance in the field of hydrogen energy systems (Bargiacchi et al., 2021). Some methodological aspects were also considered, as the choice of two different functional units (1 kg of SNG and 1 MJ of SNG) and different approaches to address the multifunctionality of such a cogeneration system. Results show that, with all the considered system layouts, SNG production from biomass and renewable electricity presents a significantly lower impact than that of conventional natural gas under GWP and CEDnr, whereas higher impacts than natural gas ones are obtained under the acidification footprint. The cause of this high impact in AP is mainly associated with the alkaline electrolyser infrastructure and in particular is ascribable to the mining of catalyst metals (nickel and rhodium) for electrodes manufacturing. When total energy in output (SNG + electricity) is considered as the functional unit, these trends of comparison with natural gas are confirmed, with the layout based on hydro-gasification showing an impact lower than the other systems under comparison. Finally, it was found that different methodological choices to deal with the system multifunctionality can affect the ranking between the different technologies under comparison but they did not overturn the overall results in terms of comparison with natural gas.

Overall, although most of the authors proved that SNG has the potential of reducing the overall environmental footprint in the gas sector rather than its fossil alternative, the environmental benefits must be carefully assessed on a case-by-case basis. Some precautions are essential to minimise the environmental impact of an SNG, such as the use of a renewable energy mix for electricity production, a suitable choice of the CO_2 source and a careful choice of the materials to avoid burden-shifts.

Nomenclature

AP	acidification potential
CEDnr	non-renewable cumulative energy demand
CSTR	continuously stirred tank reactor
F	flow in standard condition
GHSV	gas hourly space velocity
GWP	global warming potential
GWI	global warming impact
LCA	life cycle assessment
MFR	methane formation rate
SNG	substitute natural gas
V	volume

Subscripts

g	gas
R	reactor
V	volumetric

References

Assettati, A., et al. (2010). *Sperimentazione di siste mi di trattamento e conversione del syngas*. Rome: ENEA.

Bailera, M., et al. (2017). Power to gas projects review: Lab, pilot and demo plants for storing renewable energy and CO2. *Renewable and Sustainable Energy Reviews*, *69* (October 2016), 292–312. Available from https://doi.org/10.1016/j.rser.2016.11.130, Elsevier.

Bailera, M., et al. (2020). *Energy storage - Hybridization of power-to-gas technology and carbon capture*. Cham: Springer. Available from https://doi.org/10.1007/978-3-030-46527-8.

Bargiacchi, E., et al. (2021). Life cycle assessment of substitute natural gas production from biomass and electrolytic hydrogen. *International Journal of Hydrogen Energy*. (in press).

Bargiacchi, E., Frigo, S., and Spazzafumo, G. (2018). Energetic and exergetic analysis of an innovative plant for the production of electricity and substitute natural gas. *Energy Procedia*, *148*, 312–319. Available from https://doi.org/10.1016/j.egypro.2018.08.083.

Bargiacchi, E., Frigo, S., and Spazzafumo, G. (2019). From biomass and electrolytic hydrogen to substitute natural gas and power: The issue of intermediate gas storages. *International Journal of Hydrogen Energy*, *44*(38), 21045–21054. Available from https://doi.org/10.1016/j.ijhydene.2019.03.028.

Bartholomew, C. H. (1982). Carbon deposition in steam reforming and methanation. *Catalysis Reviews: Science and Engineering*, *24*(1), 67–112. Available from http://doi.org/10.1080/03602458208079650.

Bartholomew, C. H. (2001). Mechanisms of catalyst deactivation. *Applied Catalysis A: General*, *212*, 17–60. Available from http://doi.org/10.1016/S0926-860X(00)00843-7.

Bartholomew, C. H., Agrawal, P. K., and Katzer, J. R. (1982). Sulfur poisoning of metals. *Advances in Catalysis*, *31*(C), 135–242. Available from https://doi.org/10.1016/S0360-0564(08)60454-X.

Brunner, C. and Thomas, A. (2014) Arbeitspaket 6: Gasnetzanalysen und Wirt- schaftlichkeitsbetrachtung, Energie I wasser-praxis, (November). Available at: http://www.dvgw-innovation.de/fileadmin/dvgw/angebote/forschung/innovation/pdf/1411brunner.pdf.
Buchholz, O. S., et al. (2014). Power-to-gas: Storing surplus electrical energy. A design study. *Energy Procedia*, *63*, 7993−8009. Available from https://doi.org/10.1016/j.egypro.2014.11.836.
Burkhardt, M., and Busch, G. (2013). Methanation of hydrogen and carbon dioxide. *Applied Energy*, *111*, 74−79. Available from https://doi.org/10.1016/j.apenergy.2013.04.080.
Burkhardt, M., Koschack, T., and Busch, G. (2014). Biocatalytic methanation of hydrogen and carbon dioxide in an anaerobic three-phase system. *Bioresource Technology*, *178*, 330−333. Available from https://doi.org/10.1016/j.biortech.2014.08.023.
Collet, P., et al. (2017). Techno-economic and Life Cycle Assessment of methane production via biogas upgrading and power to gas technology. *Applied Energy*, *192*, 282−295. Available from https://doi.org/10.1016/j.apenergy.2016.08.181.
E&E Consultant, HESPUL, and Solagro (2014) *Etude portant sur l'hydrogène et la méthanation comme procédé de valorisation de l'électricité excédentaire*, p. 238. Available at: http://www.grtgaz.com/fileadmin/engagements/documents/fr/Power-to-Gas-etude-ADEME-GRTgaz-GrDF-complete.pdf.
Eigenberger, G. (1992) Fixed bed reactors. In *Ullmann's Encyclopedia of Industnal Chemistry*. Vol B4. VCH. Available at: https://elib.uni-stuttgart.de/bitstream/11682/1848/1/eig16.pdf.
Felder, R., and Dones, R. (2007). Evaluation of ecological impacts of synthetic natural gas from wood used in current heating and car systems. *Biomass and Bioenergy*, *31*(6), 403−415. Available from https://doi.org/10.1016/j.biombioe.2006.08.005.
Foscolo, P. U., Gallucci, K., and Micheli, F. (2011) Studio della valorizzazione dell'impianto di gassificazione di biomasse a letto fluidizzato internamente ricircolante per la produzione di metano. Parte A: Studio di configurazioni impiantistiche e variabili processuali per massimizzare la resa di metano. ENEA.
Frigo, S., and Spazzafumo, G. (2020). Comparison of different system layouts to generate a substitute of natural gas from biomass and electrolytic hydrogen. *International Journal of Hydrogen Energy*, *45*(49), 26166−26178. Available from https://doi.org/10.1016/j.ijhydene.2020.03.205.
Ganendra, G., et al. (2015). Methane biofiltration using autoclaved aerated concrete as the carrier material. *Applied Microbiology and Biotechnology*, *99*(17), 7307−7320. Available from https://doi.org/10.1007/s00253-015-6646-6.
Gao, J., et al. (2012). A thermodynamic analysis of methanation reactions of carbon oxides for the production of synthetic natural gas. *RSC Advances*, *2*(6), 2358−2368. Available from https://doi.org/10.1039/c2ra00632d.
Gao, J., et al. (2015). Recent advances in methanation catalysts for the production of synthetic natural gas. *RSC Advances*, *5*(29), 22759−22776. Available from https://doi.org/10.1039/c4ra16114a.
Gerber, L., Gassner, M., and Maréchal, F. (2011). Systematic integration of LCA in process systems design: Application to combined fuel and electricity production from lignocellulosic biomass. *Computers and Chemical Engineering*, *35*(7), 1265−1280. Available from https://doi.org/10.1016/j.compchemeng.2010.11.012.
González, M. I., and Schaub, G. (2015). Fischer-Tropsch synthesis with H2/CO2—Catalyst behavior under transient conditions. *Chemie-Ingenieur-Technik*, *87*(6), 848−854. Available from https://doi.org/10.1002/cite.201400137.

Götz, M., et al. (2016). Renewable Power-to-Gas: A technological and economic review. *Renewable Energy*, *85*, 1371–1390. Available from https://doi.org/10.1016/j.renene.2015.07.066.

Held, M., et al. (2020). Power-to-Gas: CO2 methanation concepts for SNG production at the Engler-Bunte-Institut. *Chemie-Ingenieur-Technik*, *92*(5), 595–602. Available from https://doi.org/10.1002/cite.201900181.

Hoppe, W., Bringezu, S., and Thonemann, N. (2016). Comparison of global warming potential between conventionally produced and CO2-based natural gas used in transport versus chemical production. *Journal of Cleaner Production. Elsevier Ltd*, *121*, 231–237. Available from https://doi.org/10.1016/j.jclepro.2016.02.042.

Infomine (2020) Metal prices. Available at: www.infomine.com/investment/metal-prices/ Accessed 18.11.20.

Jee, H. S., Nishio, N., and Nagai, S. (1988). Continuous CH4 production from H2 and CO2 by Methanobacterium thermoautotrophicum in a fixed-bed reactor. *Journal of Fermentation Technology*, *66*(2), 235–238. Available from https://doi.org/10.1016/0385-6380(88)90054-4.

Jensen, J. H., Poulsen, J. M., and Andersen, N. U. (2011). From coal to clean energy. *Nitrogen + Syngas*, *310*(March-April), 34–38.

Köppel, W., Götz, M., and Graf, F. (2009). Biogas upgrading for injection into the gas grid - Quality aspects, technological and ecological consideration. *gwf-Gas Erdgas*, *150*, 26–35, International Issue.

Kopyscinski, J., Schildhauer, T. J., and Biollaz, S. M. A. (2010). Production of synthetic natural gas (SNG) from coal and dry biomass - A technology review from 1950 to 2009. *Fuel*, *89*(8), 1763–1783. Available from https://doi.org/10.1016/j.fuel.2010.01.027.

Kopyscinski, J., Schildhauer, T. J., and Biollaz, S. M. A. (2011). Methanation in a fluidized bed reactor with high initial CO partial pressure: Part I-Experimental investigation of hydrodynamics, mass transfer effects, and carbon deposition. *Chemical Engineering Science*, *66*(5), 924–934. Available from https://doi.org/10.1016/j.ces.2010.11.042.

Lefebvre, J., et al. (2015). Improvement of three-phase methanation reactor performance for steady-state and transient operation. *Fuel Processing Technology*, *132*, 83–90. Available from https://doi.org/10.1016/j.fuproc.2014.10.040.

Limbri, H., et al. (2014). Coal-packed methane biofilter for mitigation of green house gas emissions from coal mine ventilation air. *PLoS One*, *9*(4). Available from https://doi.org/10.1371/journal.pone.0094641.

Liu, Z., et al. (2012). Total methanation of syngas to synthetic natural gas over Ni catalyst in a micro-channel reactor. *Fuel*, *95*, 599–605. Available from https://doi.org/10.1016/j.fuel.2011.12.045.

Luo, G., and Angelidaki, I. (2012). Integrated biogas upgrading and hydrogen utilization in an anaerobic reactor containing enriched hydrogenotrophic methanogenic culture. *Biotechnology and Bioengineering*, *109*(11), 2729–2736. Available from https://doi.org/10.1002/bit.24557.

Luo, G., et al. (2012). Simultaneous hydrogen utilization and in situ biogas upgrading in an anaerobic reactor. *Biotechnology and Bioengineering*, *109*(4), 1088–1094. Available from https://doi.org/10.1002/bit.24360.

Martin, M. R., et al. (2013). A single-culture bioprocess of methanothermobacter thermautotrophicus to upgrade digester biogas by CO2-to-CH4 conversion with H2. *Archaea*, *2013*(August 2014), 11. Available from https://doi.org/10.1155/2013/157529.

Martínez, J., et al. (2019). High selectivity and stability of nickel catalysts for CO2 Methanation: Support effects. *Catalysts*, *9*(1). Available from https://doi.org/10.3390/catal9010024.

Müller-Syring, G. et al. (2013) Entwicklung von modularen Konzepten zur Erzeugung, Speicherung und Einspeisung von Wasserstoff und Methan ins Erdgasnetz, DVGW Forschung, 350. Available at: https://www.dvgw.de/medien/dvgw/forschung/berichte/g1_07_10.pdf.

Mutz, B., et al. (2015). Methanation of CO2: Structural response of a Ni-based catalyst under fluctuating reaction conditions unraveled by operando spectroscopy. *Journal of Catalysis*, *327*, 48–53. Available from https://doi.org/10.1016/j.jcat.2015.04.006.

Nikiema, J., et al. (2005). Biofiltration of methane: An experimental study. *Chemical Engineering Journal*, *113*((2–3), 111–117. Available from https://doi.org/10.1016/j.cej.2005.04.005.

Nishimura, N., et al. (1992). Cultivation of thermophilic methanogen KN-15 on H2-CO2 under pressurized conditions. *Journal of Fermentation and Bioengineering*, *73*(6), 477–480. Available from https://doi.org/10.1016/0922-338X(92)90141-G.

Pavlostathis, S. G., and Giraldo-Gomez, E. (1991). Kinetics of anaerobic treatment: A critical review. *Critical Reviews in Environmental Control*, *21*(5–6), 411–490. Available from https://doi.org/10.1080/10643389109388424.

Peillex, J. P., et al. (1988). Growth of Methanococcus thermolithotrophicus in batch and continuous culture on H2 and CO2: influence of agitation. *Applied Microbiology and Biotechnology*, *29*(6), 560–564. Available from https://doi.org/10.1007/BF00260985.

Perry, R. H., and Green, D. W. (1998). *Perry's chemical engineers' hanbook*. McGraw-Hill.

Sabatier, P. and Senderens, J. B. (1902) Nouvelles synthèses du méthane, Comptes rendus de l'Académie des Sciences. Paris.

Schaaf, T., Graf, F., et al. (2014). Abschlussbericht: Technoökonomische Studie von Power-to-Gas-Konzepten. *Deutscher Verein des Gas- und Wasserfaches e. V. (DVGW)*, *287* (November). Available at. Available from http://www.dvgw-innovation.de/fileadmin/dvgw/angebote/forschung/innovation/pdf/g3_01_12_tp_b_d.pdf.

Schaaf, T., Grünig, J., et al. (2014). Methanation of CO2 - Storage of renewable energy in a gas distribution system. *Energy, Sustainability and Society*, *4*(1), 1–14. Available from https://doi.org/10.1186/s13705-014-0029-1.

Seifert, A. H., et al. (2013). Method for assessing the impact of emission gasses on physiology and productivity in biological methanogenesis. *Bioresource Technology*, *136*, 747–751. Available from https://doi.org/10.1016/j.biortech.2013.03.119.

Seifert, A. H., Rittmann, S., and Herwig, C. (2014). Analysis of process related factors to increase volumetric productivity and quality of biomethane with Methanothermobacter marburgensis. *Applied Energy*, *132*, 155–162. Available from https://doi.org/10.1016/j.apenergy.2014.07.002.

Singlitico, A., Goggins, J., and Monaghan, R. F. D. (2019). The role of life cycle assessment in the sustainable transition to a decarbonised gas network through green gas production. *Renewable and Sustainable Energy Reviews*, *99*(September 2018), 16–28. Available from https://doi.org/10.1016/j.rser.2018.09.040.

Skorek-Osikowska, A., Martín-Gamboa, M., and Dufour, J. (2020). Thermodynamic, economic and environmental assessment of renewable natural gas production systems. *Energy Conversion and Management: X*, *7*, 100046. Available from https://doi.org/10.1016/j.ecmx.2020.100046.

Stangeland, K., et al. (2017). CO2 methanation: the effect of catalysts and reaction conditions. *Energy Procedia*, *105*(1876), 2022–2027. Available from https://doi.org/10.1016/j.egypro.2017.03.577.

Sternberg, A., and Bardow, A. (2016). Life Cycle Assessment of power-to-gas: syngas vs methane. *ACS Sustainable Chemistry and Engineering*, *4*(8), 4156–4165. Available from https://doi.org/10.1021/acssuschemeng.6b00644.

Stone, K. A., et al. (2017). A mini review on bioreactor configurations and gas transfer enhancements for biochemical methane conversion. *Biochemical Engineering Journal*, *128*, 83–92. Available from https://doi.org/10.1016/j.bej.2017.09.003.

Tasca, A. L., et al. (2019). Biomethane from short rotation forestry and microalgal open ponds: System modeling and life cycle assessment. *Bioresource Technology*, *273*, 468–477. Available from https://doi.org/10.1016/j.biortech.2018.11.038.

Thema, M., et al. (2019). Biological CO2-methanation: An approach to standardization. *Energies*, *12*(1670), 1–32.

Ullmann. (1989). *Ullmann's encyclopedia of industrial chemistry* (5th completely revised ed.). Weinheim, Federal Republic of Germany: Wiley-VCH Verlagsgesell schaft GmbH, D-6940.

Uniongas (2020) Typical composition of natural gas. Available at https://www.uniongas.com/about-us/about-natural-gas/chemical-composition-of-natural-gas Accessed 22.11.20.

Vandewalle, J., Bruninx, K., and D'Haeseleer, W. (2015). Effects of large-scale power to gas conversion on the power, gas and carbon sectors and their interactions. *Energy Conversion and Management*, *94*, 28–39. Available from https://doi.org/10.1016/j.enconman.2015.01.038.

Wang, W., et al. (2013). Performance and microbial community analysis of the anaerobic reactor with coke oven gas biomethanation and in situ biogas upgrading. *Bioresource Technology*, *146*, 234–239. Available from https://doi.org/10.1016/j.biortech.2013.07.049.

Yentekakis, I. V., and Goula, G. (2017). Biogas management: Advanced utilization for production of renewable energy and added-value chemicals. *Frontiers in Environmental Science*, *5*(Feb). Available from https://doi.org/10.3389/fenvs.2017.00007.

Zhang, X., et al. (2017). Life cycle assessment of power-to-gas: Approaches, system variations and their environmental implications. *Applied Energy*, *190*, 326–338. Available from https://doi.org/10.1016/j.apenergy.2016.12.098.

Power to methanol

5

Daniele Candelaresi and Giuseppe Spazzafumo
Department of Civil and Mechanical Engineering, University of Cassino and Southern Lazio, Cassino, Italy

Methanol is a very important raw material in chemical industry. As Fig. 5.1 shows, several other products can be obtained starting from methanol. It is also a very flexible liquid fuel as it can be used in spark-ignited internal combustion engines, pure or blended with gasoline and ethanol, as well as in fuel cells. Methanol exhibits favourable combustion properties and its combustion efficiency is higher than that of gasoline thanks to the higher octane number, whereas the lower heating value is about the half (Bromberg and Cheng, 2007). It is also safer and no more toxic than petroleum-based fuels and less dangerous in case of spill because it is water soluble and highly biodegradable (Malcolm Pirnie Inc. and American Methanol Institute, 1999).

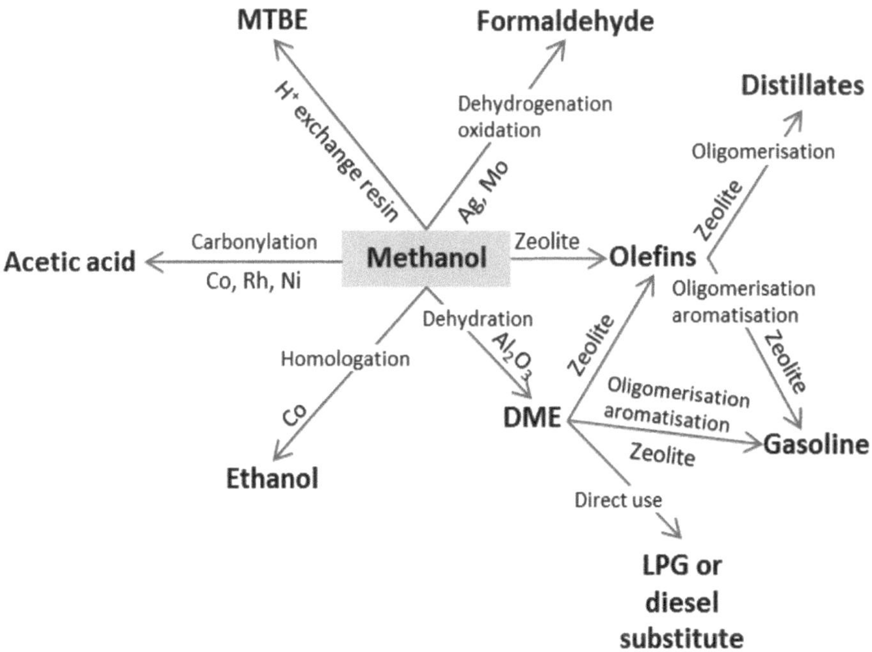

Figure 5.1 Main products obtainable from methanol (Bjorck, Dobson, and Pandhal, 2018).

Power to Fuel. DOI: https://doi.org/10.1016/B978-0-12-822813-5.00002-3

Methanol production has always been based on synthesis gas obtained from fossil fuels via natural gas reforming or coal gasification; however, in the next future, renewable methanol can be produced starting from green hydrogen and biomass or captured CO_2.

5.1 Methanol production from syngas

The first process for industrial-scale synthesis of methanol dates back to one century ago. It operated at high pressures (250–350 bar) and temperatures (320°C–450°C) with a ZnO/Cr_2O_3 catalyst (Wender, 1996). That process was replaced about 60 years ago by a lower pressure (50–100 bar) and lower temperature (200°C–300°C) process with a copper-based catalyst (Studt et al., 2015; Supp, 1990). The commercially used catalysts for the synthesis of methanol are currently based on CuO and ZnO with the addition of some stabilizing elements such as Zr, Cr, Mg (Ali, Abdullah, and Mohamed, 2015; Jadhav et al., 2014; Wang et al., 2011).

The reactions involved in methanol synthesis are as follows:

$$CO + 2H_2 \Rightarrow CH_3OH \quad \Delta H_{298K} = -90.7\ \text{kJ mol}^{-1} \tag{5.1}$$

$$CO_2 + 3H_2 \Rightarrow CH_3OH + H_2O \quad \Delta H_{298K} = -49.5\ \text{kJ mol}^{-1} \tag{5.2}$$

$$CO + H_2O \Rightarrow CO_2 + H_2 \quad \Delta H_{298K} = -41.2\ \text{kJ mol}^{-1} \tag{5.3}$$

The composition of a syngas, whether it is obtained from fossil fuels or biomass, has a hydrogen content that does not meet the stoichiometry required by methanol production, thus requiring an additional supply of hydrogen. This additional hydrogen can be derived from renewable energy sources through an electrolysis process and, therefore, it is globally a power to methanol process even in the case of fossil fuels as a carbon source. Obviously, starting from biomass, the produced methanol is totally renewable and the carbon dioxide emissions due to the final use of methanol are offset by the preventive carbon absorption by the biomass used.

Methanol synthesis is exothermic and is therefore favoured by low temperatures. Moreover, it involves a decrease in the number of moles and is therefore favoured by elevated pressures, according to Le Châtelier principle. This can be seen in Fig. 5.2 which shows the chemical equilibrium for reaction (5.1) and (5.2).

In the past, it was believed that the preferred carbon source for the production of methanol was carbon monoxide, while today it is believed that the main reaction is that of hydrogenation of carbon dioxide (Grabow and Mavrikakis, 2011; Lee et al., 1993). The reason why carbon dioxide hydrogenation experiments have given lower results than syngas hydrogenation (e.g. Pontzen et al., 2011) is that the production of water due to reaction (5.2) shifts the equilibrium of the reaction itself towards reactants and causes catalysts inactivation (Wu et al., 2001).

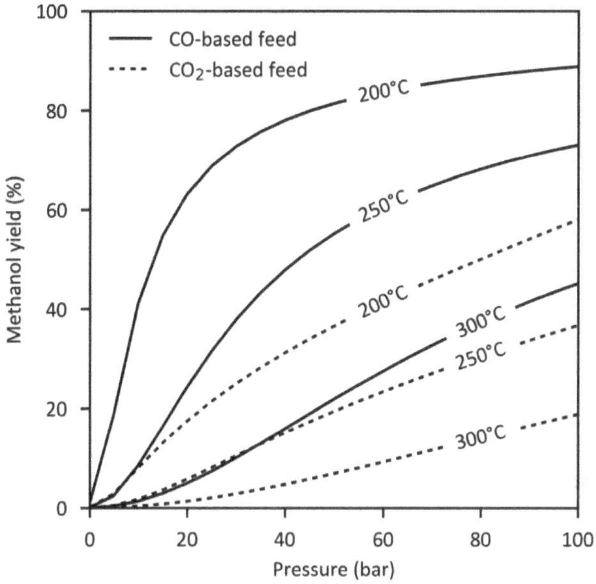

Figure 5.2 Equilibrium conversion of methanol formation (Dieterich et al., 2020).

Moreover, reaction (5.2) generates a lower amount of methanol at equilibrium. It is interesting to observe that when CO is used as a feed gas the activation energy of reaction (5.1) with a Cu catalyst is much lower than with a Cu/ZnO catalyst, whereas for a syngas the activation energy of reaction (5.2) with a Cu/ZnO is about a quarter lower than with a Cu catalyst, as shown in Fig. 5.3 (Studt et al., 2015). Moreover, the presence of carbon monoxide reduces the presence of water thanks to the water–gas shift reaction (5.3).

On the other hand, the reaction kinetics improves with higher temperatures at the expense of a lower final conversion. It is therefore necessary to recycle unconverted carbon oxides and hydrogen after cooling and steam condensation and separation (Fig. 5.4). Thanks to such a recycle, it is possible to obtain a very high conversion level, limited in practice only by the need to periodically purge the system to avoid the accumulation of unwanted substances contained into the fed gas or produced by competing reactions. To promote carbon conversion a slight excess of hydrogen can be used. This excess will be recycled as unreacted gas and therefore it is not lost, apart from a minimal amount with the purge gas.

Despite the elimination of unwanted substances by purging, the produced methanol is not entirely pure. According to the required purity, the produced methanol must therefore be subjected to a purification process. Dissolved gases are removed by flashing at a pressure between 5 and 10 bar. The next step is distillation, which can be carried out in a single column or in several columns depending on the purity required (Dieterich et al., 2020; Supp, 1990). A single column is suitable for fuel grade methanol, for which the water content is the only restriction, whereas higher

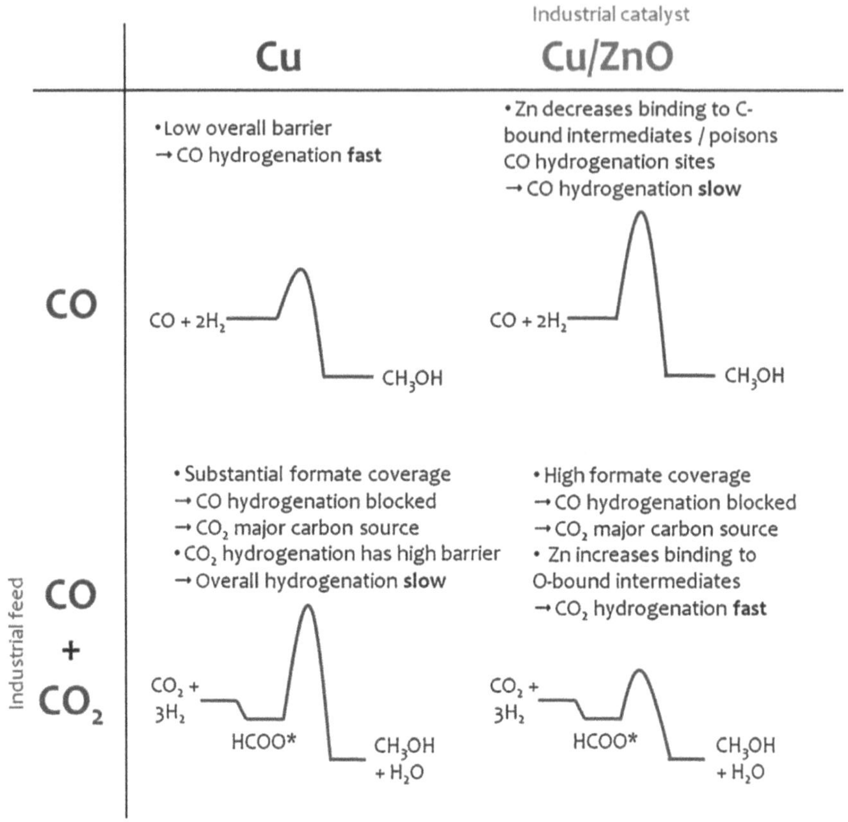

Figure 5.3 Energy barriers and activation energies in methanol synthesis reactions (Studt et al., 2015).

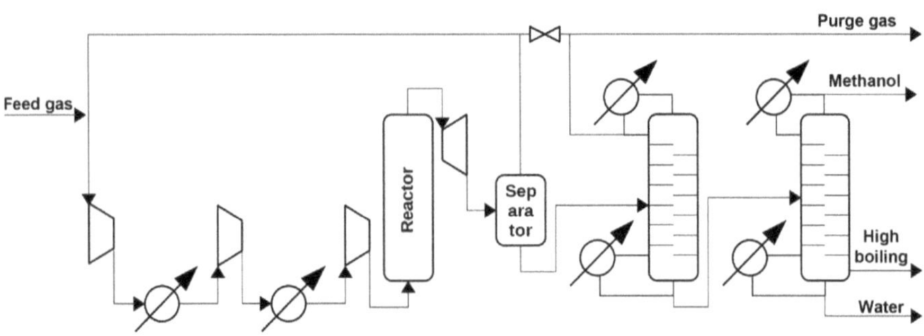

Figure 5.4 Flowsheet for a methanol production plant.

alcohols or liquid hydrocarbons do not pose problems. Commercial methanol is generally classified, according to the American Society for Testing and Materials, in A and AA purity grades (ASTM, 2012). Chemical methanol is normally grade AA. Two distillation columns are required for grade A methanol and three or four distillation columns are required for grade AA methanol (Andika et al., 2018).

In the case of a single column the high-boiling products and the process water are extracted from the bottom of the column, while liquid methanol and low-boiling products in the gaseous state are extracted from the top. In the case of two columns the first separates the low-boiling products, while the second separates methanol from the high-boiling products (Supp, 1990).

The selectivity of currently used catalysts is very high, over 99% (Twigg and Spencer, 2001). The impurities present in the final product [e.g. ethanol, methyl formate, dimethyl ether (DME), acetone] are thermodynamically favoured with respect to methanol. The formation of these by-products is favoured by the impurities of the catalyst, by high pressures and by the strong presence of carbon monoxide compared to the other reactants (Dieterich et al., 2020; Hansen and Nielsen, 2008).

The catalyst can be deactivated by poisoning or sintering (Fig. 5.5), and however, a duration of more than 4 years was demonstrated (Filippi and Badano, 2007; Hirotani, Nakamura, and Shoji, 1998). Impurities are usually eliminated upstream of the process until the typical purity requirements for the supply gas are met (Table 5.1). The compounds of sulphur and, above all, of chlorine are the most harmful. Sulphur can be eliminated thanks to the zinc oxide present in the catalysts (Kung, 1992).

Guard beds can be provided to protect the catalyst. The guard bed can be placed in a separate reactor or in the same reactor as the catalyst, obviously upstream of the catalysis bed. This guard bed can be constituted by alkalised alumina or by the catalyst itself (Moreo, 2011; Nielsen, 2009; Twigg and Spencer, 2001). It follows that sintering remains the only real cause of catalyst deactivation.

Compared to past ones, modern catalysts show a longer useful life and are able to maintain higher activity and selectivity even at the end of their life. For instance, Haldor Topsøe reported noticeable catalyst improvements in moving from MK-101,

Table 5.1 Gas purity requirements for Cu/ZnO based catalysts (Dieterich et al., 2020).

Component	Purity required
H_2S	<0.05 ppm
HCl	1 ppb
Metal carbonyl	Few ppb
Particles	<0.1 mg Nm^{-3}
Tar	<1.0 mg Nm^{-3}
Alkalis	<0.25 mg Nm^{-3}

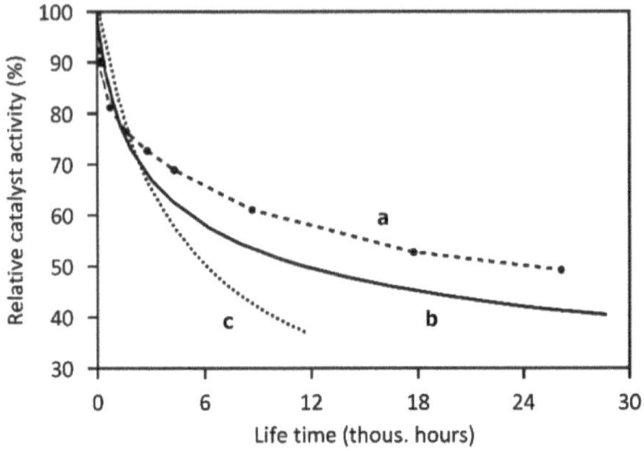

Figure 5.5 Deactivation curves of commercial methanol catalysts: (a) typical deactivation curve, (b) deactivation model fitted to a 295 t day^{-1} Lurgi plant, (c) Haldor Topsøe MK-101, industrial plant.
Source: Adapted from Dieterich, V. et al. (2020). Power-to-liquid via synthesis of methanol, DME or Fischer-Tropsch-fuels: A review. *Energy and Environmental Science*, *13*(10), 3207–3252. https://doi.org/10.1039/d0ee01187h.

to MK-121, to their recent product MK-151 Fence (Hansen, 2015; Nielsen, 2009). In general, a reduction of the catalyst activity over time is however inevitable.

To maintain a constant production of methanol along the time, the operating temperature can be gradually increased (Heydorn, Diamond, and Lilly, 2003).

Obviously, an important aspect in the development of this technology is that of costs and, consequently, of the economy of scale. Going from a capacity of 2000 to 5000 t day^{-1}, the plant cost increases by only 75%, whereas a 10,000 t day^{-1} plant costs only about the triple (Dieterich et al., 2020).

Reactor employed for methanol synthesis is similar to those utilised for Fischer–Tropsch and DME synthesis and can be classified in three different categories: fixed bed reactors (FBRs), fluidised bed reactors and liquid-phase reactors (Table 5.2). There are different FBR types:

- adiabatic, which requires a series of reactors with an intercooling system to keep the temperature at a proper level;
- polytropic
  - quench, which requires two separate bed sections and injection of cold feed reactant gas in between and
  - gas-cooled converter, in which cold gas refrigerates the catalyst; and
- steam raising converter (SRC), in which water evaporates on the opposite side of the catalyst.

Fluidised bed reactors and liquid-phase reactors are both isothermal. Fluidised bed reactors have not yet reached a commercial level, while among the liquid-phase reactors, only one has been industrialised. There are other alternative concepts as

Table 5.2 Kinds of reactors for methanol synthesis (Dieterich et al., 2020).

	FBR					
	Adiabatic FBR	**Quench**	**SRC**	**GCC**	**Fluidised bed reactor**	**Liquid-phase reactor**
Operation mode	**Adiabatic**	**Polytropic**	**Isothermal**	**Polytropic**	**Isothermal**	**Isothermal**
State of catalyst	Catalyst bed	Catalyst bed divided into sections	Catalyst filled into shell or tube side	Catalyst filled into shell or tube side	Catalyst in fluidised bed	Catalyst suspended in liquid hydrocarbons
Cooling concept	Series of reactors with intercooling	Cold feed gas is injected between sections	Cooled by water evaporation on opposing side of the catalyst	Cooled by preheating cold gas passing through the catalyst bed in tubes	Submerged coil heat exchanger	Liquid hydrocarbons as heat transfer medium; heat removed in internal heat exchanger
Advantages	Simple scale-up, no mechanical stress of catalyst, defined residence time	No mechanical stress of catalyst, improved temperature control	Very efficient heat recovery, good heat transfer and temperature control		High heat transfer coefficients, uniform temperature distribution	Excellent heat transfer performance, isothermal reaction profile
Disadvantages	Lower heat transfer, danger of hotspots, several pressure vessels and piping	Large catalyst volume	Large number of tubes, expensive	Limited heat transfer	Nonuniform residence time (bubble formation), attrition of catalyst, erosion of internals, difficult scale-up	High mechanical stress of catalyst

FBR, Fixed bed reactor; *GCC*, gas-cooled converter; *SRC*, steam raising converter.

well, such as membrane reactors, drip bed reactors and suspension reactors, but all require further development (Dieterich et al., 2020).

The most common reactor is nowadays the steam raising FBRs (or SRC) which is quasiisothermal (see Fig. 5.6) thanks to the heat removed by the water evaporation.

Fig. 5.7 shows several kinds of commercial reactors:

1. Lurgi tubular reactor, in which the gas flows inside a series of tubes containing the catalyst, whereas the boiling water is outside, in the vessel, generating medium pressure steam.
2. Linde Variobar, in which the boiling water circulates inside a coil, while the catalyst is distributed outside.
3. Toyo Multistage Radial Flow, which is a reactor developed in collaboration by Toyo Engineering Corporation and Mitsui Toatso Chemicals and differs from the previous two for the gas flow, which is radially distributed from the outside and is collected centrally after passing through the catalytic bed; cooling is achieved by means of coaxial pipes with water supply from the inside and steam production in the external annular section.
4. Mitsubishi Superconverter, which was developed in collaboration by Mitsubishi Gas Chemical and Mitsubishi Heavy Industries. Also, this reactor has a series of coaxial tubes, but they are crossed by the gas which preheats through the inner tube and then reacts by crossing the catalyst in the outer annular section; the cooling water is contained in the external vessel as in the Lurgi tubular reactor.
5. Methanol Casale isothermal methanol converter (IMC), which is characterised by the presence of hollow plates inside the catalysis bed; the bed is cooled by the circulation of a fluid inside the plates even with different flow rates at different heights, to maintain an almost isothermal temperature profile; the refrigerant fluid can be either the supply gas itself or boiling water.

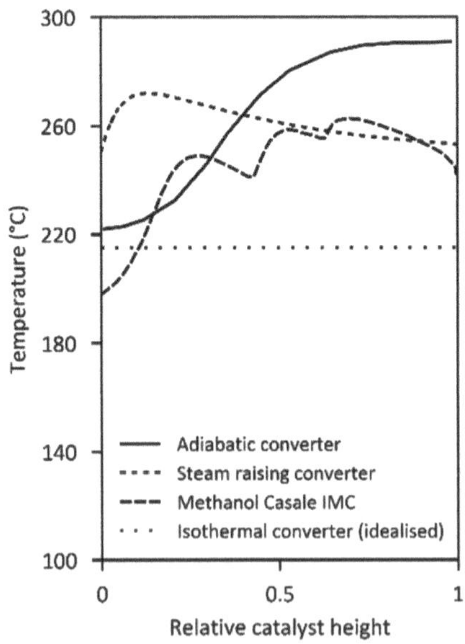

Figure 5.6 Temperature profiles along the reactor length of some reactors (Dieterich et al., 2020).

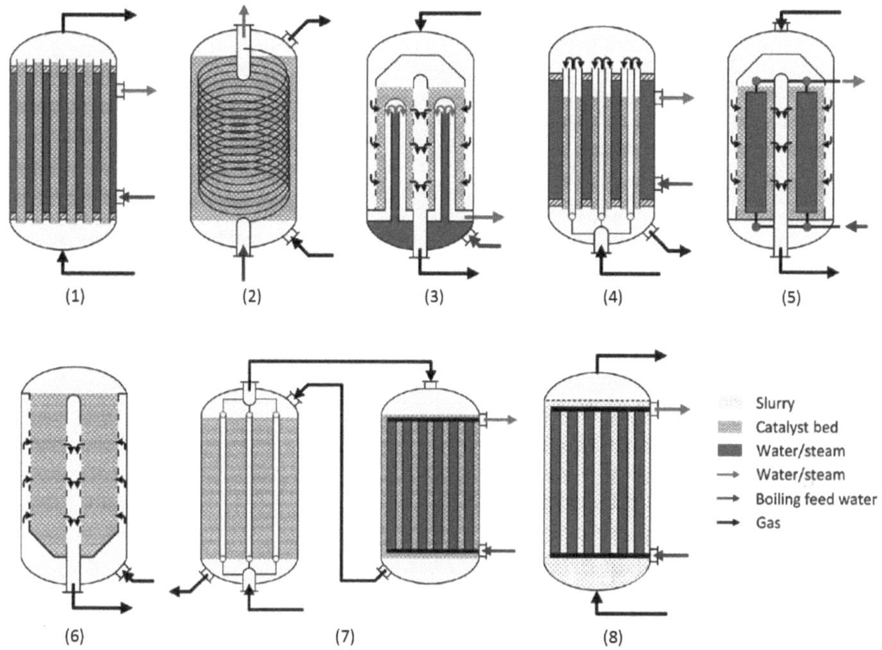

Figure 5.7 Simplified reactor layouts (Dieterich et al., 2020).

6. Haldor Topsøe adiabatic reactor, which is a multistage adiabatic reactor with intercooling.
7. Lurgi MegaMethanol, which is a two-stage reactor consisting of a first bed cooled with boiling water and a second bed cooled with the feed gas that is thus preheated.
8. Air Products LPMeOH (liquid-phase methanol), which, unlike the others, is not a FBR but formed by a mineral oil in which the catalyst particles are suspended; this solution achieves an excellent heat transfer, but at the expense of a low catalyst density inside the bed.

Several data about these kinds of reactors are shown in Tables 5.3 and 5.4 in which they are respectively grouped on the basis of gas flow direction: axial or radial and axial-radial (for Methanol Casale IMC). This is an important difference since it affects the pressure drop, which is significantly lower for radial flow reactors. As a consequence, also the power needed for recycled gas recompression is lower for radial flow reactors. Moreover, for the same reason, radial flow reactor could be higher than axial flow one.

As previously mentioned, reactions that lead to the production of methanol are exothermic and therefore favoured by low temperature which, on the other hand, is unfavourable in terms of the efficiency of the process. The high temperatures therefore speed up the process, but the final carbon conversion is lower.

Therefore the reactor design is important not only for its ability to retain temperature within the ideal range for reactions but also for the efficiency of recovery of the generated heat. In general, higher outlet/peak temperature allows to produce steam at higher pressure. For most industrial reactors in operation the ranges of

Table 5.3 Characteristics and operating parameters of main commercial axial flow reactors.

	Lurgi	**Superconverter**	**MegaMethanol**	**LPMeOH**
Licensor	Lurgi	MGC and MHI	Lurgi	Air products
Type	SRC	GCC/SRC	GCC and SRC	Slurry
Catalyst location	Tube side	Double pipes	Shell side	Shell side
Heat exchanger	Tubular	Tubular	Tubular	Tubular
Stages	1	1	2	1
T_{out} (°C)	255	190	220	215
T_{peak} (°C)	270	270	270	215
Pressure (bar)	50–100	55–100	75	30–50
Recycle ratio	3–4	2–3	2–2.7	1–5
MeOH yield (mol%)	6–7	10–15	11	8–12
Max capacity (t day^{-1})	1500–2200	Not available	5000–10,000	Low
Steam pressure (bar)	29–43	19–45	50–60	16–25
Plants/projects	>55	9	>10	Demo

GCC, Gas-cooled converter; *MGC*, Mitsubishi gas chemical; *MHI*, Mitsubishi heavy industries; *SRC*, steam raising converter.
Source: Adapted from Dieterich, V. et al. (2020). Power-to-liquid via synthesis of methanol, DME or Fischer-Tropsch-fuels: A review. *Energy and Environmental Science*, *13*(10), 3207–3252. https://doi.org/10.1039/d0ee01187h.

operating temperatures and pressures are, respectively, 200°C–300°C and 50–100 bar, while the pressure range for the steam obtained is 16–45 bar, with the sole exception of the Lurgi which allows to obtain steam between 50 and 60 bar (see Table 5.3).

Another parameter that differentiates the reactor types is the recycle factor (f_r) required to obtain high overall conversion values:

$$f_r = \frac{F_{V,rg}}{F_{V,fg}} \quad (5.4)$$

with $F_{V,rg}$ and $F_{V,fg}$ representing the volumetric flow of recycled gas and feed gas, respectively.

Table 5.4 Characteristics and operating parameters of main commercial radial flow reactors.

	IMC	**Variobar**	**MRF-Z**	**Haldor Topsøe**
Licensor	Methanol Casale	Linde	Toyo (TEC)	Haldor Topsøe
Type	SRC	SRC	SRC	Adiabatic
Catalyst location	Shell side	Shell side	Shell side	Fixed bed
Heat exchanger	Plate	Tubular	Bayonet	Intercooler
Stages	1	1	1	2–4
T_{out} (°C)	225	Not available	240	290
T_{peak} (°C)	280	Not available	280	Not available
Pressure (bar)	65–80	50–150	80–100	50–100
Recycle ratio	3	Not available	Not available	3–5
MeOH yield (mol%)	10.1–13.3	Not available	10	7
Max capacity ($t\ day^{-1}$)	7000–10,000	4000	5000	10,000
Steam pressure (bar)	25–32	40	Not available	Not available
Plants/projects	9	8	6	>40

IMC, Isothermal methanol converter; *SRC*, steam raising converter.
Source: Adapted from Dieterich, V. et al. (2020). Power-to-liquid via synthesis of methanol, DME or Fischer-Tropsch-fuels: A review. *Energy and Environmental Science, 13*(10), 3207–3252. https://doi.org/10.1039/d0ee01187h.

As already written, a total conversion is not possible due to the need for purge. As the conversion increases in the single passage inside the reactor, the recycle factor decreases: Table 5.3 shows f_r values ranging from a minimum of 1 to a maximum of 5 and allow to obtain a carbon conversion in a range of 93%–98%. Reducing the amount of recycled gas makes it possible to adopt a smaller reactor and to consume less energy for gas recompression.

When the conversion is carried out in more than one stage, it is possible to condense and separate the methanol between one stage and the other, thus favouring the conversion in the next stage and reducing the final amount of unreacted gas.

This can be done for adiabatic reactors, while among SRCs, only the Lurgi MegaMethanol features a double stage (Bertau et al., 2014).

A parameter used to normalise the gas flow to the reactor size is the gas hourly space velocity (GHSV):

$$GHSV = \frac{F_{V,g,in}}{V_R} \tag{5.5}$$

where $F_{V,g,in}$ is the volumetric flow in standard condition of the feed gas and V_R is the volume of the reactor. In the case of SRCs, its value ranges from about 6000 to 12,000 h^{-1} for each stage, whereas LPMeOH shows GHSV lower than 5000 h^{-1} (Dieterich et al., 2020).

Other types of reactors have been developed but have not been built on an industrial scale (Arora et al., 2018; Gallucci and Basile, 2007; Pass et al., 1990; Rahimpour and Ghader, 2003; Terreni et al., 2019).

5.2 Methanol production from carbon dioxide

As can be seen from the comparison between reactions (5.1) and (5.2), this second reaction is less favourable because it requires an extra mole of hydrogen for each mole of methanol obtained. However, the methanol production starting from carbon dioxide is interesting today because it is connected to the possibility of reducing climate-altering gases emissions into the atmosphere by capturing the carbon dioxide directly from the sources with the greatest concentration (chimneys of power plants and industries).

Although catalysts similar to those for syngas conversion have been used (Doss, Ramos, and Atkins, 2009; Pontzen et al., 2011), these are not ideal for this process. In fact, in this case the presence of water in the reactor is much greater (30%–40% vs 10%–15% for production from syngas) and, since it is also a product, it hinders the progress of the reaction. Against this disadvantage, carbon dioxide production offers a much lower by-product content (see Table 5.5) and a lower consumption for methanol distillation as a consequence.

Due to the large water production, it is necessary to develop specific catalysts. Information on these new catalysts can be found in two recent reviews (Dang et al., 2019; Guil-López et al., 2019).

The study and application of specific catalysts for the production of methanol starting from carbon dioxide began in the 1990s. Examples and information on both such catalysts and reactors are reported by Dieterich et al. (2020). They are mostly small fixed bed pilot reactors, with only three of them exceeding 100 t $year^{-1}$. The temperatures are all around 250°C, whereas the pressures have a much wider range, between 30 and 100 bar. The GHSV and the conversion rate are also very variable, respectively, from 1 to 18 h^{-1} and from 40% to 96.5% (the latter obtained thanks to a two-stage process).

Table 5.5 Raw methanol compositions of CO and CO_2-based feed gas.

	CO-syngas	CO_2-based feed gas (H_2:CO_2 = 3)
Temperature (°C)	250	230–270
Main components (wt%)		
Methanol	84.5	63.0–63.7
Water	15.4	35.6–36.9
Impurities (wt ppm)		
n-Paraffins	78	0
Higher alcohols	626	28–148
Esters (mainly methyl formate)	582	129–450
Ketones	24	0
Dimethyl ether	61	14–24
Total impurities (wt ppm)	1371	248–478
Methanol selectivity, except water (%)	99.84	99.92–99.96

Source: Adapted from Dieterich, V. et al. (2020). Power-to-liquid via synthesis of methanol, DME or Fischer-Tropsch-fuels: A review. *Energy and Environmental Science*, *13*(10), 3207–3252. https://doi.org/10.1039/d0ee01187h.

5.3 Innovative processes

Alternative processes to thermochemical synthesis are also being studied:

- electrochemical conversion (coelectrolysis);
- biological conversion; and
- photocatalytic conversion, in which photosynthesis is emulated to perform the reduction of carbon dioxide thanks to sunlight and suitable catalysts and produce synthetic fuels.

Obviously, they have not yet reached technical maturity and economic convenience. Moreover, photocatalytic conversion could store solar energy but is not a power to fuel technology and will be not considered.

5.3.1 Coelectrolysis

An interesting alternative process which is gaining increasing interest is coelectrolysis.

A coelectrolyser (Fig. 5.8) is a solid oxide electrolyser fed with steam and carbon dioxide at the cathode where H_2O and CO_2 reduction reactions (5.6) and (5.7) occur together with reaction (5.3):

$$H_2O + 2\,e^- \Rightarrow H_2 + O^{2-} \tag{5.6}$$

$$CO_2 + 2\,e^- \Rightarrow CO + O^{2-} \tag{5.7}$$

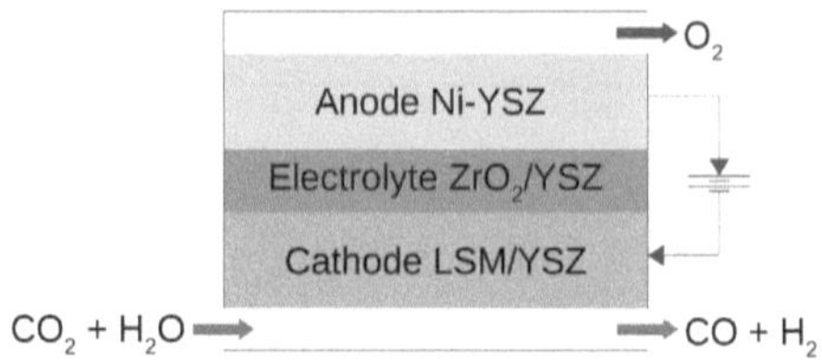

Figure 5.8 Coelectrolyser.

The oxygen evolution reaction takes place at the anode:

$$2\,O^{2-} \Rightarrow O_2 + 4\,e^- \tag{5.8}$$

The most common electrolyte is a ceramic layer of ZrO_2 doped with Y_2O_3 (YSZ, yttria-stabilized zirconia), stable and capable to conduct ions at high temperatures (800°C–1000°C). The most common materials used for electrodes are lanthanum strontium manganite/YSZ for anode and nickel–YSZ for cathode.

The direct electrochemical reduction of carbon dioxide to methanol was tried by some researchers along the last decades. Semiconductor electrodes were tested by Freese and Leach with the goal to store solar energy in methanol: GaAs and InP showed a good selectivity but required a high potential and suffered from corrosion (Canfield and Frese, 1983). Subsequent experiments with metal electrodes such as Pd, Pt, Co and Fe did not give good results because the main product was carbon monoxide. Using molybdenum electrodes, methanol can be obtained as the major carbon-containing product with small amounts of carbon monoxide and traces of methane (Summers, Leach, and Frese, 1986). Cu-based catalysts, in particular Cu/ZnO, appear more promising for the conversion of CO_2 and CO to methanol in the presence of H_2 (Frese, 1991). Ruthenium was also tested with some good results (Popić, Avramov-Ivić, and Vuković, 1997; Spataru et al., 2003).

However, the current technology is cost competitive only for carbon monoxide production (Durst et al., 2015) so that it is necessary for a subsequent synthesis process to obtain methanol.

5.3.2 *Biological oxidation of methane*

There is not a direct biological conversion from carbon dioxide to methanol, but methanol could be obtained from methane. Therefore a two-step process from power to methane and methane to methanol can be considered when the desired final product is a liquid fuel.

Just as methanogenic bacteria exist, there are many others (methanotrophic) which feed on methane and are capable to generate methanol as an intermediate compound in their metabolism. Methanogenic bacteria live in swamps, rivers, oceans, ponds and sewage sludge producing methane through the anaerobic decay of organic matter. Therefore most methanotrophic bacteria live in the same environments, exploiting the oxygen of other compounds, or at the aerobic–anaerobic interface by oxidising the methane produced by the former. They are also mesophilic bacteria too, but they adapt well to other temperatures.

Biological methanol production is catalysed by methane monooxygenase (MMO) enzymes in a single step at atmospheric pressure, 25°C–35°C, and pH in the range 7.0–7.65 (Hanson and Hanson, 1996). Methanotrophic bacteria metabolism is more complex and also based on other enzymes: methanol dehydrogenase (MDH), formaldehyde dehydrogenase (FADH) and formate dehydrogenase (FDH). The chain of subsequent transformations is represented in Fig. 5.9. The methanol produced by MMO is sequentially converted to formaldehyde by MDH, to formate by FADH and finally to carbon dioxide by FDH.

In the initial reaction the MMO breaks the O–O bond of the oxygen molecule using substances that change from the reduced to the oxidised form (e.g. NADH $\rightarrow NAD^{+}$). One of the oxygen atoms binds to the methane molecule to generate methanol, while the other oxygen atom combines with hydrogen and produces water (Lipscomb, 1994).

It could be a valid alternative to the traditional conversion from methane to methanol, which is an energy-intensive and expensive process because it requires high temperatures and pressures. Whereas the commercial process proceeds in two steps, namely methane reforming (5.9), highly endothermic:

$$CH_4 + H_2O \Rightarrow CO + 3H_2 \quad \Delta H_{298K} = +206.1\ \text{kJ mol}^{-1} \tag{5.9}$$

followed by methanol synthesis (5.1), the direct conversion from methane to methanol is an exothermic reaction:

$$CH_4 + \frac{1}{2} O_2 \Rightarrow CH_3OH \quad \Delta H_{298K} = -126.2\ \text{kJ mol}^{-1} \tag{5.10}$$

Another advantage is that this process is suitable for small-scale operations, that is connected to a distributed production of methane, and requires a low investment. Unfortunately, the transition from laboratory experiments to industrialisation of the process requires, as always, the overcoming of some obstacles.

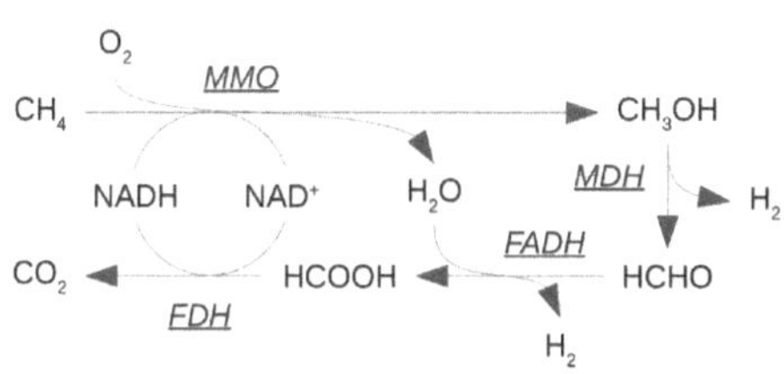

Figure 5.9 Simplified pathway for oxidation of methane in methanotrophic bacteria.

The main options for achieving the partial oxidation of methane to methanol using the knowledge developed about the enzyme MMO are the following:

- whole-cell methanotrophic culture
- genetically modified organisms
- isolated MMO enzymes
- synthetic MMO analogues
- ammonia oxidising bacteria

The direct culture of methanotrophic bacteria is certainly the cheapest solution since the bacteria themselves take care of all the biological apparatus necessary for the process as well as their own reproduction. However, it is evident that the metabolism of whole cells is not devoted to the production of methanol and therefore problems arise related to the interruption of the natural process which is feasible by inhibition of the MDH enzyme. Furthermore, a splitting of the process must be provided in a methanol production section and a mother culture section intended for cell reproduction.

An alternative is to try to optimise the reaction processes using whole cells by genetic engineering. However, in this sector, we are still at the patent level (e.g. Silverman and Regitsky, 2014) without applications that can demonstrate its feasibility.

The only component of methanotrophic bacteria useful for the conversion of methane to methanol is the MMO enzyme. Therefore it is possible to think of using it in isolation, immobilised on artificial matrices, thus avoiding the other reactions of the metabolism of these bacteria. The advantage of avoiding unnecessary reactions is contrasted by the need to isolate and purify the enzyme (Hakemian and Rosenzweig, 2007) and, ultimately, the activity and duration that impact on the operating cost and do not make this option currently advantageous over whole-cell culture.

Wanting to overcome these limits, one could resort to the creation of catalysts that emulate the behaviour of MMO enzymes, to obtain a stable thermochemical process aimed only at the production of methanol. This is an option that could also bring other advantages in terms of optimisation (selectivity, efficiency and speed of conversion), even if it is still an immature technology (Que and Tolman, 2008).

Even the bacteria that oxidise ammonia, through the enzyme ammonia monooxygenase, are able to oxidise methane to form methanol (Hyman, Murton, and Arp, 1988; Taher and Chandran, 2013), but also in this case the technology is not yet mature.

Nomenclature

F	flow in standard condition
f	factor
FADH	formaldehyde dehydrogenase
FBR	fixed bed reactor

FDH	formate dehydrogenase
GCC	gas-cooled converter
GHSV	gas hourly space velocity
LSM	lanthanum strontium manganite
MDH	methanol dehydrogenase
MMO	methane monooxygenase
NAD	nicotinamide adenine dinucleotide
SRC	steam raising converter
V	volume
YSZ	yttria-stabilized zirconia

Subscripts

f	feed
g	gas
R	reactor
r	recycled
V	volumetric

References

Ali, K. A., Abdullah, A. Z., and Mohamed, A. R. (2015). Recent development in catalytic technologies for methanol synthesis from renewable sources: A critical review. *Renewable and Sustainable Energy Reviews*, *44*, 508–518. Available from https://doi.org/10.1016/j.rser.2015.01.010.

Andika, R., et al. (2018). Co-electrolysis for power-to-methanol applications. *Renewable and Sustainable Energy Reviews*, *95*, 227–241. Available from https://doi.org/10.1016/j.rser.2018.07.030.

Arora, A., et al. (2018). Optimal methanol production via sorption-enhanced reaction process. *Industrial and Engineering Chemistry Research*, *57*(42), 14143–14161. Available from https://doi.org/10.1021/acs.iecr.8b02543.

ASTM. (2012). *ASTM D1152-06*. Standard specification for methanol (methyl alcohol). American Society for Testing and Materials.

Bertau, M., et al. (2014). In M. Bertau, et al. (Eds.), *Methanol. The basic chemical and energy feedstock of the future: Asinger's vision today* (1st ed.). Berlin, Heidelberg: Springer-Verlag. Available from https://doi.org/10.1007/978-3-642–39709-7.

Bjorck, C. E., Dobson, P. D., and Pandhal, J. (2018). Biotechnological conversion of methane to methanol: Evaluation of progress and potential. *AIMS Bioengineering*, *5*(1), 1–38. Available from https://doi.org/10.3934/bioeng.2018.1.1.

Bromberg, L., and Cheng, W. K. (2007). *Methanol as an alternative transportation fuel in the U.S.: Options for sustainable and/or energy-secure transportation*. Cambridge: Massachusetts Institute of Technology, Sloan Laboratories for Automotive and Aircraft Engines.

Canfield, D., and Frese, K. W. (1983). Reduction of carbon dioxide to methanol on n-and p-GaAs and p-InP. Effect of crystal face, electrolyte and current density. *Journal of the Electrochemical Society*, *130*(8), 1772–1773. Available from https://doi.org/10.1149/1.2120090.

Dang, S., et al. (2019). A review of research progress on heterogeneous catalysts for methanol synthesis from carbon dioxide hydrogenation. *Catalysis Today*, *330*, 61–75. Available from https://doi.org/10.1016/j.cattod.2018.04.021.

Dieterich, V., et al. (2020). Power-to-liquid *via* synthesis of methanol, DME or Fischer-Tropsch-fuels: A review. *Energy and Environmental Science*, *13*(10), 3207–3252. Available from https://doi.org/10.1039/d0ee01187h.

Doss, B., Ramos, C., and Atkins, S. (2009). Optimization of methanol synthesis from carbon dioxide and hydrogen: Demonstration of a pilot-scale carbon-neutral synthetic fuels process. *Energy and Fuels*, *23*(9), 4647–4650. Available from https://doi.org/10.1021/ef900466u.

Durst, J., et al. (2015). Electrochemical CO_2 reduction – A critical view on fundamentals, materials and applications. *Chimia*, *69*(12), 769–776. Available from https://doi.org/10.2533/chimia.2015.769.

Filippi, E., and Badano, M. (2007). Methanol converters and syncope designs for gasification plants. In *World methanol conference*. Orlando, FL: Methanol Casale S.A., Lugano, Switzerland. Available at https://www.researchgate.net/publication/276292511_Methanol_converters_and_syncope_designs_for_gasification_plants.

Frese, K. W. (1991). Electrochemical reduction of CO_2 at intentionally oxidized copper electrodes. *Journal of the Electrochemical Society*, *138*(11), 3338–3344. Available from https://doi.org/10.1149/1.2085411.

Gallucci, F., and Basile, A. (2007). A theoretical analysis of methanol synthesis from CO_2 and H_2 in a ceramic membrane reactor. *International Journal of Hydrogen Energy*, *32* (18), 5050–5058. Available from https://doi.org/10.1016/j.ijhydene.2007.07.067.

Grabow, L. C., and Mavrikakis, M. (2011). Mechanism of methanol synthesis on Cu through CO_2 and CO hydrogenation. *ACS Catalysis*, *1*(4), 365–384. Available from https://doi.org/10.1021/cs200055d.

Guil-López, R., et al. (2019). Methanol synthesis from CO_2: A review of the latest developments in heterogeneous catalysis. *Materials*, *12*(23). Available from https://doi.org/10.3390/ma122333902.

Hakemian, A. S., and Rosenzweig, A. C. (2007). The biochemistry of methane oxidation. *Annual Review of Biochemistry*, *76*, 223–241. Available from https://doi.org/10.1146/annurev.biochem.76.061505.175355.

Hansen, J. B. (2015). Methanol production technology: Todays and future renewable solutions. In *Methanol workshop*. Lund: Haldor Topsøe. Available at http://www.lth.se/fileadmin/mot2030/filer/9._Bogild_Hansen_-_Methanol_Production_Technology_Todays_and_Future.pdf.

Hansen, J. B., and Nielsen, P. E. H. (2008). 13.13 Methanol synthesis. In G. Ertl, et al. (Eds.), *Handbook of heterogeneous catalysis – Part 13. Energy-related catalysis* (1st ed., pp. 2920–2949). Wiley-VCH Verlag GmbH & Co. Available from https://doi.org/10.1002/9783527610044.hetcat0148.

Hanson, R. S., and Hanson, T. E. (1996). Methanotrophic bacteria. *Microbiological Reviews*, *60*(2), 439–471. Available from https://doi.org/10.1128/mmbr.60.2.439-471.1996.

Heydorn, E. C., Diamond, B. W., and Lilly, R. D. (2003). *Commercial-scale demonstration of the liquid phase methanol (LPMEOH) process. Final report, Volume 2: Project performance and economics*. U.S. Department of Energy.

Hirotani, K., Nakamura, H., and Shoji, K. (1998). Optimum catalytic reactor design for methanol synthesis with TEC MRF-Z® reactor. *Catalysis Surveys From Japan*, *2*(1), 99–106. Available from https://doi.org/10.1023/A:1019061921451.

Hyman, M. R., Murton, I. A. N. B., and Arp, D. J. (1988). Interaction of ammonia monooxygenase from *Nitrosomonas europaea* with alkanes, alkenes and alkynes. *Applied and Environmental Microbiology*, *54*(12), 3187–3190.

Jadhav, S. G., et al. (2014). Catalytic carbon dioxide hydrogenation to methanol: A review of recent studies. *Chemical Engineering Research and Design*, *92*(11), 2557–2567. Available from https://doi.org/10.1016/j.cherd.2014.03.005.

Kung, H. H. (1992). Deactivation of methanol synthesis catalysts – A review. *Catalysis Today*, *11*(4), 443–453. Available from https://doi.org/10.1016/0920-5861(92)80037-N.

Lee, J. S., et al. (1993). A comparative study of methanol synthesis from CO_2/H_2 and CO/H_2 over a $Cu/ZnO/Al_2O_3$ catalyst. *Journal of Catalysis*, *144*(2), 414–424. Available from https://doi.org/10.1006/jcat.1993.1342.

Lipscomb, J. D. (1994). Biochemistry of the soluble methane monooxygenase. *Annual Review of Microbiology*, *48*, 371–399. Available from https://doi.org/10.1146/annurev.mi.48.100194.002103.

Malcolm Pirnie Inc. and American Methanol Institute. (1999). *Evaluation of the fate and transport of methanol in the environment*. Washington, DC: American Methanol Institute.

Moreo, P. (2011). Casale coal-based ammonia and methanol synloops. In *Nitrogen & syngas conference*. Duesseldorf, Germany: Casale Group, Lugano, Switzerland. Available at https://dokumen.tips/documents/casale-coal-based-ammonia-and-methanol-coal-based-ammonia-and-methanol-synloops.html.

Nielsen, C. S. (2009). *Topsøe methanol technology for coal based plants*. Available at http://topsoe.ru/sites/default/files/topsoe_methanol_coal_based_plants.ashx_.pdf.

Pass, G., et al. (1990). Methanol synthesis in a trickle-bed reactor. *AIChE Journal*, *36*(7), 1054–1060. Available from https://doi.org/10.1002/aic.690360712.

Pontzen, F., et al. (2011). CO_2-based methanol and DME – Efficient technologies for industrial scale production. *Catalysis Today*, *171*(1), 242–250. Available from https://doi.org/10.1016/j.cattod.2011.04.049.

Popić, J. P., Avramov-Ivić, M. L., and Vuković, N. B. (1997). Reduction of carbon dioxide on ruthenium oxide and modified ruthenium oxide electrodes in 0.5 M $NaHCO_3$. *Journal of Electroanalytical Chemistry*, *421*(1–2), 105–110. Available from https://doi.org/10.1016/S0022-0728(96)04823-1.

Que, L., and Tolman, W. B. (2008). Biologically inspired oxidation catalysis. *Nature*, *455* (7211), 333–340. Available from https://doi.org/10.1038/nature07371.

Rahimpour, M. R., and Ghader, S. (2003). 'Theoretical investigation of a Pd-membrane reactor for methanol synthesis. *Chemical Engineering and Technology*, *26*(8), 902–907. Available from https://doi.org/10.1002/ceat.200301717.

Silverman, J., and Regitsky, D. (2014). *Genetically engineered microorganisms for biological oxidation of hydrocarbons – World patent WO 2014/062703 A1 – PCT/US2013/065087*.

Spataru, N., et al. (2003). Electrochemical reduction of carbon dioxide at ruthenium dioxide deposited on boron-doped diamond. *Journal of Applied Electrochemistry*, *33*(12), 1205–1210. Available from https://doi.org/10.1023/B:JACH.0000003866.85015.b6.

Studt, F., et al. (2015). The mechanism of CO and CO_2 hydrogenation to methanol over Cu-based catalysts. *ChemCatChem*, *7*(7), 1105–1111. Available from https://doi.org/10.1002/cctc.201500123.

Summers, D. P., Leach, S., and Frese, K. W. (1986). The electrochemical reduction of aqueous carbon dioxide to methanol at molybdenum electrodes with low overpotentials. *Journal of Electroanalytical Chemistry*, *205*(1–2), 219–232. Available from https://doi.org/10.1016/0022-0728(86)90233-0.

Supp, E. (1990). *How to produce methanol from coal*. Berlin, Heidelberg: Springer-Verlag. Available from http://doi.org/10.1007/978-3-662-00895-9.

Taher, E., and Chandran, K. (2013). High-rate, high-yield production of methanol by ammonia-oxidizing bacteria. *Environmental Science and Technology*, *47*(7), 3167–3173. Available from https://doi.org/10.1021/es3042912.

Terreni, J., et al. (2019). Sorption-enhanced methanol synthesis. *Energy Technology*, *7*(4). Available from https://doi.org/10.1002/ente.201801093.

Twigg, M. V., and Spencer, M. S. (2001). Deactivation of supported copper metal catalysts for hydrogenation reactions'. *Applied Catalysis A: General*, *212*(1−2), 161−174. Available from https://doi.org/10.1016/S0926-860X(00)00854-1.

Wang, W., et al. (2011). Recent advances in catalytic hydrogenation of carbon dioxide. *Chemical Society Reviews*, *40*(7), 3703−3727. Available from https://doi.org/10.1039/c1cs15008a.

Wender, I. (1996). 'Reactions of synthesis gas. *Fuel Processing Technology*, *48*(3), 189−297. Available from https://doi.org/10.1016/S0378-3820(96)01048-X.

Wu, J., et al. (2001). The stability of Cu/ZnO-based catalysts in methanol synthesis from a CO_2-rich feed and from a CO-rich feed. *Applied Catalysis A: General*, *218*(1−2), 235−240. Available from https://doi.org/10.1016/S0926-860X(01)00650-0.

Power-to-DME: a cornerstone towards a sustainable energy system

6

Malte Gierse[1], Ramy E. Ali[1], Mohamed Ouda[1], Achim Schaadt[1], Jörg Sauer[2] and Christopher Hebling[1]
[1]Fraunhofer Institute for Solar Energy Systems ISE, Freiburg, Germany, [2]Institute of Catalysis Research and Technology (IKFT), Karlsruhe Institute of Technology (KIT), Eggenstein-Leopoldshafen, Germany

6.1 Introduction

Dimethyl ether (DME) is the simplest of all ethers and a well-known chemical entity. Due to its thermophysical and chemical properties as well its combustion behaviour, it offers a wide range of applications and, thus, is a promising target molecule for power-to-liquid processes. DME is nontoxic, noncarcinogenic, nonteratogenic and nonmutagenic. It is a volatile compound; however, liquefaction can be achieved by cooling to $-24.9°C$ at atmospheric pressure or by pressurising to an absolute pressure of 0.6 MPa at 25°C (Maus, 2019; Semelsberger, Borup, and Greene, 2006). The vapour pressure is comparable to propane and butane, making DME a suitable substitute or blending component for liquefied petroleum gas (LPG).

Furthermore, DME can be used as a green solvent, refrigerant and propellant. In addition to its application as final product, DME plays a key role as an intermediate in many subsequent synthesis routes for fuels and chemicals. Particular focus is directed towards the use of DME as an alternative fuel and energy carrier. Besides the mentioned potential as LPG substitute, DME has very beneficial characteristics for the use in internal combustion diesel engines. In addition, DME can be used for fuel cell application either as a hydrogen source (Ledesma et al., 2019) in hydrogen fuel cells or as a fuel in direct DME fuel cells (Kashyap, Teller, and Schechter, 2019). To date, the major application of DME is as an LPG replacement and blending component. Especially countries with limited access to gas reserves could profit from a domestic DME production to limit LPG imports while at the same time fulfilling stricter environmental regulations (Grand View Research Inc., 2020). The rising energy demand in Asia Pacific is expected to boost the overall DME market in the upcoming years. China is responsible for more than 85% of the worldwide DME demand (Pulidindi and Pandey, 2018).

The current world production of DME is approximately 5 million tonnes per annum (Methanol Institute, 2016) and is based on the feedstock methanol, which is being produced from synthesis gas (syngas, CO/H_2). Furthermore, the direct

Power to Fuel. DOI: https://doi.org/10.1016/B978-0-12-822813-5.00010-2

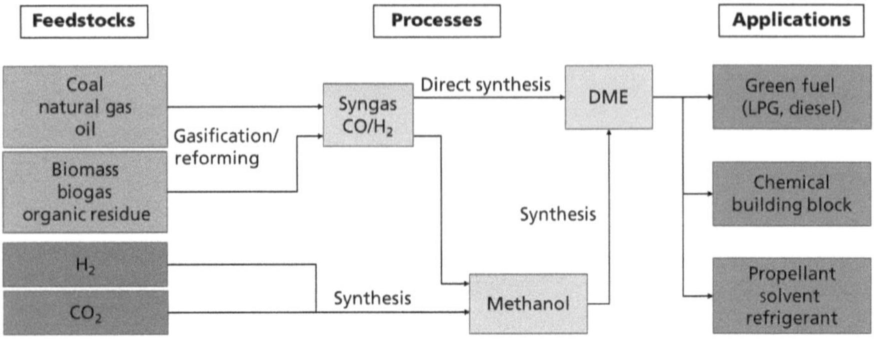

Figure 6.1 Overview of DME production pathways for different feedstocks. *DME*, Dimethyl ether.

synthesis of DME from syngas in a single step is possible. Syngas can be produced from conventional feedstock such as coal, oil or natural gas but also from renewable sources including biomass, biogas or residues from agriculture or paper industry (Dahmen et al., 2017; Maus, 2019). Moreover, methanol can be synthesised from CO_2 and H_2. By using this methanol as a building block for the DME synthesis the production of carbon-neutral DME is possible. Fig. 6.1 illustrates feedstocks, production pathways and applications of DME.

6.1.1 Dimethyl ether as a green fuel

DME is discussed intensively as a green fuel. Besides a great deal of attention in research, many companies such as Volvo (BioAge Group LLC, 2013), Mack Trucks (AB Volvo, 2017) and Ford (Ford-werke GmbH, 2015) promote DME as a fuel due to various ecological, technical and economic aspects. A major benefit of DME compared to conventional fuels is the absence of carbon-to-carbon bonds, leading to a practically soot-free combustion (Matzen and Demirel, 2016). This additionally entails the advantage, that DME engines can be operated with higher exhaust gas recirculation rates to reduce NO_x emissions regardless of the NO_x-soot-trade-off present at conventionally fuelled engines (Arcoumanis et al., 2008).

DME exhibits a high cetane number (CN) in the range of 55–60 which is slightly higher than that of conventional diesel (CN = 51–54), indicating further optimisation potential for diesel engines (Maus, 2019; Schemme et al., 2017). For this reason, DME has been promoted as an alternative diesel fuel since the mid-1990s (Semelsberger et al., 2006). Due to its oxygen content, DME has a lower energy density compared to conventional diesel. However, compared to other green fuels such as methane or methanol, the volumetric energy density of DME is superior. Table 6.1 compares the physical properties of DME to that of conventional diesel.

The apparent differences compared to conventional diesel lead to the necessity for technical adaptions of the tank and injection system. Owing to the vapour pressure similar to that of LPG, comparable tank systems can be used to store DME at

Table 6.1 Physical properties of DME and conventional diesel.

		Diesel	DME	
Molecular formula	–	$C_{14}H_{30}$	CH_3OCH_3	Semelsberger et al. (2006)
Oxygen content	wt.%	<1	34.8	Willems et al. (2020)
Density	$kg\ L^{-1}$	0.856	0.661[a]	Semelsberger et al. (2006)
Lower heating value	$MJ\ kg^{-1}$	41.66	28.62	Semelsberger et al. (2006)
Cetane number	–	51–54	55–60	Willems et al. (2020)
Kinematic viscosity (40°C)	$mm^2\ s^{-1}$	3	0.184	Willems et al. (2020)
Normal boiling point	°C	125–400	−24.9	Semelsberger et al. (2006)

DME, Dimethyl ether.
[a]Density in liquid state at p = 1.013 bar and T = −25°C.

saturated state between 2 and 7 bar. To account for the lower energy density of DME, an enlargement of the injector nozzle's flow cross section is necessary. The poor lubrication behaviour of DME requires an oil lubrication of the injection pump (Willems et al., 2020). Furthermore, precaution must be taken concerning the chemical stability of the used rubber seals since DME can lead to swelling and dissolution of various rubber materials. Ethylene propylene diene monomer and perfluoroelastomer have been identified as suitable seal materials (Sorenson, 2001; Willems et al., 2020; Wu, Zhang, and Huang, 2008).

A major criterion for the potential evaluation of alternative fuels is the required effort to create an infrastructure for the fuel. This comprises not only refuelling stations, but also tank trucks, harbour terminals and tank ships for ocean transport. In this regard, DME benefits from its vapour pressure comparable to that of propane and butane, allowing the use of existing LPG infrastructure (Azizi et al., 2014; Semelsberger et al., 2006).

Compared to other green fuels such as Fischer–Tropsch (FT) diesel, biodiesel, methanol, ethanol and methane, DME exhibits the lowest well-to-wheel greenhouse gas emissions (Semelsberger et al., 2006). Furthermore, considering a hydrogen-based production of green fuels, DME shows the highest conversion efficiency and the lowest cost in comparison to the other examined liquid fuels polyoxymethylene DMEs (OME), FT diesel, iso-octane, ethanol and gasoline from the methanol-to-gasoline (MTG) process (Schemme et al., 2019).

Another application of DME is the substitution of LPG as residential fuel for heating and cooking. Gas stove systems running on pure DME require modifications of the burner tip, vaporiser and a slightly increased tank volume to account for the reduced energy density compared to LPG (Semelsberger et al., 2006). However, using DME for blending with LPG is the major commercial application, representing 90% of China's DME demand (Fleisch, Basu, and Sills, 2012; Saraswat and Bansal, 2017). Recently, the Californian renewable DME producer

Oberon Fuels partnered with the world's largest propane distributor SHV Energy, to reduce the carbon footprint of conventional propane gas by blending it with Oberon's renewable produced DME (Williams, 2020).

6.1.2 Dimethyl ether in fuel cells

Besides its use in internal combustion engines, DME is also a suitable energy carrier for fuel cell applications. So-called direct DME fuel cells use DME at the anode and oxygen at the cathode of Proton Exchange Membrane Fuel Cells (Du et al., 2019). Most research groups used a platinum-based electrode on carbon support as anode catalyst (Ong, Kamarudin, and Basri, 2017). In comparison to other fuels such as methanol, ethanol and formic acid, DME shows low crossover through Nafion membranes (Du et al., 2019; Kashyap et al., 2019). However, low reaction kinetics is the major drawback of this technology and the reason for ongoing research addressing the development of new catalysts such as platinum-based PtRu (Li et al., 2014; Zhang et al., 2008) and PtCu (Gavriel, Sharabi, and Elbaz, 2017).

Besides the direct use in fuel cells, DME is also discussed as a hydrogen carrier, allowing easy and efficient energy storage in the liquid phase. By evaporation and catalytic reforming, hydrogen can be formed on-board for further use in a fuel cell. Due to the absence of C–C bonds, catalytic reforming can take place at moderate temperatures of 300°C–600°C (Ledesma et al., 2019). In the reforming process, CO is formed which is critical to the fuel cell. For this reason, a hydrogen purification step needs to be included such as a Pd–Ag hydrogen permeation membrane (Zhang et al., 2018).

6.1.3 Dimethyl ether as chemical building block

In the chemical industry, DME is of importance as a chemical building block for the production of value-added chemicals. Besides its use for the production of the methylating agent dimethyl sulphate or the solvent methyl acetate, DME is also a potential feedstock for the production of acetic acid and ethanol (Liu et al., 2013; Müller and Hübsch, 2000; Park, Shin, and Bae, 2016; Qian et al., 2018).

Particularly in the field of methanol-to-chemicals and methanol-to-fuels, DME plays a key role as an intermediate. It was shown that the methanol-based processes – MTG, methanol-to-olefins (MTO) and methanol-to-propylene – are actually based on DME as intermediate, which is synthesised by methanol dehydration as the primary reaction step (Keil, 1999; Lee, Gogate, and Kulik, 1995; Pérez-Uriarte, Ateka, et al., 2016; Pérez-Uriarte, Gamero, et al., 2016). Thus it is possible to use DME directly as feedstock. In fact, it was shown that the DME-to-olefins reaction allows higher yields of olefins at lower reaction temperatures compared to the MTO process, while at the same time reducing the degree of exothermicity and the water content of the reaction medium (Pérez-Uriarte, Ateka, et al., 2016). The same applies to a DME-to-gasoline process, where higher yields are reported compared to the MTG process. Furthermore, it was shown that the lower water concentrations and lower degree of exothermicity are beneficial for the catalyst (Lee et al., 1995).

Another development is the Mobil olefins-to-gasoline/distillate process, which can be employed subsequent to the MTO process to convert the produced olefins to final fuels, including jet fuel (Keil, 1999; Tabak and Yurchak, 1990).

A recently emerging green fuel is OMEs, which – as well as DME – have no C–C bonds, thus enabling a soot-free combustion (Vertin et al., 1999). A product mixture of OME_{3-5} has similar thermophysical and intrinsic combustion properties as diesel. Contrarily to DME, OMEs are liquid at ambient conditions. Furthermore, OME/diesel blends show advantages regarding tail emissions. For these reasons, OMEs have the potential to use the existing infrastructure and diesel engines without extensive changes (Oestreich et al., 2018). However, an optimised and efficient production of OMEs still presents a major challenge. The use of DME as an intermediate presents a promising strategy for the production of OMEs, as DME is the source of two methyl groups as well as a formaldehyde source through oxidation, dehydration and oxo-dehydrogenation of DME. The combination of integrated formaldehyde synthesis and OME synthesis – referred to as oxidative OME synthesis from DME – presents the opportunity of a one-step OME process, referred to as oxidative OME synthesis from DME. However, this synthesis route is still under early developments (Gao et al., 2018; Wang et al., 2016, 2019; Zhang et al., 2007, 2014, 2016) and the formation of the industrially relevant higher OME chains was not yet successful (Arnold et al., 2019). In contrast, the nonoxidative OME synthesis from DME requires an external formaldehyde source. While the synthesis based on DME and monomeric formaldehyde has not been realised yet, the reaction with trioxane as formaldehyde source has been patented (Ströfer et al., 2006) and published in literature (Breitkreuz et al., 2018; Haltenort et al., 2018).

6.1.4 Other uses

DME demonstrated good performance as a green solvent in liquid extraction for a variety of compounds (Bauer and Kruse, 2019). Due to its high vapour pressure, it can easily be removed from the final product by depressurisation. Furthermore, DMEs partial miscibility in water allows the use for the extraction of highly moist products (Boonnoun et al., 2019).

In addition, DME was studied as a solvent for enhanced oil recovery (EOR) and Shell developed a chemical EOR technique called DME-enhanced waterflood. In this process, DME is injected into an oil reservoir, leading to swelling and viscosity reduction and consequently an increased yield of crude oil (Cho, Kim, and Lee, 2018).

When released to the atmosphere, DME shows low global warming potential and an ozone depletion potential of zero. Therefore DME is discussed as a green refrigerant. To overcome the drawback of DME's flammability, mixtures of DME and other refrigerants can be used (Kim, 2016; Onaka, Miyara, and Tsubaki, 2010). On top of that, DME can be used as pesticide, polishing agent and antirust agent (Azizi et al., 2014).

6.2 Dimethyl ether production pathways

Until 1975 DME was a by-product of the high-pressure methanol synthesis. However, the introduction of the highly selective low-pressure methanol synthesis necessitated the development of a dedicated DME-synthesis route (Müller and Hübsch, 2000). This route is referred to as indirect route, in which DME is synthesised from methanol via dehydration and under corresponding water formation. A more recent alternative synthesis route is the so-called direct route which allows a direct synthesis of DME from synthesis gas. A process comparison of direct and indirect route is shown in Fig. 6.2. In the following subsections, both production routes are described in detail. In the first instance, the synthesis technology is presented in general. In a second step, the specific conditions, opportunities and limitations of a CO_2-based production are discussed.

6.2.1 *Indirect route*

The indirect route represents the conventional process with all existing commercial-scale DME plants being based on that route. The technology is offered by several licensors such as Haldor Topsøe, Toyo Engineering Corporation, Lurgi, Mitsubishi Gas Chemical and ThyssenKrupp Industrial Solutions AG (Bertau et al., 2014). In this approach, DME is synthesised in two steps. In the first step, methanol is synthesised and subsequently purified via distillation. The second step is the actual DME synthesis by dehydration of methanol according to the exothermic reaction below (Arnold et al., 2019):

$$2CH_3OH \Leftrightarrow CH_3OCH_3 + H_2O \quad \Delta H^0_{298K} = -23.5\ \text{kJ mol}^{-1} \tag{6.1}$$

The reaction is catalysed by acidic groups, typically using solid weakly acidic catalysts such as γ-Al_2O_3 (Bakhtyari and Rahimpour, 2018; Maus, 2019). Due to the exothermic and equimolar reaction, pressure has no significant influence on equilibrium conversion, but low temperatures are favourable for high methanol conversion. All commercial plants are based on a gas-phase reaction. In practice, depending on the water content of the feed, reaction temperatures range between 220°C and 360°C (Bakhtyari and Rahimpour, 2018; Pontzen et al., 2011) at pressures up to 20 bar (Ahlers et al., 2014; Bakhtyari and Rahimpour, 2018; Catizzone,

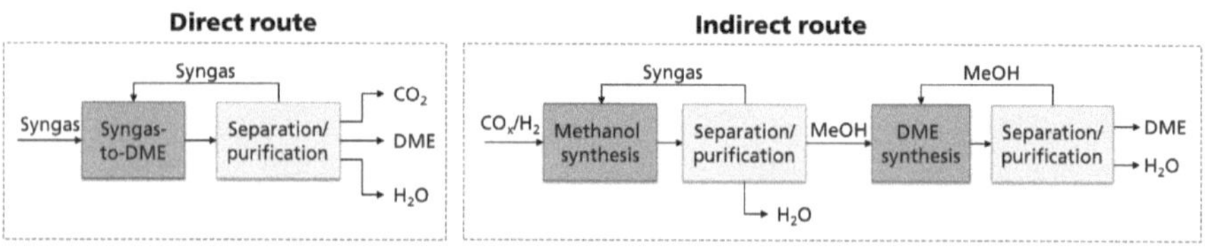

Figure 6.2 Simplified comparison of direct and indirect routes.

Bonura, et al., 2017; Shoji and Terai, 2004). The reaction is equilibrium limited and reaches methanol conversions up to 70%–85% (Pontzen et al., 2011). Consequently, the reactor outlet gas consists mainly of the reaction products DME and water but also considerable amounts of unreacted methanol. Although methanol dehydration is highly selective towards DME, the formation of by-products such as hydrocarbons as well as CO and H_2 has been reported (Bakhtyari and Rahimpour, 2018; Liu et al., 2010; Yaripour et al., 2005, 2015).

6.2.1.1 Process description

The indirect route of DME production starting from crude methanol is depicted in Fig. 6.3. For the DME-synthesis pure methanol is required. The purification of crude methanol is conducted by a series of distillation columns. In a first forerun distillation column, low-boiling by-products such as CO, CO_2, CH_4 and hydrocarbons are removed. The water–methanol mixture is then charged to the methanol column, where water is removed as bottoms product. Purified methanol is discharged from the column head, preheated and subsequently fed to the DME-synthesis reactor. The DME product mixture is then purified in a first step by removing water in a mixture column. The methanol and DME containing distillate

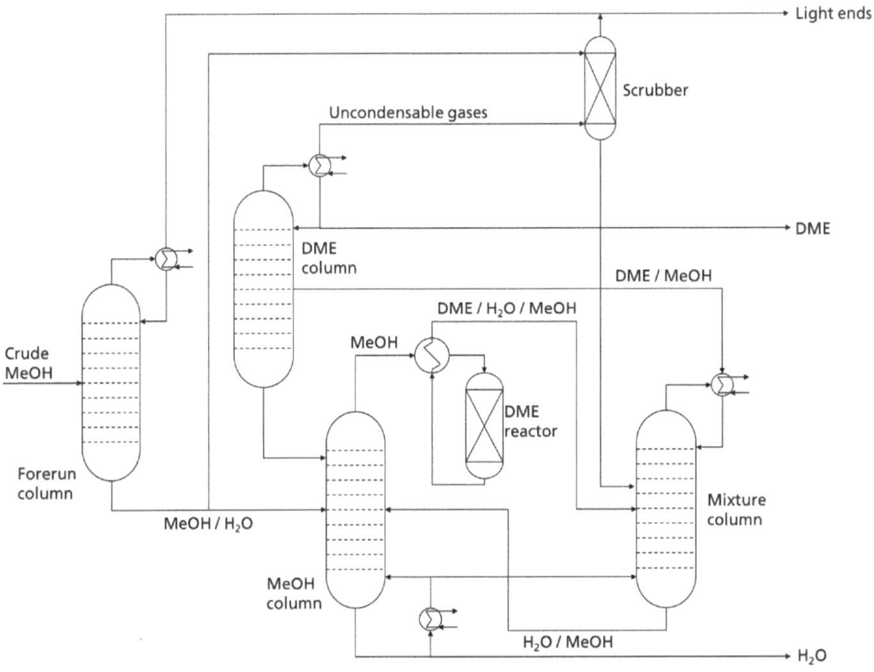

Figure 6.3 Indirect route of DME production based on crude methanol feedstock, according to the Lurgi MegaDME process (Ahlers et al., 2014). *DME*, Dimethyl ether.

is charged to the DME column and separated into a methanol stream (bottom) and a DME stream (distillate) representing purified DME product. The methanol stream is recycled to the top of the methanol column. Incondensable gases in the DME stream are fed into a scrubber to recover residual DME.

6.2.1.2 Catalysts

A large amount of research was conducted regarding the catalysis of methanol dehydration reaction. As the reaction is catalysed with an acidic function, a high concentration of acid sites is generally favourable in terms of catalyst activity. However, there is a consensus that the formation of hydrocarbons as an undesired by-product can be attributed to the abundance of acid sites with high acidic strength (Khandan, Kazemeini, and Aghaziarati, 2008; Ramos et al., 2005; Sun et al., 2014; Xu et al., 1997; Yaripour et al., 2005; Zhu et al., 2008). Thus a high amount of acid sites of low and medium acid strength is desired to maintain high activity while concurrently suppressing the formation of by-products.

Researchers agree that the surface reaction is the rate-controlling step. However, two positions exist regarding the mechanism, namely the associative pathway and the dissociative pathway. The associative pathway assumes coadsorption of two methanol molecules which react on the catalyst surface directly. In contrast, the dissociative pathway involves adsorption of one methanol molecule followed by water elimination and subsequent reaction with another methanol molecule. There is a disagreement considering the appropriate description of the reaction kinetics, with a variety of different kinetic models published. Both Eley-Rideal (An, Chuang, and Sanger, 2004; Lei et al., 2011) and Langmuir-Hinshelwood (Bandiera and Naccache, 1991; Hosseininejad, Afacan, and Hayes, 2012) kinetics have been reported.

As already mentioned, γ-Al_2O_3 is the most investigated solid acid catalyst for methanol dehydration. However, a wide range of catalysts including zeolites, heteropoly acids, ion exchange resins, Nafion and acid-activated carbon have been tested and developed specifically for this reaction (see Table 6.2). A detailed overview of catalysis in methanol dehydration is given in the review of Bateni et al. (Bateni and Able, 2019).

Advantages of alumina are its low cost, high selectivity up to 400°C, good performance, catalyst life-time and mechanical and thermal stability (Keshavarz et al., 2010; Stiefel et al., 2011; Yaripour et al., 2015). However, the activity of alumina is strongly affected by the presence of water, which adsorbs on the Lewis acid active sites (Bateni and Able, 2019; Migliori et al., 2014). Zeolites possess both Brønsted and Lewis acid sites. This results in methanol adsorption being preferred over water, thus significantly increasing the stability in the presence of water. A major drawback of zeolites is a reduced selectivity, resulting from the formation of hydrocarbons as a by-product, especially at higher temperatures. Furthermore, deactivation by coking can be critical and needs to be addressed in the selection of the reaction conditions. Lowering the silica/alumina ratio increases the acid strength of the zeolite and leads to a higher activity but also a reduced selectivity (Bateni and Able, 2019).

Table 6.2 Overview of most commonly researched catalysts for methanol dehydration.

Catalyst	Source
γ-Al_2O_3	Barbarossa et al. (2020), Boon et al. (2019), Catizzone, Migliori, et al. (2019), Keshavarz, Rezaei, and Yaripour (2010), Khandan et al. (2008), Yaripour et al. (2015)
Zeolites	Aloise et al. (2020), Bandiera and Naccache (1991), Catizzone, Aloise, et al. (2017), Catizzone, Migliori, et al. (2019), Catizzone, van Daele, et al. (2019), Ha et al. (2011), Hosseini, Taghizadeh, and Eliassi (2012), Khandan et al. (2008), Masih et al. (2017), Migliori et al. (2014), Migliori, Aloise, and Giordano (2014), Moradi, Yaripour, and Vale-Sheyda (2010), Ortega et al. (2018), Ramos et al. (2005), Sabour et al. (2014), Xu et al. (1997), Zhu et al. (2008)
Nafion	Barbarossa et al. (2020), Ciftci, Sezgi, and Dogu (2010), Kolesnichenko et al. (2011), Varışlı and Doğu (2008)
Heteropoly acids	Alharbi, Kozhevnikova, and Kozhevnikov (2015), Liuzzi and Peinado (2020)
Ion exchange resins	An et al. (2004), Hosseininejad et al. (2012), Lei et al. (2011)
Acid-activated carbon	Jasińska, Krzyżyńska, and Kozłowski (2011), Moreno-Castilla et al. (2001), Valero-Romero et al. (2019)

6.2.1.3 CO_2-based production

As DME synthesis in the indirect route is based on methanol as feedstock, the DME-synthesis step itself remains unchanged when using CO_2 instead of CO in methanol production. However, regarding the process steps prior to the DME reactor in Fig. 6.3, a CO_2-based production positively influences the raw methanol purification step. For conventional methanol synthesis based on CO-comprising feedstock such as syngas from steam reforming, impurities with higher vapour pressure are present in the raw methanol. Thus, the purification of the yielded raw methanol is based on the two-step distillation process discussed above. The CO_2-based methanol production in contrast shows a higher selectivity and a strongly reduced amount of impurities (Goehna and Koenig, 1994; Marlin, Sarron, and Sigurbjörnsson, 2018). Thus the yielded raw methanol can be purified without a dedicated light ends column. Overall, purification can be simplified to a single heavy ends column including a stripper at the column head to vent dissolved CO_2 (Marlin et al., 2018).

6.2.2 Direct route

The direct route of DME production presents a more recent development and has been studied thoroughly in literature. Although it has not been implemented in commercial scale, several companies offer the technology including Toyo Engineering Corporation, Mitsubishi Gas Chemical Ltd. and Air Liquide S.A. (Arnold et al., 2019). The direct route is based on the combined synthesis of methanol and the subsequent dehydration to DME in a single reactor. Thus, in this so-called syngas-to-DME process, DME is synthesised from syngas in a single step. It is important to note, however, that this single step consists of two separate reactions, namely the methanol synthesis and the methanol dehydration. Both reactions occur on their dedicated catalyst sites and the catalyst bed consists of either a physical mixture of methanol catalyst and dehydration catalyst or a bifunctional catalyst, including active centres for both reactions (Maus, 2019).

Methanol synthesis can either be based on CO or CO_2, although recent research suggests that methanol synthesis is mainly due to CO_2 hydrogenation, whereas the hydrogenation of CO scarcely occurs (Nestler et al., 2018).

$$CO + 2\,H_2 \Leftrightarrow CH_3OH \quad \Delta H^0_{298\ K} = -90.4\ kJ\ mol^{-1} \tag{6.2}$$

$$CO_2 + 3\,H_2 \Leftrightarrow CH_3OH + H_2O \quad \Delta H^0_{298\ K} = -49.4 kJ/mol \tag{6.3}$$

The DME-synthesis step is represented by the methanol dehydration reaction equation as already depicted in Eq. (6.1). The reactions for methanol and DME synthesis are furthermore complemented by the water–gas shift (WGS):

$$CO + H_2O \Leftrightarrow H_2 + CO_2 \quad \Delta H^0_{298\ K} = -41.0\ kJ\ mol^{-1} \tag{6.4}$$

Thus, taking into account methanol synthesis, methanol dehydration and WGS, the DME synthesis from syngas can be represented by the overall reaction:

$$3\,CO + 3\,H_2 \Leftrightarrow CH_3OCH_3 + CO_2 \quad \Delta H^0_{298\ K} = -90.4\ kJ\ mol^{-1} \tag{6.5}$$

Besides the integration of two separate reactors in one, this concept has two fundamental advantages due to the synergy of the involved reactions as illustrated in Fig. 6.4. First, formed methanol is converted in situ to DME which enhances the reaction conversion by overcoming the equilibrium limitation of methanol synthesis

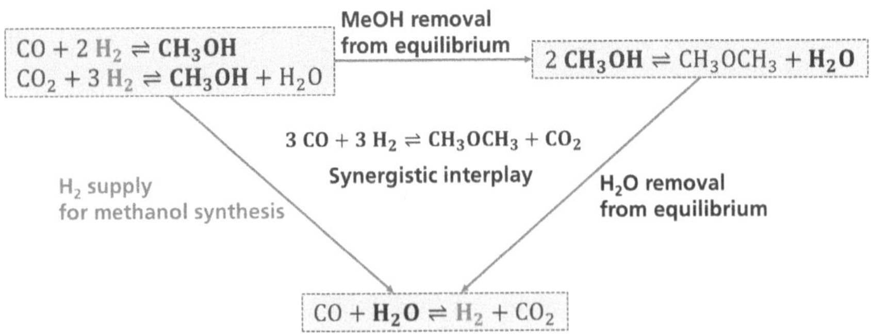

Figure 6.4 Illustration of the synergistic interplay of methanol synthesis, methanol dehydration and WGS. *WGS*, Water–gas shift.

(Gogate, 2018). Second, the water produced in the methanol dehydration step is removed from the reaction equilibrium by the WGS when CO-rich syngas is used (Peng et al., 1997). Overall, syngas conversion and volumetric reactor productivity can be increased this way. Furthermore, compared to the indirect route, the required hydrogen supply is reduced which can also be traced back to the WGS, where hydrogen is formed from the reaction of water and CO (Mondal and Yadav, 2019).

A drawback of the direct route is the complex purification of the product mixture exiting the reactor. Whereas in the indirect route methanol, water and small amounts of hydrocarbons are the only by-products, the product mixture in the direct route additionally includes unreacted syngas and CO_2. Due to the lower vapour pressure of DME, a separation of CO_2 and syngas from DME in a flash column is not possible. In addition, the presence of methanol hampers the separation of CO_2 and DME. For this reason, in a suggested purification process by Peng et al. (2002), methanol and water are condensed in a flash column and subsequently fed in a separate methanol dehydration reactor as depicted in Fig. 6.5. The gas phase consisting mainly of DME, syngas and CO_2 is then fed to an absorber. Here, DME and CO_2 are absorbed, while

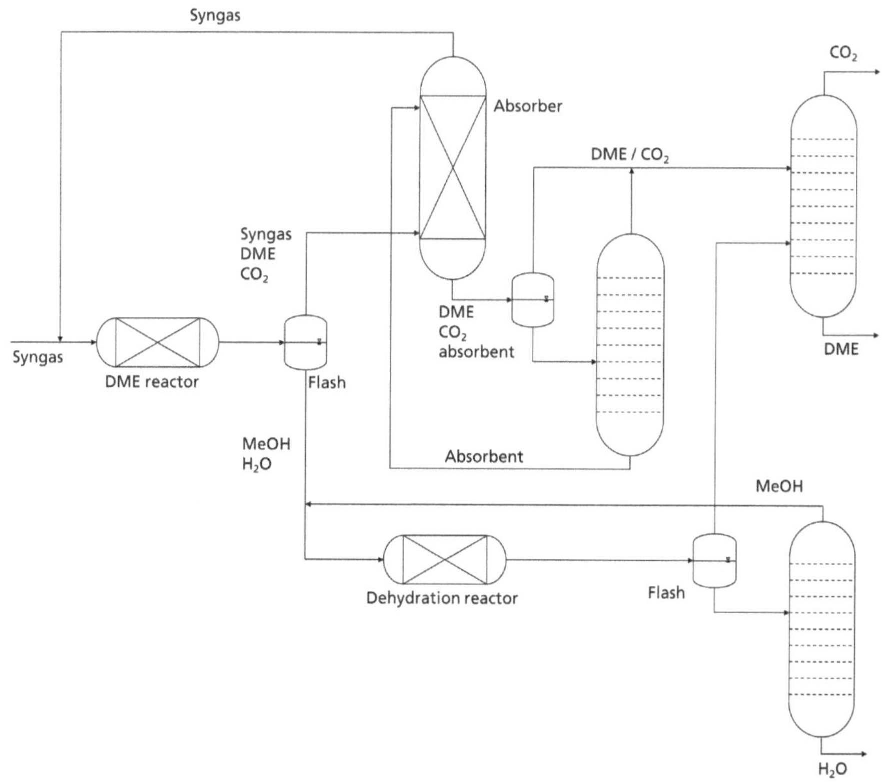

Figure 6.5 Process flowsheet for the direct route including syngas recycle and purification according to Patent US6458856B1 (Peng et al., 2002).

syngas leaves the absorber and is subsequently recycled into the reactor. The absorbed DME and CO_2 are diluted in the absorbent and consequently need to be purified. In a flash column the absorbent is regenerated and recycled to the absorber. DME and CO_2 exit the flash column in the gas stream and are charged to a distillation column, where pure DME is finally obtained as bottom stream.

Due to the exothermic nature of methanol synthesis, methanol dehydration and WGS, the direct DME synthesis from syngas is highly exothermic. This leads to challenges in the reactor design to avoid hot spots, which promote deactivation of the methanol catalyst due to sintering (Ateka et al., 2017). Another disadvantage of the direct route is the generation of CO_2 when using CO as carbon source. Due to the water generation in the methanol dehydration, the WGS reaction leads to the oxidation of CO to CO_2 according to Eq. (6.1). Thus the direct route of DME production from syngas is a net CO_2 producing process.

6.2.2.1 CO_2-based production

As discussed earlier, the direct DME-synthesis route is premised on the synergetic interplay of methanol synthesis, methanol dehydration and WGS. Thus replacing CO with CO_2 as carbon source has a significant influence on the syngas-to-DME process. Regardless of the carbon source, the thermodynamic inhibition from methanol can be overcome due to the in situ conversion to DME.

However, when using pure CO_2 and hydrogen as feedstock, the reverse WGS (rWGS) dominates compared to the WGS, thus producing water. The water from the rWGS adds to the water generated in the methanol synthesis and methanol dehydration, leading to a thermodynamic inhibition of all involved reactions, where water is a by-product (compare Eqs. 6.1 and 6.3). The absence of CO in the feed gas eliminates the possibility of water removal from the chemical equilibrium through WGS. The synergistic interplay between the reactions as illustrated earlier is thus not present when using CO_2 rather than CO as carbon source. As a result, the equilibrium DME yield strongly decreases when gradually replacing CO with CO_2. Fig. 6.6 illustrates the influence of CO_2 on equilibrium yield and water formation in the direct DME-synthesis reaction. The carbon oxide ratio (COR) is a measure for the CO_2 content of the feed and is defined as follows (Nestler et al., 2018):

$$\mathrm{COR} = \frac{y_{\mathrm{CO_2}}}{y_{\mathrm{CO_2}} + y_{\mathrm{CO}}} \tag{6.6}$$

Due to the exothermic nature of the involved reactions, lower reaction temperatures are beneficial for the equilibrium yield. However, the low reaction rate below 250°C due to kinetic limitation should be noted (Ateka et al., 2017). Increasing the reaction pressure is beneficial for the DME yield; however, high pressure promotes catalyst deactivation by coking (Ereña et al., 2011).

Besides thermodynamic limitations, the CO_2-induced water generation furthermore affects catalysis. Due to competitive adsorption, water molecules block not only the metallic sites of the methanol catalyst but also the acid sites of the

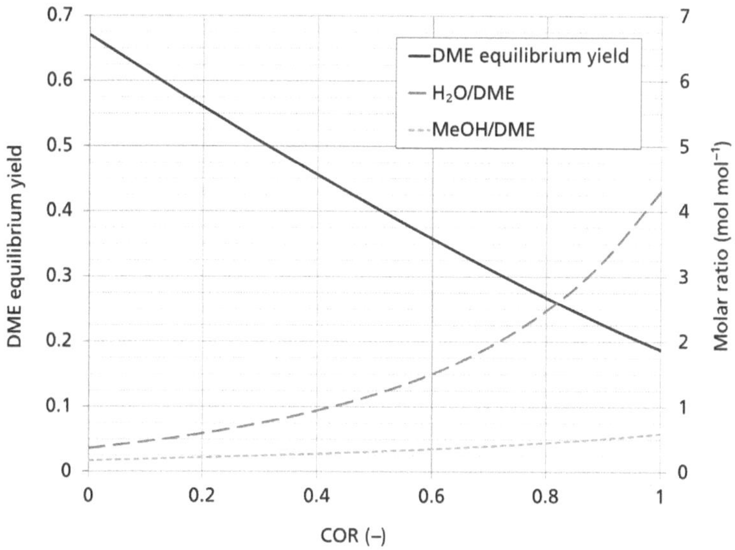

Figure 6.6 Influence of COR on the DME equilibrium yield and moles of water/methanol produced per mole of DME (T = 275°C, p = 50 bar, $H_2/CO_x = 3$). *COR*, Carbon oxide ratio; *DME*, Dimethyl ether.

methanol dehydration catalyst (Ateka et al., 2019; Ereña et al., 2011). Therefore, catalysts with high acid strength like zeolite FER are preferred as dehydration catalysts (Frusteri et al., 2017). This phenomenon further reduces the performance of the CO_2-based direct DME synthesis. There is a consensus that process intensification techniques for in situ water removal are required to overcome the thermodynamic inhibition by water and increase DME yield (Ateka et al., 2019; De Falco, Capocelli, and Centi, 2016; Mondal and Yadav, 2019).

In principle, a preceding rWGS reactor could be used to convert CO_2 to CO and thus to allow high yields in the direct route. However, the addition of a separate reversible reaction step in a process that already involves multiple steps and catalysts significantly would have a critical influence on the performance (De Falco et al., 2016).

6.2.3 Emerging pathways

To improve the conventional production pathways and to reduce costs, significant effort has been made to develop alternative production pathways. A central focus has been put on the in situ removal of water in DME synthesis. In the indirect route, water removal has the potential to increase the methanol conversion to nearly full conversion. This in turn leads to a simplification of the downstream purification process. In the direct route, in situ water removal is considered a necessity to commercialise a CO_2-based production, since this would displace the WGS equilibrium. The studied concepts hereby are reactive distillation, sorption-based water removal and membrane-based water removal, which are presented in the following sections.

6.2.3.1 Reactive distillation

Reactive distillation is often seen as the front-runner in process intensification and is based on the synergistic combination of reaction and separation in a single unit operation (Harmsen, 2007; Kiss, 2018). As the product of the reaction is continuously removed from the reaction equilibrium, a full conversion of otherwise equilibrium-limited reactions can be achieved. In addition to the simultaneous removal and purification of the product, this averts the requirement of a separate reagent recovery and reagent recycle step. Furthermore, exothermic reactions provide the opportunity of in situ heat utilisation for evaporation of the liquid, thus reducing the external heat demand in the reboiler.

Fig. 6.7 shows the process design for DME production based on methanol, when using reactive distillation compared to the conventional reactor-separator-recycle sequence. As can be seen, reactive distillation can replace one reactor and two distillation columns.

However, the implementation of a reactive distillation requires the compliance of the conditions of reaction and distillation. In addition, products and reactants have to be in different phases (liquid or gaseous). The research so far has been focused on reactive distillation processes using ion exchange resins at column pressures below 11 bar and temperatures below 150°C (An et al., 2004; Bîldea et al., 2017;

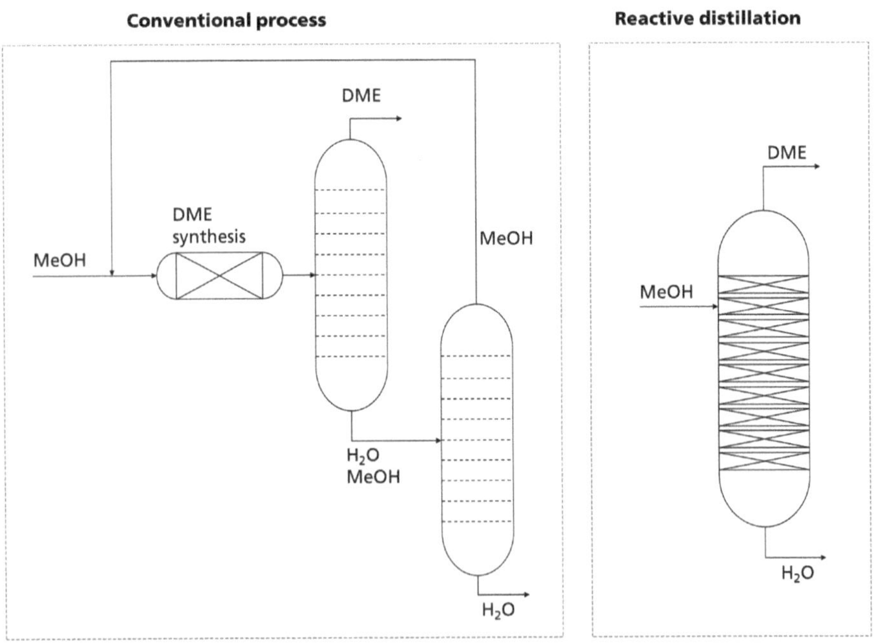

Figure 6.7 Simplified process flowsheet for conventional DME production based on methanol feed compared to process flowsheet for a process with reactive distillation. *DME*, Dimethyl ether.

Lei et al., 2011). Lei et al. (2011) simulated a process concept with a fixed-bed reactor prior to the reactive distillation and concluded that this concept reduces the energy demand compared to the conventional process. Bîldea et al. (2017) investigated the full conversion of methanol in a single reactive distillation column and compared the process via simulation to the conventional indirect route, reporting lower capital expenditures (CAPEX) and operational expenditures (OPEX) for the reactive distillation. Besides conventional reactive distillation configurations, Kiss et al. (Kiss and Suszwalak, 2012) studied reactive dividing-wall column for DME production. In this concept, a distillation column is separated into two thermally coupled sections by a middle wall, one of which is loaded with heterogeneous catalyst and serves as reactive zone. In this configuration, DME won as distillate, water as bottoms product and methanol as side stream that can be recycled. The authors report lower capital cost and 12% lower energy cost compared to the conventional process.

6.2.3.2 Sorption based

The aforementioned thermodynamic limitation through water generation in the CO_2-based direct route has led to research focusing on the in situ removal of water. One solution may be the adsorption of water in the reactor. Iliuta, Iliuta, and Larachi (2011) proposed a fixed-bed reactor that was packed with a homogeneous mixture of bifunctional catalyst for DME production and zeolite particles for water adsorption. The simulation results indicated that the CO_2 conversion is strongly increased in the sorption-enhanced reactor compared to a conventional fixed-bed reactor. This is due to the shift of the WGS equilibrium. Furthermore, the amount of unreacted methanol is decreased, as the thermodynamic limitation of the methanol dehydration reaction is reduced by the water removal. The enhancement of CO_2 conversion and DME selectivity increases with increasing CO_2-level in the feed.

It is important to note that the adsorption process results in dynamic operation of the reactor/adsorber, leading to the practical challenge of realizing an overall continuous production process. Hamidi, Samimi, and Rahimpour (2015) established a simulation model of a fixed-bed reactor for methanol dehydration with continuous adsorbent regeneration. Hereby zeolite adsorbent is continuously fed to the top of the reactor and withdrawn from the lower end. In a separate regenerator the water is removed from the adsorbent and fresh adsorbent can be recycled into the reactor. An alternative approach is the cyclic regeneration of the adsorbent as soon as the capacity limit is reached (Iliuta et al., 2011). The adsorbent regeneration is typically conducted by pressure swing, temperature swing, concentration swing and combinations thereof. Since the regeneration process consists of several steps and usually takes longer than the reaction step, van Kampen et al. (2020) simulated a sorption-enhanced DME-synthesis process based on three reactors. Each reactor is operated independently and in periodic order. This way, a continuous process can be established based on three discontinuously operated reactors. One reactor is in the reaction step, while the other two reactors are in the regeneration phase consisting of a blowdown step for pressure reduction, purging, heating and repressurisation. A reaction term of 112.5 min and a total regeneration time of 225 min were assumed for the simulation.

The sorption-enhanced DME synthesis was demonstrated experimentally and proofed to allow a DME yield of more than 80%, even when utilising pure CO_2 as carbon source (van Kampen et al., 2019). The major drawback, however, is the cyclic discontinuous operation of the reactors and the high technical effort required to achieve a continuous synthesis. Furthermore, the purge gas used in the purge step is still present in the reactor at the beginning of the reaction step. When using an external purge gas, this adds an additional component to the process, thus complicating the purification process. Using syngas for purging in contrast would unfavourably impact the material balance of the process. Another issue that needs to be addressed is the blowdown gas, emerging in the pressure reduction step. This gas consists mainly of syngas and water at low pressure. Thus a separate drying and repressurisation step would be required not to send this stream to waste.

6.2.3.3 Membrane based

An alternative principle for the in situ removal of water is the membrane separation. This approach is based on a semipermeable membrane, selectively allowing water to be transported out of the reactor while retaining all other components. To increase the trans-membrane driving force, the membrane is swept at low pressure.

Iliuta, Larachi, and Fongarland (2010) examined the effect of in situ water removal in a membrane reactor on the direct DME synthesis theoretically in a simulation model. It was confirmed, that DME yield and selectivity increase with an increasing membrane permeance. This effect was more distinct for higher COR. The authors pointed out that the implementation of such membrane reactors requires membranes with good thermal and mechanical stability, high membrane permeances and permselectivity towards water (Iliuta et al., 2010).

Lee, Youn, and Sea (2006) prepared alumina–silica composite membranes and tested the membranes in a reactor for DME synthesis from methanol. The membranes proved to be stable at 250°C and successfully enhanced the conversion of methanol.

Diban, Urtiaga, et al. (2013) studied the influence of membrane properties on the reactor performance in the direct DME synthesis from CO_2-containing syngas. Real properties of zeolite membranes found in literature were used to provide realistic simulation results. It was found that the use of zeolite membranes indeed enhances CO_2 conversion due to the in situ removal of water from the equilibrium. However, DME yields are reduced in the membrane reactor compared to a conventional packed bed reactor. This was attributed to the low water permselectivity of the studied zeolite membranes leading to a considerable mass transfer of methanol from the reactor side to the sweep gas side. Although this in situ methanol removal enhanced methanol yields, the removal of methanol prevented the further dehydration to DME, thus reducing the DME yield. The highest CO_2 conversions were achieved for a membrane with high water permselectivity and low permeance (Diban, Urtiaga, et al., 2013).

Ateka et al. (2019) simulated direct DME synthesis in a membrane reactor and reported a 25% increase in DME yield for the hydrogenation of CO_2 compared to a conventional packed bed reactor. Different sweeping modes and agents were furthermore examined. Nonetheless, it was concluded that the influence is minor. Despite the promising results, the authors pointed out that obtaining membranes that are stable at the conditions of direct DME synthesis presents a tremendous challenge.

Membrane reactors are an interesting emerging pathway to realise the direct DME synthesis in the prospect of a CO_2-based production as was demonstrated in the mentioned simulation studies. However, for a practical implementation the high demands on the membranes need to be fulfilled. Direct DME synthesis presents a challenging environment for the implementation of membranes due to high temperatures and pressures and the presence of water inducing the requirement of hydrothermal stability. Zeolite membranes seem to be the most adequate material for these demands. Yet, the reproducible fabrication presents the limiting challenge for a successful commercialisation (Diban, Aguayo, et al., 2013).

6.3 Techno-economic insights

In the previous sections the technical background of DME production routes has been presented and discussed. It was shown that a carbon-neutral production of DME based on CO_2 and H_2 is technically feasible. However, to penetrate the market, this renewable production path has to be economically viable. This section provides insights into the economic aspects of DME production. Since the direct route has not yet been commercialised, the following analysis will focus on the indirect route. As stated previously, the influence of the feedstock on the process design is marginal. However, as shown in the upcoming sections, the influence of the feedstock on the overall production cost is far more pronounced. Fig. 6.8 shows the CAPEX and OPEX breakdown as well as the total production cost for a DME production facility based on CO_2 and H_2. The facility includes hydrogen production via electrolysis and CO_2 capture from a cement plant via membrane separation. The DME synthesis is based on the indirect route described in the previous section and

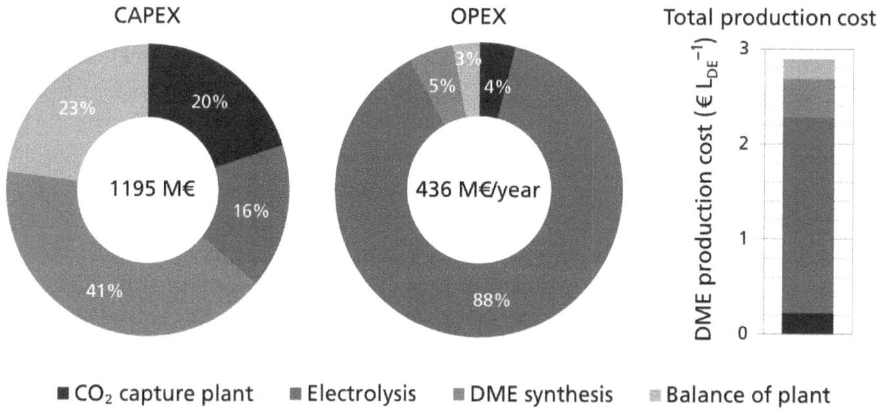

Figure 6.8 Key economic indicators of a power-to-DME process based on Michailos et al. (2019). *DME*, Dimethyl ether.

thus includes the CO_2-based methanol synthesis, methanol dehydration and all purification steps. The balance of plant accounts for the peripheral processes and consists of wastewater treatment, cooling towers and a combined heat and power plant for the utilisation of the purge stream from methanol synthesis. For a detailed description of the process, the reader is referred to Michailos et al. (2019).

The overall CAPEX of the entire production facility with an annual production capacity of 246.7 kt DME sums up to 1195 million €. It can clearly be seen that the CAPEX is dominated by the synthesis unit itself, accounting for 41% of the overall CAPEX. The residual CAPEX is allocated to electrolysis, CO_2-capture unit and the balance of plant. The OPEX in contrast is nearly entirely dominated by the electricity costs for the electrolysis. Together with the operational costs for CO_2 capture, the supply of feedstock accounts for a total of 92% of the operational expenses. Resulting from both CAPEX and OPEX, the total production cost of green DME is broken down in the stacked column diagram. Electrolysis distinctly represents the lion's share of the power-to-DME process, accounting for 71% of the total DME production cost. This emphasises the importance of cheap renewable energy resources on the one hand and an efficient production process on the other. The underlying electricity price reported in this study is 87.7 € MWh^{-1}. In regions with high potential for renewable energy, levelised cost of electricity below 20 € MWh^{-1} have been reported to date (Maus, 2019) and are expected to drop down to 16 € MWh^{-1} (Ram et al., 2017). Such a favourable electricity price would in turn result in a much lower overall production cost. The influence of CO_2 capture is far less pronounced than hydrogen production when it comes to both OPEX and overall production cost. It should be noted that CO_2 capture from a cement plant is considered in the aforementioned study. Though being an economically viable option, this necessitates the availability of such a source in place. If no CO_2 source is available, direct air capture can be an option. However, this technology is to date a very expensive option which would make CO_2-feedstock a considerable cost driver of a power-to-DME process. Specific CO_2-capture cost from direct air capture is targeted to reach 75 € t CO_2^{-1} (Fasihi, Efimova, and Breyer, 2019), which would pave the way for an economic and unlimited CO_2 source.

The second largest contributor to the total production cost is the DME synthesis with a share of 14%. Considering the little influence on OPEX (5%), the contribution of DME synthesis can be majorly attributed to the CAPEX of the DME-synthesis plant. It should further be noted that with lower electricity prices, the contribution of electrolysis to the total production cost declines, thus increasing the contribution of DME synthesis and consequently raising the significance of efficient and economic DME production processes.

Fig. 6.9 shows the total production cost for DME production from CO_2 and hydrogen reported in different literature sources and for different hydrogen prices. Production cost is quoted per litre diesel equivalent (L_{DE}), incorporating the lower heating value of diesel of 9.97 kWh L^{-1} (Edwards et al., 2014). The market price of DME – which is produced from fossil feedstock – is given for comparison as a dashed line. The effect of the change from fossil to renewable feedstock on the overall production cost can clearly be seen. While fossil-based DME has a market price of 0.60 € L_{DE}^{-1} (China Petroleum and Chemical Industry Federation, 2020),

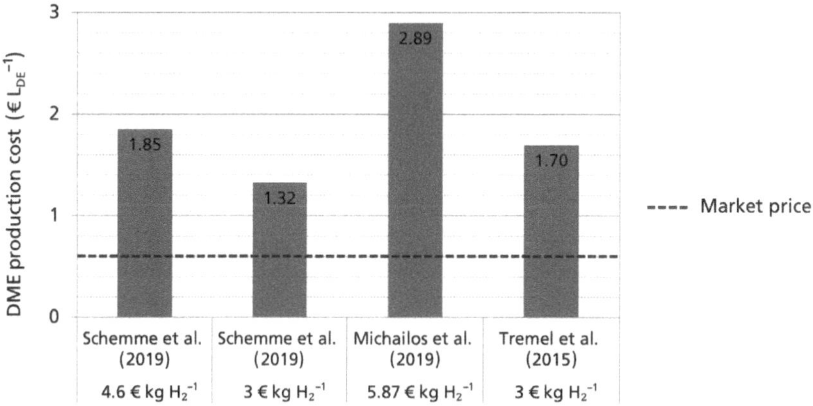

Figure 6.9 DME total production cost based on CO_2 and hydrogen from different sources. Current market price of fossil DME as a reference (China Petroleum and Chemical Industry Federation, 2020). *DME*, Dimethyl ether.

DME from CO_2 and hydrogen is in the range between 1.32 and 2.89 € L_{DE}^{-1}. However, once again it needs to be emphasised that the production price is strongly dependent on the assumptions regarding the hydrogen cost. With costs for renewable energy generation dropping in the future, electrolysis at locations with great renewable energy potential could lead to hydrogen production costs below 1.6 USD kg^{-1} (Hydrogen Council, 2020; IEA, 2019).

Furthermore, a variation of the production costs can be observed in the different studies. This can first be attributed to the assumptions regarding the hydrogen cost. While in Schemme et al. (2019) a hydrogen price of 4.6 and 3 € kg H_2^{-1} is assumed, Tremel et al. (2015) use a hydrogen price of 3 € kg^{-1}. In contrast, in the analysis of Michailos et al. (2019), hydrogen is not considered an input stream but electrolysis is integrated in the production plant. The electricity cost and CAPEX of the electrolyser translate to a hydrogen cost of 5.36 € kg H_2^{-1}. Second, the overall process efficiency varies between the authors. While (Schemme et al., 2019) calculate a power-to-fuel efficiency of 60%, Michailos calculates a value of 44%, which can mainly be attributed to a lower assumed electrolyser efficiency. Last but not least, the scale of the production plants varies strongly between the presented studies. Schemme et al. (2019) consider a production capacity of 328.5, a capacity of 246.7 (Michailos et al., 2019) and a capacity of 24.9 kt DME $year^{-1}$ (Tremel et al., 2015). Hence, the economy of scale effect is one additional criterion to account for in future analyses of economic efficient DME syntheses.

6.4 Summary and outlook

The transport sector and the chemical industry largely rely on carbonaceous substances, whose complete replacement is not expected in the future. Thus, the reduction of net carbon dioxide emissions can be achieved by a defossilisation rather than a decarbonisation. In this context, DME is gaining considerable attention as a power-to-liquids product due to its multifarious applications and the possibility to produce DME from various feedstocks.

To date, DME is primarily used as propellant and LPG replacement or blending component. Furthermore, DME proved to be suitable for use as refrigerant and solvent for extraction processes. Due to its molecule structure and high CN, DME allows an efficient and soot-free combustion in diesel engines. The DME-imposed challenges such as the stability have been successfully solved in demonstration vehicles. The use of DME as diesel substitute presents a possible new application and bears the potential to dramatically increase the global DME market volume. Besides the direct utilisation of DME as fuel, it is furthermore an intermediate for the production of other fuels and chemicals such as OME, jet fuel, gasoline and olefins.

Current DME production is based on fossil syngas from coal and natural gas. DME is formed by dehydration of methanol, thus releasing water. The conventional production pathway is the indirect route, which comprises the synthesis and purification of methanol, the subsequent dehydration of methanol and the purification of the formed DME. All existing commercial plants are based on this production route. The production of DME directly from CO-rich syngas is thermodynamically favoured due to the synergistic interplay between WGS, methanol synthesis and methanol dehydration. However, this production route has not been commercialised yet.

To comply with CO_2-emission goals and to "defossilise" the transport and chemical sector, closed carbon cycles are required. This reinforces the need for CO_2-based production processes. In this regard, the direct route is distinctly less suitable than for a CO-based process due to the intensive water formation. To commercialise the direct route for a CO_2-based process, in situ water removal is seen as a prerequisite. In contrast, the CO_2-based methanol synthesis is already state-of-the art and thus allows a CO_2-based indirect DME production. Within the power-to-liquids concept, methanol is discussed as an energy carrier for the global transport of renewable energy (Hank et al., 2020). In this context, the indirect route might maintain being the prevalent production route.

The use of hydrogen and CO_2 as a feedstock implies higher production cost than the use of fossil feedstocks. However, the production cost is chiefly dominated by the hydrogen price. The key to an economically viable Power-to-DME process is the availability of cheap renewable hydrogen or cheap renewably produced methanol. Electrolysis at geographical sites with high potential for renewable energy is expected to provide hydrogen at reasonably low cost. In conjunction with political instruments such as a carbon price, DME has the potential to become an economically competitive and carbon-neutral fuel or chemical.

Nomenclature

CN	cetane number
COR	carbon oxide ratio
DME	dimethyl ether

EOR	enhanced oil recovery
FT	Fischer–Tropsch
LPG	liquefied petroleum gas
MTG	methanol-to-gasoline
MTO	methanol-to-olefins
OME	polyoxymethylene dimethyl ether
rWGS	reverse water–gas shift
Syngas	synthesis gas
WGS	water–gas shift

Symbols

ΔH^0 **(kJ mol^{-1})**	standard enthalpy of reaction
y_i (–)	Molar fraction of component i
p **(bar)**	pressure
T (°C)	temperature

References

AB Volvo. (2017). *Mack Trucks tests alternative fuel DME | Volvo Group*. Available at https://www.volvogroup.com/en-en/news/2017/jan/mack-trucks-tests-alternative-fuel-dme.html (Accessed 11 March 2020).

Ahlers, B., et al. (2014). *Method and device for producing dimethyl ether from methanol. US 8,884,074 B2*.

Alharbi, W., Kozhevnikova, E. F., and Kozhevnikov, I. V. (2015). Dehydration of methanol to dimethyl ether over heteropoly acid catalysts: The relationship between reaction rate and catalyst acid strength. *ACS Catalysis*, *5*(12), 7186–7193. Available from https://doi.org/10.1021/acscatal.5b01911.

Aloise, A., et al. (2020). Desilicated ZSM-5 zeolite: Catalytic performances assessment in methanol to DME dehydration. *Microporous and Mesoporous Materials*, *302*, 110198. Available from https://doi.org/10.1016/j.micromeso.2020.110198.

An, W., Chuang, K. T., and Sanger, A. R. (2004). Dehydration of methanol to dimethyl ether by catalytic distillation. *The Canadian Journal of Chemical Engineering*, *82*(5), 948–955. Available from https://doi.org/10.1002/cjce.5450820510.

Arcoumanis, C., et al. (2008). The potential of di-methyl ether (DME) as an alternative fuel for compression-ignition engines: A review. *Fuel*, *87*(7), 1014–1030. Available from https://doi.org/10.1016/j.fuel.2007.06.007.

Arnold, U., et al. (2019). Die Rolle von Dimethylether (DME) als Schlüsselbaustein synthetischer Kraftstoffe aus erneuerbaren Rohstoffen. In W. Maus (Ed.), *Zukünftige Kraftstoffe* (pp. 532–561). Berlin, Heidelberg: Springer Vieweg. Available from https://doi.org/10.1007/978-3-662-58006-6_22.

Ateka, A., et al. (2017). A comparative thermodynamic study on the CO_2 conversion in the synthesis of methanol and of DME. *Energy*, *120*, 796–804. Available from https://doi.org/10.1016/j.energy.2016.11.129.

Ateka, A., et al. (2019). Strategies for the intensification of CO_2 valorization in the one-step dimethyl ether synthesis process. *Industrial and Engineering Chemistry Research*, *59*(2), 713–722. Available from https://doi.org/10.1021/acs.iecr.9b05749.

Azizi, Z., et al. (2014). Dimethyl ether: A review of technologies and production challenges. *Chemical Engineering and Processing: Process Intensification*, *82*, 150–172. Available from https://doi.org/10.1016/j.cep.2014.06.007.

Bakhtyari, A., and Rahimpour, M. R. (2018). Methanol to dimethyl ether. In A. Basile, and F. Dalena (Eds.), *Methanol* (pp. 281–311). Elsevier. Available from https://doi.org/10.1016/B978-0-444-63903-5.00010-8.

Bandiera, J., and Naccache, C. (1991). Kinetics of methanol dehydration on dealuminated H-mordenite: Model with acid and basic active centres. *Applied Catalysis*, *69*(1), 139–148. Available from https://doi.org/10.1016/S0166-9834(00)83297-2.

Barbarossa, V., et al. (2020). Kinetic parameter estimation for methanol dehydration to dimethyl ether over sulfonic and polymeric acid catalysts. *Journal of Chemical Technology and Biotechnology*, *95*(6), 1739–1747. Available from https://doi.org/10.1002/jctb.6372.

Bateni, H., and Able, C. (2019). Development of heterogeneous catalysts for dehydration of methanol to dimethyl ether: A review. *Catalysis in Industry*, *11*(1), 7–33. Available from https://doi.org/10.1134/S2070050419010045.

Bauer, M. C., and Kruse, A. (2019). The use of dimethyl ether as an organic extraction solvent for biomass applications in future biorefineries: A user-oriented review. *Fuel.*, *254*, 115703. Available from https://doi.org/10.1016/j.fuel.2019.115703.

Bertau, M., et al. (2014). *Methanol: The basic chemical and energy feedstock of the future: Asinger's vision today*. Berlin, Heidelberg: Springer-Verlag. Available from https://doi.org/10.1007/978-3-642-39709-7.

Bîldea, C. S., et al. (2017). Optimal design of intensified processes for DME synthesis. *Computers and Chemical Engineering*, *105*, 142–151. Available from https://doi.org/10.1016/j.compchemeng.2017.01.004.

BioAge Group LLC. (2013). *Volvo Trucks to begin limited production of DME heavy-duty trucks in NA in 2015; customer trials; Oberon partnership – Green Car Congress*. Available at https://www.greencarcongress.com/2013/06/dme-20130607.html (Accessed 11 March 2020).

Boon, J., et al. (2019). Reversible deactivation of γ-alumina by steam in the gas-phase dehydration of methanol to dimethyl ether. *Catalysis Communications*, *119*, 22–27. Available from https://doi.org/10.1016/j.catcom.2018.10.008.

Boonnoun, P., et al. (2019). Optimization of rubber seed oil extraction using liquefied dimethyl ether. *Chemical Engineering Communications*, *206*(6), 746–753. Available from https://doi.org/10.1080/00986445.2018.1522502.

Breitkreuz, C. F., et al. (2018). Design of a production process for poly(oxymethylene) dimethyl ethers from dimethyl ether and trioxane. *Chemie Ingenieur Technik*, *90*(10), 1489–1496. Available from https://doi.org/10.1002/cite.201800038.

Catizzone, E., Aloise, A., et al. (2017). The effect of FER zeolite acid sites in methanol-to-dimethyl-ether catalytic dehydration. *Journal of Energy Chemistry*, *26*(3), 406–415. Available from https://doi.org/10.1016/j.jechem.2016.12.005.

Catizzone, E., Bonura, G., et al. (2017). CO_2 recycling to dimethyl ether: State-of-the-art and perspectives. *Molecules (Basel, Switzerland)*, *23*(1), 1–28. Available from https://doi.org/10.3390/molecules23010031.

Catizzone, E., Migliori, M., et al. (2019). Ferrierite vs. γ-Al_2O_3 : The superiority of zeolites in terms of water-resistance in vapour-phase dehydration of methanol to dimethyl ether. *Journal of Energy Chemistry*, *30*, 162–169. Available from https://doi.org/10.1016/j.jechem.2018.05.004.

Catizzone, E., van Daele, S., et al. (2019). Catalytic application of ferrierite nanocrystals in vapour-phase dehydration of methanol to dimethyl ether. *Applied Catalysis B: Environmental*, *243*, 273–282. Available from https://doi.org/10.1016/j.apcatb.2018.10.060.

China Petroleum and Chemical Industry Federation. (2020). *China CN: Market price: Monthly avg: Organic chemical material: Dimethyl ether: 99.0% or above.* Available at https://www.ceicdata.com/en/china/china-petroleum--chemical-industry-association-petrochemical-price-organic-chemical-material/cn-market-price-monthly-avg-organic-chemical-material-dimethyl-ether-990-or-above (Accessed 9 June 2020).

Cho, J., Kim, T. H., and Lee, K. S. (2018). Compositional modeling and simulation of dimethyl ether (DME)-enhanced waterflood to investigate oil mobility improvement. *Petroleum Science, 15*(2), 297−304. Available from https://doi.org/10.1007/s12182-017-0212-z.

Ciftci, A., Sezgi, N. A., and Dogu, T. (2010). Nafion-incorporated silicate structured nanocomposite mesoporous catalysts for dimethyl ether synthesis. *Industrial and Engineering Chemistry Research, 49*(15), 6753−6762. Available from https://doi.org/10.1021/ie9015667.

Dahmen, N., et al. (2017). The bioliq process for producing synthetic transportation fuels. *Wiley Interdisciplinary Reviews: Energy and Environment, 6*(3), e236. Available from https://doi.org/10.1002/wene.236.

De Falco, M., Capocelli, M., and Centi, G. (2016). Dimethyl ether production from CO_2 rich feedstocks in a one-step process: Thermodynamic evaluation and reactor simulation. *Chemical Engineering Journal, 294*, 400−409. Available from https://doi.org/10.1016/j.cej.2016.03.009.

Diban, N., Aguayo, A. T., et al. (2013). Membrane reactors for *in situ* water removal: A review of applications. *Industrial and Engineering Chemistry Research, 52*(31), 10342−10354. Available from https://doi.org/10.1021/ie3029625.

Diban, N., Urtiaga, A. M., et al. (2013). Influence of the membrane properties on the catalytic production of dimethyl ether with *in situ* water removal for the successful capture of CO_2. *Chemical Engineering Journal, 234*, 140−148. Available from https://doi.org/10.1016/j.cej.2013.08.062.

Du, L., et al. (2019). Direct dimethyl ether fuel cells with low platinum-group-metal loading at anode: Investigations of operating temperatures and anode Pt/Ru ratios. *Journal of Power Sources, 433*, 126690. Available from https://doi.org/10.1016/j.jpowsour.2019.05.096.

Edwards, R., et al. (2014). *Well-to-wheels report version 4.a: JEC well-to-wheels analysis: Well-to-wheels analysis of future automotive fuels and powertrains in the European context.* Luxembourg: Publications Office of the European Union. Available from https://doi.org/10.2790/95533.

Ereña, J., et al. (2011). Kinetic modelling of dimethyl ether synthesis from ($H_2 + CO_2$) by considering catalyst deactivation. *Chemical Engineering Journal, 174*(2−3), 660−667. Available from https://doi.org/10.1016/j.cej.2011.09.067.

Fasihi, M., Efimova, O., and Breyer, C. (2019). Techno-economic assessment of CO_2 direct air capture plants. *Journal of Cleaner Production, 224*, 957−980. Available from https://doi.org/10.1016/j.jclepro.2019.03.086.

Fleisch, T. H., Basu, A., and Sills, R. A. (2012). Introduction and advancement of a new clean global fuel: The status of DME developments in China and beyond. *Journal of Natural Gas Science and Engineering, 9*, 94−107. Available from https://doi.org/10.1016/j.jngse.2012.05.012.

Ford-werke GmbH. (2015). *Ford leitet Forschungsprojekt an Diesel-Motoren, die mit Kraftstoffen aus CO_2 nahezu rußfrei betrieben, Presseportal.de, 11 September.* Available at https://www.presseportal.de/pm/6955/3120034 (Accessed 11 March 2020).

Frusteri, F., et al. (2017). Direct CO_2-to-DME hydrogenation reaction: New evidences of a superior behaviour of FER-based hybrid systems to obtain high DME yield. *Journal of CO_2 Utilization, 18*, 353−361. Available from https://doi.org/10.1016/j.jcou.2017.01.030.

Gao, X. J., et al. (2018). Synthesis of polyoxymethylene dimethyl ethers from dimethyl ether direct oxidation over carbon-based catalysts. *ChemCatChem*, *10*(1), 273–279. Available from https://doi.org/10.1002/cctc.201701213.

Gavriel, B., Sharabi, R., and Elbaz, L. (2017). Direct electro-oxidation of dimethyl ether on Pt-Cu nanochains. *ChemSusChem*, *10*(15), 3069–3074.

Goehna, H., and Koenig, P. (1994). Producing methanol from CO_2: The keys are a catalyst that can tolerate a high CO_2 concentration and an unusual process configuration. *ChemTech*, 36–39.

Gogate, M. R. (2018). The direct dimethyl ether (DME) synthesis process from syngas I. Process feasibility and chemical synergy in one-step LPDME^{tm} process. *Petroleum Science and Technology*, *36*(8), 547–554. Available from https://doi.org/10.1080/10916466.2018.1428628.

Grand View Research Inc. (2020). *Dimethyl ether market size, share and trends analysis report by application, regional outlook, competitive strategies, and segment forecasts, 2019 to 2025*.

Ha, K. S., et al. (2011). New reaction pathways and kinetic parameter estimation for methanol dehydration over modified ZSM-5 catalysts. *Applied Catalysis A: General*, *395* (1–2), 95–106. Available from https://doi.org/10.1016/j.apcata.2011.01.025.

Haltenort, P., et al. (2018). Heterogeneously catalyzed synthesis of oxymethylene dimethyl ethers (OME) from dimethyl ether and trioxane. *Catalysis Communications*, *109*, 80–84. Available from https://doi.org/10.1016/j.catcom.2018.02.013.

Hamidi, M., Samimi, F., and Rahimpour, M. R. (2015). Dimethyl ether synthesis in a gas-solid-solid trickle flow reactor with continuous adsorbent regeneration. *Journal of the Taiwan Institute of Chemical Engineers*, *47*, 105–112. Available from https://doi.org/10.1016/j.jtice.2014.10.013.

Hank, C., et al. (2020). Energy efficiency and economic assessment of imported energy carriers based on renewable electricity. *Sustainable Energy and Fuels*, *4*(5), 2256–2273. Available from https://doi.org/10.1039/d0se00067a.

Harmsen, G. J. (2007). Reactive distillation: The front-runner of industrial process intensification. A full review of commercial applications, research, scale-up, design and operation. *Chemical Engineering and Processing: Process Intensification*, *46*(9), 774–780. Available from https://doi.org/10.1016/j.cep.2007.06.005.

Hosseini, S., Taghizadeh, M., and Eliassi, A. (2012). Optimization of hydrothermal synthesis of H-ZSM-5 zeolite for dehydration of methanol to dimethyl ether using full factorial design. *Journal of Natural Gas Chemistry*, *21*(3), 344–351. Available from https://doi.org/10.1016/S1003-9953(11)60375-7.

Hosseininejad, S., Afacan, A., and Hayes, R. E. (2012). Catalytic and kinetic study of methanol dehydration to dimethyl ether. *Chemical Engineering Research and Design. Institution of Chemical Engineers*, *90*(6), 825–833. Available from https://doi.org/10.1016/j.cherd.2011.10.007.

Hydrogen Council. (2020). *Path to hydrogen competitiveness: A cost perspective*. Available at https://hydrogencouncil.com/en/path-to-hydrogen-competitiveness-a-cost-perspective/ (Accessed 20 January 2020).

IEA. (2019). *The future of hydrogen: Seizing today's opportunities*. Available at https://www.iea.org/reports/the-future-of-hydrogen.

Iliuta, I., Iliuta, M. C., and Larachi, F. (2011). Sorption-enhanced dimethyl ether synthesis—Multiscale reactor modeling. *Chemical Engineering Science*, *66*(10), 2241–2251. Available from https://doi.org/10.1016/j.ces.2011.02.047.

Iliuta, I., Larachi, F., and Fongarland, P. (2010). Dimethyl ether synthesis with in situ H_2O removal in fixed-bed membrane reactor: Model and simulations. *Industrial and Engineering Chemistry Research*, *49*(15), 6870–6877. Available from https://doi.org/10.1021/ie901726u.

Jasińska, J., Krzyżyńska, B., and Kozłowski, M. (2011). Influence of activated carbon modifications on their catalytic activity in methanol and ethanol conversion reactions. *Central European Journal of Chemistry*, *9*(5), 925–931. Available from https://doi.org/10.2478/s11532-011-0078-7.

Kashyap, D., Teller, H., and Schechter, A. (2019). Dimethyl ether oxidation on an active SnO_2/Pt/C catalyst for high-power fuel cells. *ChemElectroChem*, *6*(9), 2407–2414. Available from https://doi.org/10.1002/celc.201900216.

Keil, F. J. (1999). Methanol-to-hydrocarbons: Process technology. *Microporous and Mesoporous Materials*, *29*(1–2), 49–66. Available from https://doi.org/10.1016/S1387-1811(98)00320-5.

Keshavarz, A. R., Rezaei, M., and Yaripour, F. (2010). Nanocrystalline gamma-alumina: A highly active catalyst for dimethyl ether synthesis. *Powder Technology.*, *199*(2), 176–179. Available from https://doi.org/10.1016/j.powtec.2010.01.003.

Khandan, N., Kazemeini, M., and Aghaziarati, M. (2008). Determining an optimum catalyst for liquid-phase dehydration of methanol to dimethyl ether. *Applied Catalysis A: General*, *349*(1–2), 6–12. Available from https://doi.org/10.1016/j.apcata.2008.07.029.

Kim, N. H. (2016). Application of the natural refrigerant mixture R-290/DME to a soft ice cream refrigerator. *International Journal of Air-Conditioning and Refrigeration*, *24*(4). Available from https://doi.org/10.1142/S2010132516500279.

Kiss, A. A. (2018). Novel catalytic reactive distillation processes for a sustainable chemical industry. *Topics in Catalysis*, *89*, 479. Available from https://doi.org/10.1007/s11244-018-1052-9.

Kiss, A. A., and Suszwalak, D. J. P. C. (2012). Innovative dimethyl ether synthesis in a reactive dividing-wall column. *Computers and Chemical Engineering*, *38*, 74–81. Available from https://doi.org/10.1016/j.compchemeng.2011.11.012.

Kolesnichenko, N. V., et al. (2011). Dehydration of methanol on the catalyst based on perfluorinated copolymer F-4SF. *Fluorine Notes*, *1*.

Ledesma, C., et al. (2019). Catalytic reforming of dimethyl ether in microchannels. *Catalysis Today*, *323*, 209–215. Available from https://doi.org/10.1016/j.cattod.2018.03.011.

Lee, K. H., Youn, M. Y., and Sea, B. (2006). Preparation of hydrophilic ceramic membranes for a dehydration membrane reactor. *Desalination*, *191*(1–3), 296–302. Available from https://doi.org/10.1016/j.desal.2005.07.026.

Lee, S., Gogate, M., and Kulik, C. J. (1995). Methanol-to-gasoline vs. DME-to-gasoline II. Process comparison and analysis. *Fuel Science and Technology International*, *13*(8), 1039–1057. Available from https://doi.org/10.1080/08843759508947721.

Lei, Z., et al. (2011). Synthesis of dimethyl ether (DME) by catalytic distillation. *Chemical Engineering Science*, *66*(14), 3195–3203. Available from https://doi.org/10.1016/j.ces.2011.02.034.

Li, Q., et al. (2014). Direct dimethyl ether fuel cell with much improved performance. *Electrocatalysis*, *5*(3), 310–317. Available from https://doi.org/10.1007/s12678-014-0196-z.

Liu, L., et al. (2010). The dehydration of methanol to dimethyl ether over a novel slurry catalyst. *Energy Sources, Part A: Recovery, Utilization and Environmental Effects*, *32*(15), 1379–1387. Available from https://doi.org/10.1080/15567030903030724.

Liu, Y., et al. (2013). Synthesis of ethanol from methanol and syngas through an indirect route containing methanol dehydrogenation, DME carbonylation, and methyl acetate hydrogenolysis. *Fuel Processing Technology*, *110*, 206–213. Available from https://doi.org/10.1016/j.fuproc.2012.12.016.

Liuzzi, D., and Peinado, C. (2020). Effects of support and reaction pressure for the synthesis of dimethyl ether over heteropolyacid catalysts. *Scientific reports*, *10*(1), S.8551. Available from https://doi.org/10.1038/s41598-020-65296-3.

Marlin, D. S., Sarron, E., and Sigurbjörnsson, Ó. (2018). Process advantages of direct CO_2 to methanol synthesis. *Frontiers in Chemistry*, *6*, 446. Available from https://doi.org/10.3389/fchem.2018.00446.

Masih, D., et al. (2017). Low-temperature methanol dehydration to dimethyl ether over various small-pore zeolites. *Applied Catalysis B: Environmental*, *217*, 247–255. Available from https://doi.org/10.1016/j.apcatb.2017.05.089.

Matzen, M., and Demirel, Y. (2016). Methanol and dimethyl ether from renewable hydrogen and carbon dioxide: Alternative fuels production and life-cycle assessment. *Journal of Cleaner Production*, *139*, 1068–1077. Available from https://doi.org/10.1016/j.jclepro.2016.08.163.

Maus, W. (2019). In W. Maus (Ed.), *Zukünftige Kraftstoffe—Energiewende des Transports als ein weltweites Klimaziel*. Berlin, Heidelberg: Springer Vieweg. Available from https://doi.org/10.1007/978-3-662-58006-6.

Methanol Institute. (2016). *DME: An emerging global fuel*. Available at http://www.methanol.org/wp-content/uploads/2016/06/DME-An-Emerging-Global-Guel-FS.pdf (Accessed 2 June 2020).

Michailos, S., et al. (2019). Dimethyl ether synthesis via captured CO_2 hydrogenation within the power to liquids concept: A techno-economic assessment. *Energy Conversion and Management*, *184*, 262–276. Available from https://doi.org/10.1016/j.enconman.2019.01.046.

Migliori, M., et al. (2014). Kinetic analysis of methanol to dimethyl ether reaction over H-MFI catalyst. *Industrial and Engineering Chemistry Research*, *53*(38), 14885–14891. Available from https://doi.org/10.1021/ie502775u.

Migliori, M., Aloise, A., and Giordano, G. (2014). Methanol to dimethylether on H-MFI catalyst: The influence of the Si/Al ratio on kinetic parameters. *Catalysis Today*, *227*, 138–143. Available from https://doi.org/10.1016/j.cattod.2013.09.033.

Mondal, U., and Yadav, G. D. (2019). Perspective of dimethyl ether as fuel: Part I. Catalysis. *Journal of CO_2 Utilization*, *32*, 299–320. Available from https://doi.org/10.1016/j.jcou.2019.02.003.

Moradi, G. R., Yaripour, F., and Vale-Sheyda, P. (2010). Catalytic dehydration of methanol to dimethyl ether over mordenite catalysts. *Fuel Processing Technology*, *91*(5), 461–468. Available from https://doi.org/10.1016/j.fuproc.2009.12.005.

Moreno-Castilla, C., et al. (2001). Dehydration of methanol to dimethyl ether catalyzed by oxidized activated carbons with varying surface acidic character. *Carbon*, *39*(6), 869–875. Available from https://doi.org/10.1016/S0008-6223(00)00192-5.

Müller, M., and Hübsch, U. (2000). *Dimethyl ether. Ullmann's Encyclopedia of Industrial Chemistry* (p. 39) Weinheim: Wiley-VCH Verlag GmbH and Co. KGaA.

Nestler, F., et al. (2018). Methanol synthesis – Industrial challenges within a changing raw material landscape. *Chemie Ingenieur Technik*, *90*(10), 1409–1418. Available from https://doi.org/10.1002/cite.201800026.

Oestreich, D., et al. (2018). Production of oxymethylene dimethyl ether (OME)-hydrocarbon fuel blends in a one-step synthesis/extraction procedure. *Fuel*, *214*, 39–44. Available from https://doi.org/10.1016/j.fuel.2017.10.116.

Onaka, Y., Miyara, A., and Tsubaki, K. (2010). Experimental study on evaporation heat transfer of CO_2/DME mixture refrigerant in a horizontal smooth tube. *International Journal of Refrigeration*, *33*(7), 1277–1291. Available from https://doi.org/10.1016/j.ijrefrig.2010.06.014.

Ong, B. C., Kamarudin, S. K., and Basri, S. (2017). Direct liquid fuel cells: A review. *International Journal of Hydrogen Energy*, *42*(15), 10142–10157. Available from https://doi.org/10.1016/j.ijhydene.2017.01.117.

Ortega, C., et al. (2018). Methanol to dimethyl ether conversion over a ZSM-5 catalyst: Intrinsic kinetic study on an external recycle reactor. *Chemical Engineering Journal*, *347*, 741–753. Available from https://doi.org/10.1016/j.cej.2018.04.160.

Park, S. Y., Shin, C. H., and Bae, J. W. (2016). Selective carbonylation of dimethyl ether to methyl acetate on ferrierite. *Catalysis Communications*, *75*, 28–31. Available from https://doi.org/10.1016/j.catcom.2015.12.006.

Peng, X. D., et al. (1997). Progress in development of LPDME process: Kinetics and catalysts. *Coal Liquefaction and Solid Fuels Contractors Review Conference* (610), 1–6. Available at. Available from http://www.fischer-tropsch.org/DOE/_conf_proc/CoalConferences/coal_liq_conf97/peng.pdf.

Peng, X.-D., et al. (2002). *Separation process for one-step production of dimethyl ether from synthesis gas*. Patent number: US6458856B1.

Pérez-Uriarte, P., Ateka, A., et al. (2016). Kinetic model for the reaction of DME to olefins over a HZSM-5 zeolite catalyst. *Chemical Engineering Journal*, *302*, 801–810. Available from https://doi.org/10.1016/j.cej.2016.05.096.

Pérez-Uriarte, P., Gamero, M., et al. (2016). Effect of the acidity of HZSM-5 zeolite and the binder in the DME transformation to olefins. *Industrial and Engineering Chemistry Research*, *55*(6), 1513–1521. Available from https://doi.org/10.1021/acs.iecr.5b04477.

Pontzen, F., et al. (2011). CO_2-based methanol and DME – Efficient technologies for industrial scale production. *Catalysis Today*, *171*(1), 242–250. Available from https://doi.org/10.1016/j.cattod.2011.04.049.

Pulidindi, K., and Pandey, H. (2018). Bio-based and synthetic dimethyl ether (DME) market size by raw material (fossil fuel based, bio-based), by application (LPG blending, aerosol propellants, transportation fuel, industrial), industry analysis report, regional outlook, growth potential, price trends, competitive market share and forecast, 2019–2025. Global Market Insights, Inc. Available at https://www.gminsights.com/methodology/detail/dimethyl-ether-dme-market.

Qian, Q., et al. (2018). Synthesis of ethanol via a reaction of dimethyl ether with CO_2 and H_2. *Green Chemistry*, *20*(1), 206–213. Available from https://doi.org/10.1039/c7gc02807e.

Ram, M., et al. (2017). *Vergleich der Stromgestehungskosten von Erneuerbaren Energien mit denen fossiler und nuklearer Kraftwerke in den G20-Ländern*. Available at https://www.greenpeace.de/g20-studie.

Ramos, F. S., et al. (2005). Role of dehydration catalyst acid properties on one-step DME synthesis over physical mixtures. *Catalysis Today*, *101*(1), 39–44. Available from https://doi.org/10.1016/j.cattod.2004.12.007.

Sabour, B., et al. (2014). Catalytic dehydration of methanol to dimethyl ether (DME) over Al-HMS catalysts. *Journal of Industrial and Engineering Chemistry*, *20*(1), 222–227. Available from https://doi.org/10.1016/j.jiec.2013.03.044.

Saraswat, V. K., and Bansal, R. (2017). *India's leapfrog to methanol economy*. Available at http://indiaenergy.gov.in/wp-content/uploads/2017/09/Indias-Leapfrog-to-Methanol-Economy.pdf.

Schemme, S., et al. (2017). Power-to-fuel as a key to sustainable transport systems – An analysis of diesel fuels produced from CO_2 and renewable electricity. *Fuel*, *205*, 198–221. Available from https://doi.org/10.1016/j.fuel.2017.05.061.

Schemme, S., et al. (2019). H_2-based synthetic fuels: A techno-economic comparison of alcohol, ether and hydrocarbon production. *International Journal of Hydrogen Energy*, *45* (8), 5395–5414. Available from https://doi.org/10.1016/j.ijhydene.2019.05.028.

Semelsberger, T. A., Borup, R. L., and Greene, H. L. (2006). Dimethyl ether (DME) as an alternative fuel. *Journal of Power Sources*, *156*(2), 497–511. Available from https://doi.org/10.1016/j.jpowsour.2005.05.082.

Shoji, K., and Terai, S. (2004). *Process for producing dimethyl ether. US2004/0034255A1.*

Sorenson, S. C. (2001). Dimethyl ether in diesel engines: Progress and perspectives. *Journal of Engineering for Gas Turbines and Power*, *123*(3), 652–658. Available from https://doi.org/10.1115/1.1370373.

Stiefel, M., et al. (2011). Direct synthesis of dimethyl ether from carbon-monoxide-rich synthesis gas: Influence of dehydration catalysts and operating conditions. *Fuel Processing Technology*, *92(*(8), 1466–1474. Available from https://doi.org/10.1016/j.fuproc.2011.03.007.

Ströfer, E., et al. (2006). *Verfahren zur herstellung von polyoxymethylendialkylethern aus trioxan und dialkylether. WO2006134081A1.*

Sun, J., et al. (2014). Catalysis chemistry of dimethyl ether synthesis. *ACS Catalysis*, *4*(10), 3346–3356. Available from https://doi.org/10.1021/cs500967j.

Tabak, S. A., and Yurchak, S. (1990). Conversion of methanol over ZSM-5 to fuels and chemicals. *Catalysis Today*, *6*(3), 307–327. Available from https://doi.org/10.1016/0920-5861(90)85007-B.

Tremel, A., et al. (2015). Techno-economic analysis for the synthesis of liquid and gaseous fuels based on hydrogen production via electrolysis. *International Journal of Hydrogen Energy*, *40*(35), 11457–11464. Available from https://doi.org/10.1016/j.ijhydene.2015.01.097.

Valero-Romero, M. J., et al. (2019). Phosphorus-containing mesoporous carbon acid catalyst for methanol dehydration to dimethyl ether. *Industrial and Engineering Chemistry Research*, *58*(10), 4042–4053. Available from https://doi.org/10.1021/acs.iecr.8b05897.

van Kampen, J., et al. (2019). Steam separation enhanced reactions: Review and outlook. *Chemical Engineering Journal*, *374*, 1286–1303. Available from https://doi.org/10.1016/j.cej.2019.06.031.

van Kampen, J., et al. (2020). Sorption enhanced dimethyl ether synthesis for high efficiency carbon conversion: Modelling and cycle design. *Journal of CO_2 Utilization*, *37*, 295–308. Available from https://doi.org/10.1016/j.jcou.2019.12.021.

Varışlı, D., and Doğu, T. (2008). Production of clean transportation fuel dimethylether by dehydration of methanol over Nafion catalyst. *Gazi University Journal of Science*, *21* (2), 37–41.

Vertin, K. D., et al. (1999). Methylal and methylal-diesel blended fuels for use in compression-ignition engines. *SAE Technical Papers* (724). Available from https://doi.org/10.4271/1999-01-1508.

Wang, W., et al. (2016). VOx modified H-beta zeolite for dimethyl ether direct oxidation to polyoxymethylene dimethyl ethers. *Chemical Sciences Journal*, *7*(2). Available from https://doi.org/10.4172/2150-3494.1000124.

Wang, W., et al. (2019). Vanadium oxide modified H-beta zeolite for the synthesis of polyoxymethylene dimethyl ethers from dimethyl ether direct oxidation. *Fuel*, *238*, 289–297. Available from https://doi.org/10.1016/j.fuel.2018.10.098.

Willems, W., et al. (2020). Sauerstoffhaltige Kraftstoffe im selbstzündenden Verbrennungsmotor. *MTZ - Motortechnische Zeitschrift, 81*(3), 28–37. Available from https://doi.org/10.1007/s35146-019-0186-z.

Williams, J. (2020) *Oberon fuels partners with SHV energy to accelerate use of renewable dme (rDME) to decarbonize transportation.* Available at http://oberonfuels.com/2020/02/11/oberon-fuels-partners-with-shv-energy-to-accelerate-use-of-renewable-dme-rdme-to-decarbonize-transportation/ (Accessed 7 April 2020).

Wu, N., Zhang, W., and Huang, Z. (2008). Impact of dimethyl ether on engine seal materials. *Frontiers of Energy and Power Engineering in China, 2*(3), 279–284. Available from https://doi.org/10.1007/s11708-008-0042-1.

Xu, M., et al. (1997). Synthesis of dimethyl ether (DME) from methanol over solid-acid catalysts. *Applied Catalysis A: General, 149*(2), 289–301.

Yaripour, F., et al. (2005). Synthesis of dimethyl ether from methanol over aluminium phosphate and silica-titania catalysts. *Catalysis Communications, 6*(8), 542–549. Available from https://doi.org/10.1016/j.catcom.2005.05.003.

Yaripour, F., et al. (2015). The effects of synthesis operation conditions on the properties of modified γ-alumina nanocatalysts in methanol dehydration to dimethyl ether using factorial experimental design. *Fuel, 139*, 40–50. Available from https://doi.org/10.1016/j.fuel.2014.08.029.

Zhang, Q., et al. (2007). $MnCl_2$ modified $H_4SiW_{12}O_{40}/SiO_2$ catalysts for catalytic oxidation of dimethy ether to dimethoxymethane. *Journal of Molecular Catalysis A: Chemical, 263*(1–2), 149–155. Available from https://doi.org/10.1016/j.molcata.2006.08.044.

Zhang, Q., et al. (2008). The electro-oxidation of dimethyl ether on platinum-based catalyst. *Electrochimica Acta, 53*(28), 8298–8304. Available from https://doi.org/10.1016/j.electacta.2008.05.044.

Zhang, Q., et al. (2014). Promotional effects of Sm_2O_3 on Mn-$H_4SiW_{12}O_{40}/SiO_2$ catalyst for dimethyl ether direct-oxidation to dimethoxymethane. *Journal of Industrial and Engineering Chemistry, 20*(4), 1869–1874. Available from https://doi.org/10.1016/j.jiec.2013.09.004.

Zhang, Q., et al. (2016). Low-temperature oxidation of dimethyl ether to polyoxymethylene dimethyl ethers over CNT-supported rhenium catalyst. *Catalysts, 6*(3), 43. Available from https://doi.org/10.3390/catal6030043.

Zhang, T., et al. (2018). Dynamic analysis of a PEM fuel cell hybrid system with an on-board dimethyl ether (DME) steam reformer (SR). *International Journal of Hydrogen Energy, 43*(29), 13521–13531. Available from https://doi.org/10.1016/j.ijhydene.2018.05.098.

Zhu, Q., et al. (2008). The study of methanol-to-olefin over proton type aluminosilicate CHA zeolites. *Microporous and Mesoporous Materials, 112*(1–3), 153–161. Available from https://doi.org/10.1016/j.micromeso.2007.09.026.

Power to ammonia and urea

7

András Perl
Hanze University of Applied Sciences, Groningen, The Netherlands

7.1 Ammonia as a bridge between agriculture, industry and energy

Globally, around 180 million tonnes of ammonia are produced and transported annually. While ammonia remains the main chemical of the fertiliser industry, its physical properties and its established infrastructure for both storage and distribution make ammonia a prominent energy carrier candidate.

Ammonia, NH_3, is a colourless gas at standard pressure and temperature, with a boiling point of $-33.33°C$ and melting point of $-77.73°C$. With a relative molecular mass of $17.031\ g\ mol^{-1}$, the hydrogen mass content of 1 kg of ammonia is around 17.6%. Therefore 1 L saturated liquid ammonia contains more hydrogen ($120\ kg\ H_2\ m^{-3}$) than 1 L liquefied hydrogen at $-265°C$ ($71\ kg\ H_2\ m^{-3}$). This fact offers to ammonia the potential to become the main carrier of renewable hydrogen.

The solid and crystalline ammonia derivative, urea, is also a possible prominent hydrogen-carrier alternative. Urea, $(NH_2)_2CO$, as a nonflammable, relatively nontoxic and colourless substance, contains 6.6% hydrogen on a weight basis or $87.9\ kg\ H_2\ m^{-3}$ pure crystal. Thus the gravimetric hydrogen content of urea is also higher than that of liquefied hydrogen. In addition, the fact that urea is solid at standard conditions gives additional safety-related and economic advantages for hydrogen storage applications.

The decomposition of ammonia into hydrogen and nitrogen and the extraction of hydrogen from urea are, however, challenging technical and economic aspects of their usage as hydrogen carrier at industrial scale. The decomposition of ammonia is an endothermic reaction whose equilibrium conversion is high enough only at 450°C or higher temperatures (Mukherjee et al., 2018). The state-of-the-art catalysts for decomposition of ammonia are ruthenium doped with potassium, barium and caesium and supported on various oxides and carbon supports. These catalysts are not sufficiently active for low-temperature ammonia decomposition and are economically not viable for large-scale applications. Transition metals like iron, cobalt or nickel also show catalytic activity for ammonia decomposition, but only at relatively high temperatures ($>600°C$) (Mukherjee et al., 2018). This high temperature not only negatively impacts the operating costs but also requires more expensive reactor materials as well. Photoelectrochemical and electrochemical methods are

Power to Fuel. DOI: https://doi.org/10.1016/B978-0-12-822813-5.00007-2

available for urea splitting (Zhu, Liang, and Zou, 2020), but all these processes are additional steps that can significantly lower the overall process efficiency.

The decomposition of ammonia and urea is not necessarily needed. Both are already heavily used in the agriculture as fertilisers. Urea is the most commonly used source of nitrogen for fertiliser usage worldwide. Ammonia can be used directly to the soil from tanks containing the liquefied gas or in the form of ammonium salts. Crop growth requires sunlight, carbon dioxide (CO_2), water and a balanced supply of the several essential plant nutrients. The essential nutrients can be divided into three main categories: nitrogen, phosphorus and potassium. They are essential for metabolic functions of the plants. The plants absorb them with water from the soil via their root system. When a plant is harvested, the nutrients it has absorbed are extracted from the local agricultural system, and without a natural cycle of local growth and decomposition, the nutrient supply in the soil is depleted. That is why these nutrients are supplemented by organic and mineral sources such as fertilisers.

The world's yearly demand of ammonia for fertiliser purposes is expected to grow from 107 million tons in 2019 to around 111 million tonnes by 2022 (FAO, 2019). On the other hand, ammonia-based fertiliser consumption is expected to slightly decrease in the European Union in the coming decade (Fertilizers Europe, 2019).

While nitrogen in the form of ammonia is a vital nutrient to plants, high concentrations are harmful to people and nature. Excess nitrogen from agricultural sources is one of the main causes of water pollution. In the European Union the Nitrate Directive and its national implementation rules promote ammonia use efficiently and limit the use of fertilisers with high ammonia emissions. The negative effect of ammonia on biodiversity might hinder the production size, but fertilisers are a very important part of our food production. Generally, it is expected for the future that urea-based fertilisers will have to be applied under special conditions or will have to be applied with inhibitors that reduce the potential for ammonia emissions (Fertilizers Europe, 2018). The production of fertilisers is also expected to face increased regulatory pressure. The society's need for a more sustainable industry and agriculture will limit the emissions from ammonia and urea production facilities and control the wastes and by-products from fertiliser production sites. The limitations of the fossil-based ammonia production promote renewable power-to-ammonia production technologies.

Besides its use as fertiliser, ammonia has several important industrial uses. Ammonia is used for the production of plastics and fibres in the form of polyamides, urea–formaldehyde–phenol resins, polyurethanes and polyacrylonitrile. It is also used for the manufacture of organic nitrogen compounds that serve as intermediate for dyes, pharmaceuticals and explosives. Ammonia is used for removing sulphur dioxide from flue gas. In the pulp and paper industry, it is used for pulping wood and as casein dispersant in the coating of paper. Ammonia is used in the rubber industry for the stabilisation of natural and synthetic latex to prevent premature coagulation. It can be used as a solvent in certain processes. The alkali metals as well as the heavier alkaline–earth metals and even some inner transition metals dissolve in liquid ammonia. Ammonia is used in the mining industry for the extraction of metals such as copper, nickel and molybdenum from their ores. Because of its low boiling point and high heat of evaporation, it is commonly used as a refrigerant (Appl, 1999).

Ammonia can be used directly to generate power. Fuel cell-based and combustion-based technologies as well are virtually available to transform the chemical energy potential of ammonia into power.

In a fuel cell the oxidation and reduction half reactions of the fuel combustion are separated with the help of an ion-conducting membrane. The membrane serves as a medium/electrolyte to transport the charge carrier between the anodic and cathodic compartments, it functions as a medium for the prevention of gas cross-over and it keeps the system compact. In a proton exchange/polymer electrolyte membrane fuel cell the polymer-based membrane transports electricity via protons (hydrogen ions) within the 5°C–95°C range and various pressures. In an alkaline fuel cell (AFC), OH^- ions cross the membrane to maintain the charge balance between the anodic and cathodic compartments. Solid oxide fuel cells (SOFCs) operate at temperatures as high as several hundred degree Celsius. The membrane, as its name suggests, is a solid oxide that uses O^{2-} ions as charge carrier (Kishimoto et al., 2020), but proton-conducting versions are also available (Valera-Medina et al., 2018).

Ammonia can be used as fuel for fuel cells. In the direct ammonia fuel cells, hydrogen is the main fuel for those fuel cells; ammonia is decomposed in situ into nitrogen and hydrogen at the anode. SOFCs are more suitable for ammonia because of their higher operating temperature that favours the ammonia decomposition process. The undesirable effects of ammonia on the cell-stack components are hindering the commercial application of direct ammonia fuel cells. Large-scale production has not been realized and the long-term stability of the catalysts has to be confirmed. Recent performance and durability tests of a 1 kW ammonia-fuelled SOFC with direct ammonia supply showed 55.9% electric efficiency. However, the nitriding of the separator materials, resulting in morphology change of the separator surface, was observed. The same SOFC runs with an operating mode based on the external decomposition of ammonia achieved 57.3% (Kishimoto et al., 2020).

Direct urea fuel cells are based on the electrochemical decomposition of urea into CO_2, N_2, H_2O and electricity in an alkaline medium. Besides the electricity generation, urea-based AFCs can be also applied to treat urea-contaminated wastewater. Ni-based electrocatalysts are in the focus of the research, but the produced CO and CO_2 act as poisoning components during urea oxidation (Sayed et al., 2019). To our best knowledge, the process is not yet applied at a commercial level.

The use of ammonia as fuel in internal combustion engines and gas turbines was intensively studied (Valera-Medina et al., 2018). Due to the corrosive nature of ammonia, components made out of brass or copper must be removed from the system to avoid operability issues. Several studies showed that in internal combustion engines, ammonia needs to be mixed with other fuels as combustion promoters owing to ammonia's low flame speed and high resistance to auto-ignition.

The air-pollutant nitrogen oxides produced during the combustion process and the emission of unburned ammonia however are slowing down the commercialisation of these technologies. News about patented ammonia engines and prototypes of ammonia-fuelled cars in Canada, Italy, Japan, South Korea and the United States have been seen in the last decade (Dimitriou and Javaid, 2020).

7.2 Ammonia and urea synthesis

The most straightforward synthesis route of ammonia is via the reaction of its two elemental components: hydrogen and nitrogen. Although the reaction is exothermic, it does not take place spontaneously. The molecular nitrogen's energy need of dissociation is much higher than that of hydrogen owing to the stronger triple bonds between the nitrogen atoms. This high dissociation energy gives nitrogen a low reactivity at room temperature. Virtually, there are no naturally occurring reactions of nitrogen at ambient conditions besides the nitrogen fixation with nitrogenase in the bacteria of the root nodules of legumes, in blue algae (Appl, 1999) and in purple nonsulphur bacteria (Sakarika et al., 2020). The homogeneous reaction of hydrogen and nitrogen in gas phase at industrially reasonable pressures would require temperatures higher than 800°C. High temperatures, however, lower the reaction constant and favour the dissociation of ammonia into its elemental components and thus decrease the conversion of the reagents.

The discovery of the heterogeneous reaction on iron-based catalyst at the beginning of the last century by Fritz Haber and Carl Bosch opened the way to high-scale industrial ammonia synthesis. Their Haber–Bosch process is used until today at large-scale ammonia production sites. The iron catalysts are used as unreduced iron oxides (mainly magnetite) with a few percent of Al, Ca and K, but other elements such as Mg and Si may also be present in small amounts (Appl, 1999). During the catalytic reaction, hydrogen and nitrogen are adsorbed to the solid catalyst surface (Fig. 7.1). The adsorption reduces their molecular dissociation energy need, lowering the activation barrier. The catalytic ammonia synthesis may be

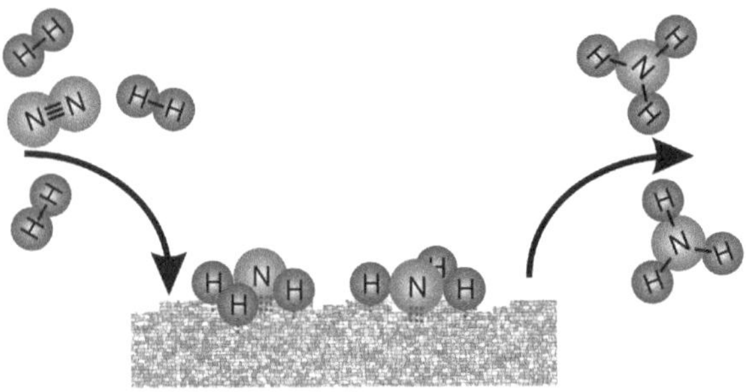

Figure 7.1 Schematic presentation of the heterogeneous reaction of nitrogen and hydrogen in a stoichiometric ratio of 1:3. The reactants are transported by diffusion and convection to the surface of the catalyst particle where they adsorb. The adsorbed species react with each other and they leave the surface.

carried out at a reasonable temperature range of 350°C–550°C. The reaction is limited by thermodynamic equilibrium and nitrogen can only achieve partial conversion (25%–35%) by a single pass through the catalyst. Unreacted reactants need to be separated and recycled back for a reasonable conversion.

Some noniron systems also show catalytic potential for the heterogeneous ammonia synthesis. Especially, metals or metal alloys from the transition-metal group for which the chemisorption energy of nitrogen is neither too high nor too low show the greatest effectiveness, but only the iron-based catalyst proved suitable for industrial use (Appl, 1999).

Some oxygen, sulphur and chlorine-based compounds can bind to the solid catalyst surface or react with it and act as a catalyst poison. While the effect of H_2O, CO, CO_2 and O_2 is reversible, and their effect can be reversed, sulphur and chlorine-based compounds may cause irreversible damage. Traces of these compounds in the reagent mixture significantly shorten the lifetime of the catalyst.

Work on alternative ammonia synthesis routes that could offer higher ammonia conversion and lower energy consumption led to electrochemical processes using either solid-state electrolytes (Amar et al., 2011) or molten salts (Yang, Weng, and Xiao, 2020). The alternative ammonia synthesis technologies did not yet reach the industrial game-changing phase, the search for highly efficient electrocatalysts with low overpotential, considerable selectivity and faradaic efficiency is still going on (Chen and Xia, 2019).

The large-scale industrial urea production is based on the Basarov reaction. The process is a two-step reaction. First, ammonium carbamate, NH_2COONH_4, is formed from ammonia and carbon dioxide, then the ammonium carbamate is dehydrated to form urea and water. The first reaction is highly exothermic and it runs almost to completion at high enough pressures where the reagents are in liquid phase. The second reaction is the limiting step that runs only in the liquid phase, and there are no known catalysts to speed up the step (Meessen, 2014). The typical conditions for the last step are 179°C–220°C and 125–250 bar.

7.3 Raw materials for ammonia and urea production

One of the two main components for ammonia synthesis, nitrogen, is abundantly present in a natural form in the atmosphere: 78% of the air around us is molecular nitrogen, N_2. Separation of nitrogen from air is a mature process. For larger scale, cryogenic separation using Linde's double column is usually used. For smaller scales, membranes or pressure swing adsorption can be used (Sánchez and Martín, 2018). Cryogenic separation techniques are based on the differences in the boiling points of the air components. Membranes and pressure swing adsorption are based on other physical properties of the gasses in air, like molecular size or adsorption affinity on surfaces. All these separation technologies are using mostly electricity to achieve separation. Due to the sensitivity of the ammonia synthesis catalyst to oxygen, the produced nitrogen often has to go through an additional deoxygenation step to ensure high enough nitrogen purity.

Hydrogen, H_2, unlike nitrogen, is not to be found naturally in a pure molecular form on our planet. It is abundantly present mostly in the form of water and in a much smaller scale in organic compounds. To have hydrogen in the pure form, it has to be extracted using energy from another source. Three main industrial hydrogen-producing methods are used nowadays: hydrocarbon reforming, coal or biomass gasification and water electrolysis. Presently, fossil energy sources dominate the industrial hydrogen production. Almost half of today's hydrogen need is produced from natural gas via steam methane reforming, while the other half is coming from gasification processes (Nikolaidis and Poullikkas, 2017). Ammonia industry almost exclusively uses the steam methane reforming or coal gasification to produce the required hydrogen. In the steam reforming process, methane is reacted with steam in the presence of mostly Ni-based catalyst at 700°C–1000°C and 3–25 bar. The heat required for the endothermic reaction is provided by external combustion of methane. The reaction products are hydrogen, carbon dioxide and carbon monoxide. After this step, carbon monoxide reacts with steam in a water–gas shift reaction to produce carbon dioxide and more hydrogen. In the final step, hydrogen is purified by removing carbon dioxide and other impurities (Bernardo et al., 2020). When analysing the total methane need of the process, a distinction can be made between the reforming feed and the fuel for energy and heat generation. The waste heat is usually used for the compression of the Haber–Bosch feed with the help of high-pressure steam.

In the gasification process, coal is reacted with steam or oxygen under high pressure and temperature. The hydrogen and carbon monoxide rich product is further involved in a water–gas shift reaction and finally the hydrogen is purified.

Production of hydrogen via the water shift steam reforming reaction of methane or coal gasification is clearly not the main technical topics here; however, their financial impact on the process economics is of great importance. The reason for these high carbon dioxide-emitting processes being favoured by industry is the very low price of today's fossil fuels. Due to their economic advantages, hydrogen production methods based on fossil methane are by some considered a necessary temporary option during the transition from a fossil-based economy to a renewable one. However, ammonia production consumes almost 1.2% of total primary energy and contributes 0.93% of greenhouse gas emissions. Approximately 1.5 tonnes of carbon dioxide are released to the environment during the production of 1 tonne of ammonia in a conventional method (Bicer and Dincer, 2018). Most of the energy demand and carbon emissions of ammonia production originate from the hydrogen production step, followed by the purification and compression steps. The successful decarbonisation of the ammonia industry could lead the energy transition with an example and speed up the process. Renewable power-to-hydrogen technologies could provide a robust decarbonisation option for the ammonia industry.

The third main hydrogen production method, electrolysis, is one of the main topics of this book. Various electrolysis technologies were described in the previous chapters. Electrolysis of liquid water can be carried out using alkaline electrolysers or proton exchange membrane electrolysers. Solid oxide electrolysers can be used to electrolyse steam. All these techniques have one common characteristic: they use electricity to dissociate a water molecule into hydrogen and oxygen gas. Electrolysis produces pure, sulphur- and chlorine-free hydrogen. Due to the possible oxygen crossover through the electrolyser membrane or diaphragm, a deoxygenation and drying step may be needed to completely remove the oxygen and water moisture from the hydrogen stream.

If the used electricity is coming from a renewable source, electrolysis of water is the most environment-friendly method of producing hydrogen. It is estimated that currently only ~4% of the global hydrogen production in the world is based on electrolysis (Bernardo et al., 2020).

The alkaline electrolyser suppliers currently dominate the industrial water electrolysis market with a broad stack capacity range and varying system energy requirements. Commercial proton exchange membrane system suppliers are on the rise, but system sizes are still below the megawatt scale. No commercial solid oxide electrolysers for industrial use are currently available to the best of our knowledge.

Several other water-splitting technologies are also on the horizon, but they are still far from a commercially competing phase. Thermochemical water splitting is based on the spontaneous decomposition of water at high temperatures. At temperatures higher than 2500°C, water brakes into its elemental components and then the separation of hydrogen gas is feasible. Reaching such high temperature with renewable energy sources is challenging; with assisting chemical reactions the decomposition temperature can significantly be lowered (Nikolaidis and Poullikkas, 2017).

In photo-electrolysis, the energy of the photons from the sunlight is used to generate an electron–hole pair on a semiconducting surface. If the energy levels are high enough, water can be split by the electron–hole pair (Nikolaidis and Poullikkas, 2017).

The reduction of carbon dioxide to carbon monoxide followed by the water–gas shift reaction is also a hydrogen-producing alternative. The carbon dioxide reduction can be carried out in nonthermal plasmas using dielectric barrier, microwave or gliding arc techniques. Plasma-based techniques show energy efficiencies in the range of 15%–50%; however, the peak energy efficiencies are observed at low conversion (Grim et al., 2020).

Besides ammonia, carbon dioxide is the second raw material for urea production. In the current ammonia industry dominated by steam methane reforming or coal gasification processes, the needed carbon dioxide is readily available.

Carbon dioxide is usually captured at a large scale based on absorption. A liquid sorbent is used to separate the CO_2 from the gas stream. The sorbent is then regenerated by heating or depressurisation. Typical sorbents are monoethanolamine, diethanolamine and potassium carbonate (Leung, Caramanna, and Maroto-Valer, 2014). Amine absorption is the most mature carbon capture technology, it can operate under a range of gas compositions and can achieve a high level of carbon dioxide removal; however, it is sensitive to amine degradation and corrosion.

Carbon capture technologies based on physisorption are also available. They use substrates such as activated carbon or metal oxides to extract carbon dioxide on the surface of their micropores. Membrane-based separation methods use polymeric membranes or ceramics to separate carbon dioxide from a gas stream (Nocito and Dibenedetto, 2020).

With the rise of renewables and the increasing social interest in decarbonisation, fossil carbon dioxide sources are likely to be replaced by renewables-based sources. Biomass is an obvious candidate; biomass fermentation, bioethanol production or other biomass upgrading processes produce high CO_2 containing waste streams. The amine adsorption technology is also suitable for CO_2 capture from these gas streams.

For a green power-to-urea process, carbon capture from air will inevitably play a relevant role in the future. The capture of CO_2 from ambient air was already commercialised in the mid-20th century. The nitrogen cryogenic separation units are producing carbon dioxide as a side product. Due to the relatively low carbon dioxide content of air, the nitrogen separation process might not provide enough carbon dioxide for a continuous urea production. Other direct air capture techniques with solid sorbents or aqueous basic solutions as the capture media are also available (Keith et al., 2018), but they are not yet competitive from an economic point of view.

7.4 Power to ammonia and urea

The increased development of the renewable power infrastructure in several countries will inevitably lead to situations where the problem of an excess electricity production could not be solved by simply transporting it to other regions. In a region with high renewable electricity penetration, new ways are needed to keep the system in balance and deal with all the excess electricity when the production outmatches the demand. Power to ammonia can be a suitable means for excess electrical storage. The possible costs of the avoided blackouts would strengthen the socio-economic performance of power to ammonia. However, ammonia plants are designed to run using pure reactants, at nearly constant pressure and temperature, almost continuously for their entire lifetime. A continuous operation can be challenging with a fluctuating renewable energy source.

The traditional Haber–Bosch process can become green and carbon-free if its hydrogen generation step is coupled to renewable electricity, and electricity-based compression and ammonia separation techniques are introduced. The water electrolysis step has to be efficient and be able to operate under varying electricity inputs for a long period of time. The Haber–Bosch process should also be available at smaller scales that require less capital, have faster response time and be easily ramped up and down (Smith, Hill, and Torrente-Murciano, 2020). The implementation of a Haber–Bosch unit that can be ramped up or down is technologically feasible, and it requires the redefinition of the conventional ammonia synthesis process with new optimisations that go beyond the current implementations based on cheap and abundant natural gas. The ammonia synthesis unit in a power-to-ammonia operation would require an autothermic operation. The reactor system, catalyst type, feed content and composition need to be customised, and a careful heat management between the inlet and outlet streams has to be implemented (Cheema and Krewer, 2018).

Power to ammonia is currently situated at the border of academic research and pilot-scale plants phase. News about planning or building power-to-ammonia demonstration plants in Australia, Japan, the United States and the United Kingdom appear more frequently in the recent years. As for scaled-down Haber–Bosch units for smaller investments, the NFUEL miniammonia plants of Proton Ventures B.V. are commercially available (Rouwenhorst et al., 2019).

Techno-economic models show that the cost of a power-to-ammonia plant is dominated by the electrolyser and hydrogen supply system (Ikäheimo et al., 2018). It is desirable in a power-to-ammonia plant to operate the electrolyser at a high as possible average operating capacity. Logically, the closer the average operating capacity is to the maximum rated power, the more profitable the electrolyser. However, the high intermittency of the renewable electricity makes it difficult to find an optimal electrolyser size. If the power-to-ammonia plant is designed to absorb all the excess renewable electricity and its peaks, the maximum rated power of the electrolyser should be as high as the maximum electricity peak. The same applies to systems where a power-to-ammonia plant is directly connected to a wind or a photovoltaic solar park.

For an electrolyser connected to a 15 MW wind farm, Fig. 7.2 shows that the higher the maximum rated power of an electrolyser, the lower the average operating capacity, and the higher the electric energy utilisation it becomes.

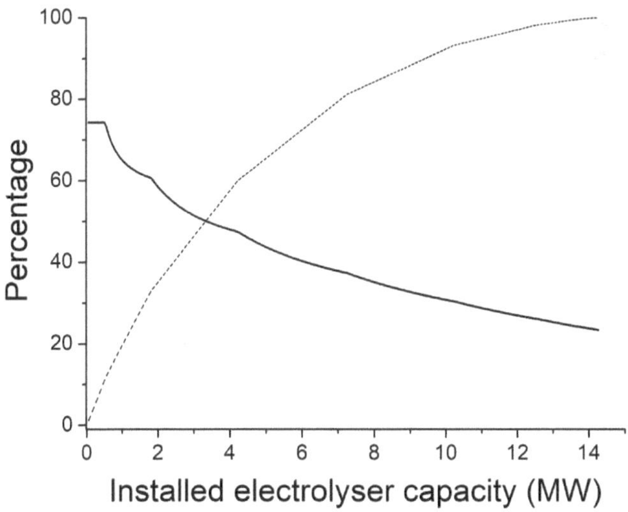

Figure 7.2 The variation of the average electrolyser operating capacity (*continuous line*) and the electric energy utilisation (*dotted line*) as a function of the maximum rated power of the electrolyser. The electricity production of a 15 MW wind park is calculated based on the hourly weather pattern for 2014 at Eelde, the Netherlands.

The electrolyser capacity depends on the main priorities of the business case. If the main focus is on ammonia production, then an electrolyser size at the lower range should be selected. The unused excess electricity will have to be transported or used for other purposes. A lower electrolyser rated power would also lower the capital expenses and increase the operating capacity. If the focus is on the storage of all the excess electricity, then the rated power of the installed electrolyser should be in the highest necessary range.

The produced hydrogen has to be stored in large enough quantity to buffer the fluctuations of the hydrogen production. The intermediate storage of the produced hydrogen is another dominating cost factor of a power-to-ammonia plant. The size of the hydrogen storage tank depends on the electricity production pattern. Fig. 7.3 shows the estimated amount of hydrogen stored during one year of operation. The stored hydrogen is the difference between the amount of hydrogen produced and consumed. The power-to-ammonia plant in this example is connected to a 15 MW wind turbine park, a 14 MW electrolyser and the Haber–Bosch ammonia production unit which has a nameplate production capacity of 38 metric tonnes day^{-1}. The calculations are performed with an electrolyser capacity of only 14 MW, because even at this nominal power practically, all the produced renewable electricity is used during the studied year. It is assumed that if the rate of the renewable electricity production requires it, the ammonia reactor can be ramped down, at the same rate as the electricity fluctuations, to the 20% of its nameplate production rate. This is however technically very challenging in real life.

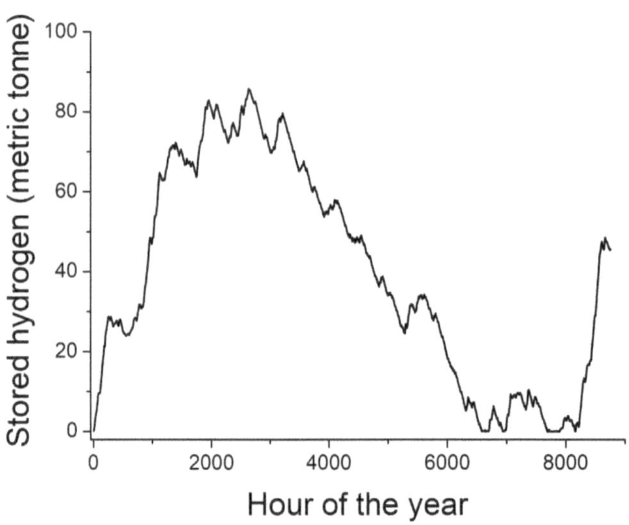

Figure 7.3 The estimated variation of the hydrogen level stored in the case of a power-to-ammonia plant powered by a 15 MW wind turbine park at the location of Eelde, the Netherlands. The electricity production of the wind park is calculated based on the hourly local weather pattern from 2014. The hydrogen level is determined by the operating power of the plant and the weather pattern.

A minimum storage tank size that would be able to accommodate the necessary amount of excess hydrogen present at any time is needed for an optimal plant design. This minimum hydrogen buffer size greatly influences the cost of the plant. The smaller the hydrogen storage size, the less expensive the plant becomes. When the hydrogen level in the storage reaches zero, the continuous ammonia production might have to be stopped. This situation should be avoided, because any restart of the ammonia synthesis unit increases the operational expenses and decreases the lifetime of the used catalysts.

It is desired to minimise the hydrogen storage size. The combination of solar and wind resources can reduce the size of the expensive buffer storage of hydrogen by producing a more uniform electricity production pattern (Armijo and Philibert, 2020).

The storage cost of hydrogen is significantly higher when it is stored for a longer period. Electricity imported from the grid during periods with no renewable electricity generation is another method to minimise the hydrogen storage infrastructure to the absolutely necessary size (Bennani et al., 2016). The elimination of the costly hydrogen storage tank from the plant design and the adaption of a strategy wherein the power-to-ammonia process is run at the lowest possible production rate during periods with no renewable energy production offers an economic advantage. However, the produced ammonia would not be carbon neutral if the purchased electricity is produced using fossil fuels. Fig. 7.4 shows the significant economic effect of the installed electrolyser capacity and hydrogen storage size. The estimated years

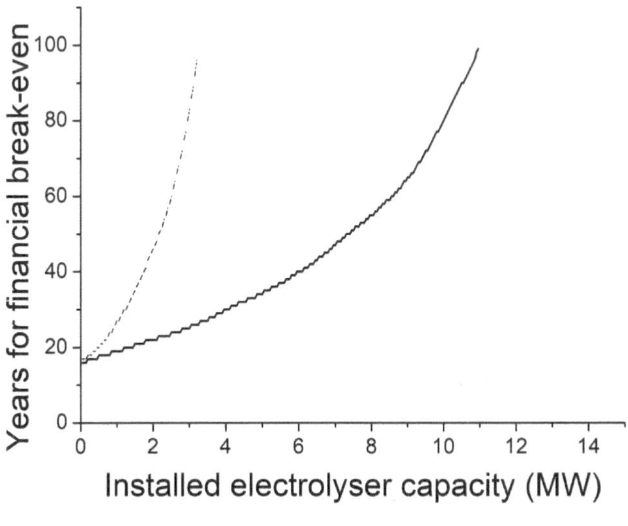

Figure 7.4 The estimated number of years needed to achieve financial break-even for an imaginary power-to-ammonia plant with hydrogen storage buffer (*dotted line*) or electricity purchase strategy (*continuous line*) as a function of the maximum rated power of the electrolyser. The electricity production of a 15 MW wind park is calculated based on the hourly weather pattern for 2014 at Eelde, the Netherlands.

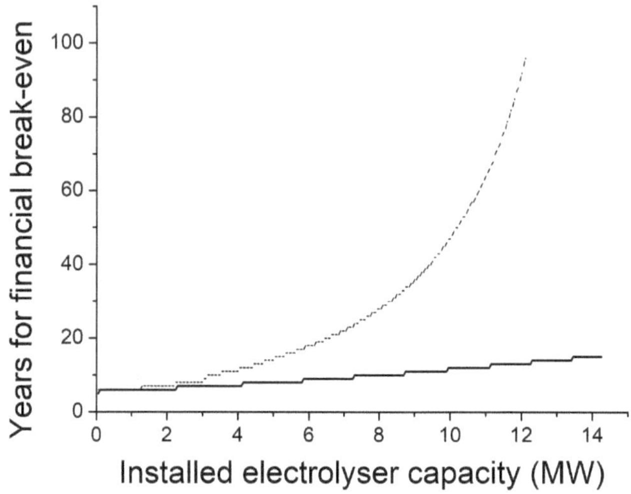

Figure 7.5 The estimated number of years needed to achieve financial break-even with hydrogen storage buffer (*dotted line*) and electricity purchase strategy (*continuous line*) as a function of the maximum rated power of the electrolyser. The electricity production of a 15 MW wind park is calculated based on the hourly weather pattern in 2014 at Hoek van Holland, the Netherlands.

needed to reach financial break-even, for both the hydrogen storage buffer and electricity purchase strategy scenarios, are increasing exponentially with the installed electrolyser capacity. The break-even point is reached when the cumulative income becomes equal with the cumulative costs and the business starts to make a profit. The curves showed on Figs. 7.4 and 7.5 are based on simple calculations using average capital and operating costs for the power-to-ammonia plant components: wind turbines, electrolyser, hydrogen storage, Haber–Bosch ammonia production unit; average selling price for ammonia and average selling and buying prices for electricity. The intention of the calculations was to visualise the relative future potential of the previously discussed scenarios, the technical details of the fast response time, the ability to easily ramp up and down, and their projected cost factors are not included in the calculations. It should also be noted that the capital cost of the onshore wind turbine park is also included in these scenarios, and the nameplate production capacity of the Haber–Bosch unit is always adjusted to the installed electrolyser size.

Power to ammonia relying on a wind park installed at Eelde, the Netherlands, under the current average electrolyser and hydrogen storage prices has no financial feasibility. A significant drop in the prices of the electrolysers and hydrogen storage would be needed for a more favourable business case.

However, the renewable energy production profile can also be a significant factor. Fig. 7.5 depicts the estimated years needed to reach financial break-even at the much windier location of Hoek van Holland, the Netherlands.

The calculations are based on the same simple financial and technical assumptions applied for Fig. 7.4. The more intensive and uniform wind pattern improves the business case significantly for both the hydrogen storage buffer and electricity purchase strategy scenarios. Hoek van Holland is a much better location than Eelde to install a power-to-ammonia plant.

7.5 Remarks about the social acceptance of power to ammonia

The current ammonia technology is safe, but even in a properly running plant, some leakages could occur. If a leakage occurs, it is likely that communities nearby the ammonia plant will be against it. The unpleasant odour and the perceived low safety when handling ammonia could cause negative public reactions. A dialogue with the public should be started at the very beginning of the planning process to create trust. The advantages and disadvantages of power to ammonia should be distributed at the community level as fairly as possible to increase its social acceptance (Bennani et al., 2016).

Industrial, large-scale power-to-ammonia plants relying on renewable electricity cannot compete financially at the moment with today's established industrial ammonia plants based on fossil resources. However, smaller and intermediate-sized power-to-ammonia plants could gain more advantages, with the rising social interest in decarbonisation and with the increasing need for local renewable electricity storage. A power-to-ammonia plant is most easily established in existing industrial areas where there is experience with chemical activity, but smaller sized plants would possibly have to be installed closer to communities. Besides the technical and economic challenges, social acceptance of power to ammonia is also an important factor for a successful implementation.

References

Amar, I. A., et al. (2011). Solid-state electrochemical synthesis of ammonia: A review. *Journal of Solid State Electrochemistry*, *15*(9), 1845–1860. Available from https://doi.org/10.1007/s10008-011-1376-x.

Appl, M. (1999). *Ammonia: Principles and industrial practice*. Wiley-VCH Verlag GmbH & Co.

Armijo, J., and Philibert, C. (2020). Flexible production of green hydrogen and ammonia from variable solar and wind energy: Case study of Chile and Argentina. *International Journal of Hydrogen Energy*, *45*(3), 1541–1558. Available from https://doi.org/10.1016/j.ijhydene.2019.11.028.

Bennani, Y., et al. (2016). *Power-to-ammonia: Rethinking the role of ammonia – From a value product to a flexible energy carrier (FlexNH3)*. Groningen: Hanzehogeschool Groningen. Available at https://research.hanze.nl/en/publications/power-to-ammonia-rethinking-the-role-of-ammonia-from-a-value-prod.

Bernardo, G., et al. (2020). Recent advances in membrane technologies for hydrogen purification. *International Journal of Hydrogen Energy*, *45*(12), 7313–7338. Available from https://doi.org/10.1016/j.ijhydene.2019.06.162.

Bicer, Y., and Dincer, I. (2018). Life cycle environmental impact assessments and comparisons of alternative fuels for clean vehicles. *Resources, Conservation and Recycling*, *132*, 141–157. Available from https://doi.org/10.1016/j.resconrec.2018.01.036.

Cheema, I. I., and Krewer, U. (2018). Operating envelope of Haber-Bosch process design for power-to-ammonia. *RSC Advances*, *8*(61), 34926–34936. Available from https://doi.org/10.1039/c8ra06821f.

Chen, A., and Xia, B. Y. (2019). Ambient dinitrogen electrocatalytic reduction for ammonia synthesis. *Journal of Materials Chemistry A*, *7*(41), 23416–23431. Available from https://doi.org/10.1039/c9ta05505c.

Dimitriou, P., and Javaid, R. (2020). A review of ammonia as a compression ignition engine fuel. *International Journal of Hydrogen Energy*, *45*(11), 7098–7118. Available from https://doi.org/10.1016/j.ijhydene.2019.12.209.

FAO. (2019). *World fertilizer trends and outlook to 2022*. Rome.

Fertilizers Europe. (2018). *Feeding for Life 2030 report*.

Fertilizers Europe. (2019). *Forecast of food, farming and fertilizer use in the European Union 2019–2029*.

Grim, R. G., et al. (2020). Transforming the carbon economy: Challenges and opportunities in the convergence of low-cost electricity and reductive CO_2 utilization. *Energy and Environmental Science*, *13*(2), 472–494. Available from https://doi.org/10.1039/c9ee02410g.

Ikäheimo, J., et al. (2018). Power-to-ammonia in future North European 100% renewable power and heat system. *International Journal of Hydrogen Energy*, *43*(36), 17295–17308. Available from https://doi.org/10.1016/j.ijhydene.2018.06.121.

Keith, D. W., et al. (2018). A process for capturing CO_2 from the atmosphere. *Joule*, *2*(8), 1573–1594. Available from https://doi.org/10.1016/j.joule.2018.05.006.

Kishimoto, M., et al. (2020). Development of 1 kW-class ammonia-fueled solid oxide fuel cell stack. *Fuel Cells*, *20*(1), 80–88. Available from https://doi.org/10.1002/fuce.201900131.

Leung, D. Y. C., Caramanna, G., and Maroto-Valer, M. M. (2014). An overview of current status of carbon dioxide capture and storage technologies. *Renewable and Sustainable Energy Reviews*, *39*, 426–443. Available from https://doi.org/10.1016/j.rser.2014.07.093.

Meessen, J. (2014). Urea synthesis. *Chemie-Ingenieur-Technik*, *86*(12), 2180–2189. Available from https://doi.org/10.1002/cite.201400064.

Mukherjee, S., et al. (2018). Low-temperature ammonia decomposition catalysts for hydrogen generation. *Applied Catalysis B: Environmental*, *226*, 162–181. Available from https://doi.org/10.1016/j.apcatb.2017.12.039.

Nikolaidis, P., and Poullikkas, A. (2017). A comparative overview of hydrogen production processes. *Renewable and Sustainable Energy Reviews*, *67*, 597–611. Available from https://doi.org/10.1016/j.rser.2016.09.044.

Nocito, F., and Dibenedetto, A. (2020). Atmospheric CO_2 mitigation technologies: Carbon capture utilization and storage. *Current Opinion in Green and Sustainable Chemistry*, *21*, 34–43. Available from https://doi.org/10.1016/j.cogsc.2019.10.002.

Rouwenhorst, K. H. R., et al. (2019). Islanded ammonia power systems: Technology review & conceptual process design. *Renewable and Sustainable Energy Reviews*, *114*. Available from https://doi.org/10.1016/j.rser.2019.109339.

Sakarika, M., et al. (2020). Purple non-sulphur bacteria and plant production: Benefits for fertilization, stress resistance and the environment. *Microbial Biotechnology*, *13*(5), 1336–1365. Available from https://doi.org/10.1111/1751-7915.13474.

Sánchez, A., and Martín, M. (2018). Scale up and scale down issues of renewable ammonia plants: Towards modular design. *Sustainable Production and Consumption*, *16*, 176–192. Available from https://doi.org/10.1016/j.spc.2018.08.001.

Sayed, E. T., et al. (2019). Direct urea fuel cells: Challenges and opportunities. *Journal of Power Sources*, *417*, 159–175. Available from https://doi.org/10.1016/j.jpowsour.2018.12.024.

Smith, C., Hill, A. K., and Torrente-Murciano, L. (2020). Current and future role of Haber-Bosch ammonia in a carbon-free energy landscape. *Energy and Environmental Science*, *13*(2), 331–344. Available from https://doi.org/10.1039/c9ee02873k.

Valera-Medina, A., et al. (2018). Ammonia for power. *Progress in Energy and Combustion Science*, *69*, 63–102. Available from https://doi.org/10.1016/j.pecs.2018.07.001.

Yang, J., Weng, W., and Xiao, W. (2020). Electrochemical synthesis of ammonia in molten salts. *Journal of Energy Chemistry*, *43*, 195–207. Available from https://doi.org/10.1016/j.jechem.2019.09.006.

Zhu, B., Liang, Z., and Zou, R. (2020). Designing advanced catalysts for energy conversion based on urea oxidation reaction. *Small*, *16*(7), 1906133. Available from https://doi.org/10.1002/smll.201906133.

Power to formic acid

8

Sudipta Chatterjee[1,2], *Indranil Dutta*[1,2] *and Kuo-Wei Huang*[1,2]
[1]KAUST Catalysis Centre, King Abdullah University of Science and Technology, Thuwal, Saudi Arabia, [2]Division of Physical Science and Engineering, King Abdullah University of Science and Technology, Thuwal, Saudi Arabia

8.1 Formic acid as an energy carrier

Formic acid (FA) is the simplest carboxylic acid. It has a chemical formula of CH_2O_2, containing 4.34 wt.% of H_2. While its gravimetric capacity is low, because of its high density (1.22 g cm^{-3}), the volumetric capacity is large (53 g H_2/L or 1.77 kWh L^{-1}). These values are comparable to those of commercial hydrogen pressure tanks at 700 bars, suggesting that the energy density of FA is suitable for similar applications of H_2. Furthermore, FA is a non-toxic and environmentally benign liquid with low flammability under ambient conditions. Even though it is corrosive, FA can be safely handled with the design of suitable containers and pipelines (Bulgarian Ministry of Environment and Water, 2017; PHE Centre for Radiation, 2015). It is thus convenient to store and transport FA. All these favourable properties make FA a strong candidate as an energy carrier (Chatterjee et al., 2021) (Table 8.1).

The potential of FA as a fuel was recognised for applications in direct formic acid fuel cells (DFAFCs) (Yu and Pickup, 2008) and 'formic acid economy' (Tsang, 2005) was proposed. Although at this stage, DFAFCs are not ready for practical use due to the unresolved electrode poisoning issues, H_2 fuel cells have become a mature technology. In the last 15 years or so, catalysis for the selective dehydrogenation of FA has been rapidly advanced. Accordingly, electricity generation devices using a FA reformer were also promisingly demonstrated with a short path to the energy market. It is clear that integrating power-to-formic acid processes will create a carbon-neutral system for energy storage and utilisation (Fig. 8.1) (Chatterjee et al., 2021).

Table 8.1 Properties of formic acid.

Formic acid: CH_2O_2			
Molecular mass	46.03	Melting point	8.4°C
Density (20°C)	1.22 g cm^3	Boiling point	100.8°C
Dielectric constant (20°C)	57.9	pK_a	3.75

Power to Fuel. DOI: https://doi.org/10.1016/B978-0-12-822813-5.00006-0

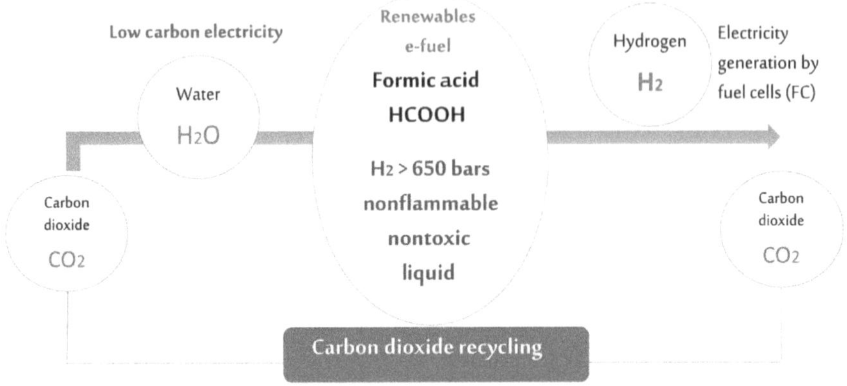

Figure 8.1 A CO_2-neutral cycle for low-carbon energy storage and utilisation.

8.2 Preparation of formic acid by hydrogenation of carbon dioxide

The process of CO_2 to HCOOH involves a phase change and hence it is entropically disfavoured [Eq. (8.1)]. However, the thermodynamics can be altered to be exergonic in the presence of aqueous medium [Eq. (8.2)]. Therefore efforts have been made to investigate various reaction conditions with tuneable parameters such as solvents, catalysts, temperature, additives etc. Because of the low equilibrium concentration of FA obtained from the reaction mixtures of CO_2 and H_2, organic and inorganic bases are often used to shift the equilibrium because of the formation of formate salts or other derivatives of FA. The continuous progress and challenges of CO_2 hydrogenation have been discussed in several reviews in the last decade (Bulushev and Ross, 2018; Jessop, 2007; Sordakis et al., 2018; Wang et al., 2011, 2015). In this section, the development of selected efficient catalytic systems is summarised and highlighted.

$$CO_{2(g)} + H_{2(g)} \Rightarrow HCOOH_{(l)} \quad \Delta G^0_{298K} = 32.9 \text{ kJ mol}^{-1} \tag{8.1}$$

$$CO_{2(aq)} + H_{2(aq)} \Rightarrow HCOOH_{(aq)} \quad \Delta G^0_{298K} = -4.0 \text{ kJ mol}^{-1} \tag{8.2}$$

8.2.1 *Noble metal-based homogeneous catalysts*

In 1976 Inoue and co-workers first reported the previously unexplored domain of homogeneous catalytic hydrogenation of CO_2 to FA using phosphine-based complexes of Ru, Rh, Pd, Ir etc. (Inoue et al., 1976). The reaction was carried out under 2.5 MPa CO_2 and 2.5 MPa H_2 to achieve the best activity [turnover number (TON) = 87, turnover frequency (TOF) $\approx 4.5\ h^{-1}$] with $Ru(H)_2(PPh_3)_4$ (**8.01**) in benzene/water containing trimethylamine (NEt_3) as the base. During the 1990s, Leitner et al. reported $[Rh(cod)Cl]_2$/dppb [cod = cyclooctadiene; dppb = 1,1-bis (diphenylphosphino)butane] (**8.02**) with a TON of 1150 in dimethyl sulfoxide (DMSO)/NEt_3 at 25°C under 40 bar H_2/CO_2 (1:1) (Graf and Leitner, 1992). The same group later demonstrated the first time water-soluble Rh complex for hydrogenation reaction in water/amine mixed solvent employing a Wilkinson-type catalyst $[RhCl(mtppms)_3]$ (mtppms = meta-trisulfonated triphenylphosphine) (**8.03**) in H_2O/dimethylamine with a TON of 3439 (Gassner and Leitner, 1993). Although initially most of the hydrogenation reactions were studied in organic solvents, it was found that the presence of water may increase the catalytic performance. The Lin and Lau group demonstrated a significant rate enhancement using $[TpRuH(PPh_3)(CH_3CN)]$ [Tp = hydrotris(pyrazolyl)borate] (**8.04**) in the presence of water (Yin et al., 2001). Experimental and theoretical studies supported that the CO_2 insertion into the Ru—H bond was promoted via a hydrogen bonding of a coordinated water molecule.

The use of supercritical CO_2 ($scCO_2$) for hydrogenation was first introduced by the Noyori group. $RuCl_2(PMe_3)_4$ (**8.05**) and $Ru(H)_2(PMe_3)_4$ (**8.06**) were used as (pre)catalysts in the presence of NEt_3 and a trace amount of water to yield formate using $scCO_2$ (Jessop, Ikariya, and Noyori, 1994). The high initial rates (TOF of $1400\ h^{-1}$) under mild conditions were attributed to the high miscibility of H_2 in $scCO_2$. Later on, Jessop et al. carried out extensive studies by varying phosphine ligands, additives and bases and found that with $[RuCl(OAc)(PMe_3)_4]$ (**8.07**) as a catalyst, a TOF of $500\ h^{-1}$ was obtained in neat NEt_3 under 100 bar H_2/CO_2 (2:3) at 50°C (Thomas et al., 2001).

In 1995 the Chen group reported TONs of up to 5000 using *cis*-$[Ru(Cl_2bpy)_2(H_2O)_2][CF_3SO_3]_2$ (bpy = 2,2′-bipyridine) catalyst (**8.08**) in EtOH in the presence of NEt_3 (Lau and Chen, 1995). In the absence of base, there was a significant reduction in the product formation. The Joó group investigated a number of rhodium and ruthenium-based catalysts viz. $[RhCl(tppms)_3]$ (**8.09**), $[RuCl_2(tppms)_2]_2$ (**8.10**) [tppms = (3-diphenylphosphinyl-benzenesulfonic acid sodium salt)], $[RuCl_2(pta)_4]$ (**8.11**) (PTA:1,3,5-triaza-7-phosphaadamantane) for the hydrogenation reaction in the aqueous medium without amines additive (Joó, 1999, 2000). Catalyst **8.09** was identified to hydrogenate 0.1 (M) aqueous solution of $CaCO_3$ to FA/formate under 100 bar of H_2/CO_2 (4:1). A maximum TOF of $9600\ h^{-1}$ was obtained using complex **8.10** at 80°C in a 0.3 M sodium bicarbonate solution. These observations laid foundation towards the use of water medium for homogeneous CO_2 hydrogenation and bicarbonate as a CO_2 like hydrogen carrier.

Himeda and co-workers developed a number of catalysts possessing hydroxyl (−OH) functionalised proton-responsive ligands (Himeda et al., 2004, 2005) (Fig. 8.2). Under the alkaline conditions, the resulting oxyanion form of the catalyst due to deprotonation was found efficient in CO_2 hydrogenation, whereas in the acidic solution, the protonated hydroxyl formed was effective in the FA dehydrogenation. Initially, Ru(II), Ir(III) and Rh(III) complexes having 4,7-dihydroxy-1, 10-phenanthroline (DHPHEN) ligand skeleton were synthesised and applied for hydrogenation of bicarbonate solution (**8.12–8.14**) (Fig. 8.2). The highest activity

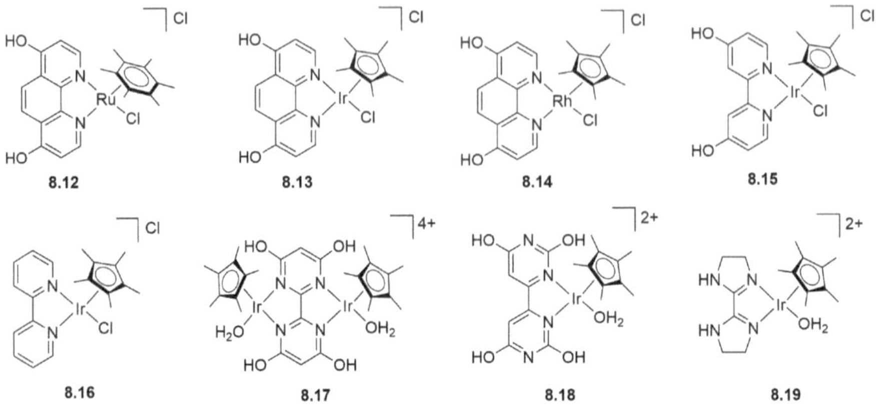

Figure 8.2 Structures of catalysts investigated for aqueous CO_2 hydrogenation in basic media.

was achieved using [Cp*Ir(DHPHEN)Cl]Cl (**8.13**, Fig. 8.2) (Cp* = 1,2,3,4,5-pentamethylcyclopentadienyl) leading to a TON of 21,000 (formate concentration of 0.42 M) and a TOF of 23,000 h^{-1} under basic conditions at 120°C. The reaction was even effective at room temperature albeit with reduced efficiency. Decreasing the catalyst concentration and keeping other parameters intact led to an improved TOF of 33,000 h^{-1}. The ruthenium analogue **8.12** showed almost threefold increase in the final formate concentration (1.2 M), albeit with a decreased TOF of 3360 h^{-1} at 120°C. When the unsubstituted 1,10-phenanthroline ligand was used, both Ir and Rh complexes did not show any reactivity. These observations strongly supported the importance of hydroxyl group and formation of oxyanionic species under alkaline conditions which enhanced the catalysts' solubility. The strong electron-donating ability of the oxyanion was understood as the reactivity of [Cp*Ir(4DHBP)Cl]Cl (**8.15**, Fig. 8.2) (4DHBP = 4,4′-dihydroxy-2,2′-bipyridine, TOF of 5100 h^{-1}) was found to be over 1000 times higher than that of the unsubstituted one [Cp*Ir(bpy)Cl]Cl (**8.16**), TOF of 4.7 h^{-1}] under the similar optimised conditions (80°C, 1 MPa, CO_2/H_2 = 1). Based on this design principle, Himeda co-workers prepared a binuclear iridium complex (**8.17**) having a bipyrimidine ligand with four hydroxyl functionalities (Fig. 8.2) (Hull, 2012). Carrying out the hydrogenation of 2 M $KHCO_3$ under 50 bar led to the formation of formate salt with TOF of 53,800 h^{-1} and TON = 79,000. The corresponding monomer analogue (**8.18**, Fig. 8.2) also showed a similar activity compared to that of **8.17**, suggesting that the bimetallic cooperativity may not be the governing factor of the excellent reactivity. Interestingly, iridium complex (**8.19**, Fig. 8.2) scaffolded on a non-functionalised imidazoline moiety also showed similar activity (Xu et al., 2015).

In 2009 Nozaki and co-workers prepared a PNP [2,6-bis(di-isopropylphosphinomethyl)pyridine] based Ir trihydride catalyst (**8.20**) and examined its catalytic activity towards hydrogenating of CO_2 in a 1 M aqueous KOH solution (Tanaka, Yamashita, and Nozaki, 2009). This system achieved a TOF of 150,000 h^{-1} at

200°C under 50 bar H_2/CO_2 (1:1) and an unprecedentedly high TON of 3,500,000 at 120°C in a total period of 48 h. The proposed reaction pathway consists of three elementary steps: (1) insertion of CO_2 into the Ir—H bond to generate species **8.21**, (2) deprotonative dearomatization with dissociation of the formate ligand to produce **8.22**, and (3) hydrogenation (Fig. 8.3, left). Step (2) was identified as the rate-determining step (RDS). Hazari et al. demonstrated a water-soluble and air-stable IrH_3(PNP) catalyst (**8.23**) (Schmeier et al., 2011). The 'saturated' ligand contains an N—H group as the hydrogen bond donor in the secondary coordination sphere and can interact with the product formate molecule. The presence of a strongly electron-donating *trans* ligand (hydride) may weaken of Ir—H bond to facilitate the CO_2 insertion process. The hydrogen-bonded [Ir]—formate species (**8.24**) was found to be active catalytic species to give an overall TON of 348,000 after 24 h. The combined experimental and computation studies suggested a plausible mechanism (Fig. 8.3, right). After the CO_2 insertion, the coordinated formate (η^1-formate species) is replaced by H_2 to generate the dihydrogen species (**8.25**), followed by the deprotonation of the coordinated H_2 to regenerate the trihydride **8.23** and complete the catalytic cycle.

Sanford and co-workers employed a dearomatized Ru-PNN pincer catalyst (**8.26**, Fig. 8.4) for CO_2 hydrogenation in the presence of K_2CO_3 under 40 bar of H_2/CO_2 (3:1) at 120°C in diglyme to afford a TON of 23,000 (Huff and Sanford, 2013). Pidko and co-workers also used a similar ruthenium PNP [2,6-bis(di-*tert*-butylphosphinomethyl)-pyridine] pincer complex (**8.27**; Fig. 8.4) and a very efficient system

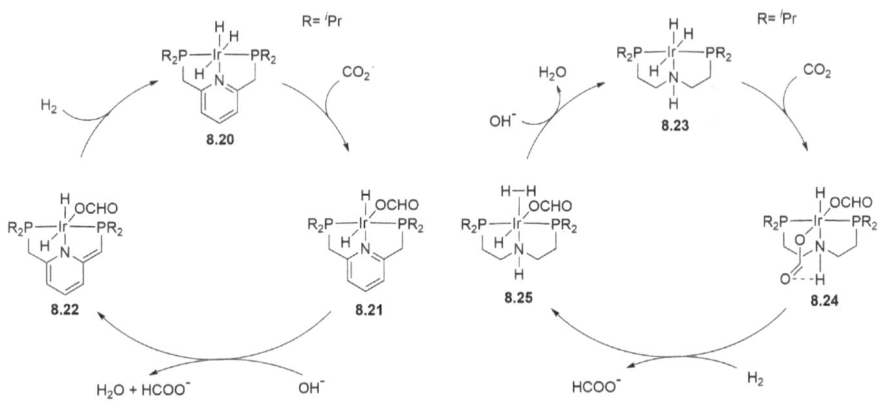

Figure 8.3 Proposed reaction mechanism for CO_2 hydrogenation using **8.20** and **8.23**.

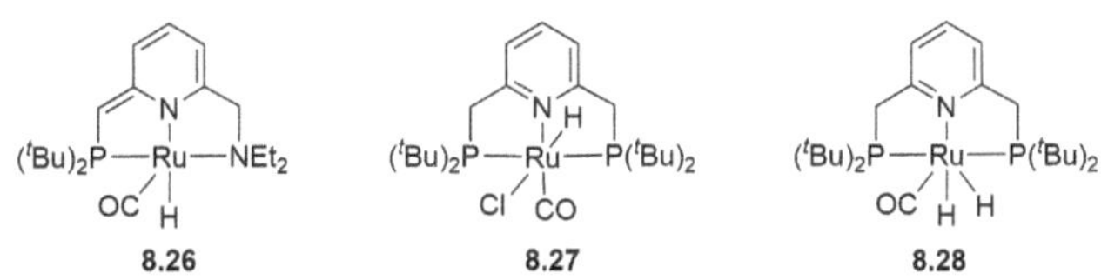

Figure 8.4 Structures of ruthenium-based pincer complexes **8.26–8.28**.

under mild conditions was identified (Filonenko et al., 2014). Using dimethylformamide (DMF) as the solvent and 1,8-diazabicyclo[5.4.0]undec-7-ene (DBU) as the base, this catalyst afforded a TOF of 1,100,000 h^{-1} under 40 bar H_2/CO_2 (3:1) at 120°C. Under a relatively milder condition of 5 bar of equimolar pressure, a TOF of 60,000 h^{-1} could still be achieved at 90°C. Combination of the experimental and theoretical studies demonstrated that the bis-hydrido ruthenium species (**8.28**; Fig. 8.4) is the active catalyst, whereas the ligand-assisted CO_2 adduct is an inactive state (Filonenko, 2013; Filonenko, Hensen, and Pidko, 2014). These results further suggested that the pathways involving metal − ligand cooperativity were not necessarily favourable.

The Huang group has explored the catalytic performance based on a novel PN^3P ligand platform (Li, 2019; Li, Zheng, and Huang, 2015; Huang, 2013) towards CO_2 hydrogenation. In 1:1 THF/H_2O (THF = tetrahydrofuran), catalyst **8.29** was found to be effective in the $NaHCO_3$ solution and a TON of 33,000 after 50 h was obtained under 1600 psi H_2 at 130°C (Fig. 8.5) (Guan, 2018). A model for CO_2 capture and utilisation was also explored by employing several amines for CO_2 capture from the air at an efficiency of 39%. Moreover, recyclability studies were carried out in a biphasic 2-$MeTHF/H_2O$ system using PMDTA (*N,N,N′,N″,N″*-pentamethyldiethylenetriamine) as the CO_2 absorber leading to an easy product separation and catalyst recycling. On another study on Ir − hydride PN^3P pincer complexes (**8.30** and **8.31**) for direct hydrogenation of CO_2 to formate (Pan et al., 2019), a maximum TON of 5100 was achieved using equimolar 120 psi pressure of H_2/CO_2 in 1:4 THF/H_2O at 130°C (Fig. 8.5). However, further increase in temperature did not improve the yield and suggested possible deactivation of the catalyst. Detailed stoichiometric study and spectroscopic characterisation confirmed a dearomatized PN^3P^*−Ir(I)−CO species (**8.32**, Fig. 8.5) was formed as a possible key pathway responsible for the loss of catalytic activity in the CO_2 hydrogenation.

In 2010 Peris group first explored the transfer hydrogenation methodology for CO_2 toward formate synthesis (Sanz, Azua, and Peris, 2010; Sanz, Benítez, and Peris, 2010). Several Ru and Ir complexes (**8.33**−**8.36**) containing N-heterocyclic carbene (NHC) ligands were tested for the FA synthesis from CO_2 in the presence of either ^{i}PrOH as the hydrogen donor or H_2 gas (Fig. 8.6). Employing Ir catalyst **8.33** resulted in a TON of 150 (50 bar CO_2, 0.5 M KOH, 110°C) under the transfer hydrogenation condition. Using bis NHC-based Ir complex **8.34**, a TON of 1600 was achieved under the equimolar 60 bar pressure of H_2/CO_2 in a 2 M KOH

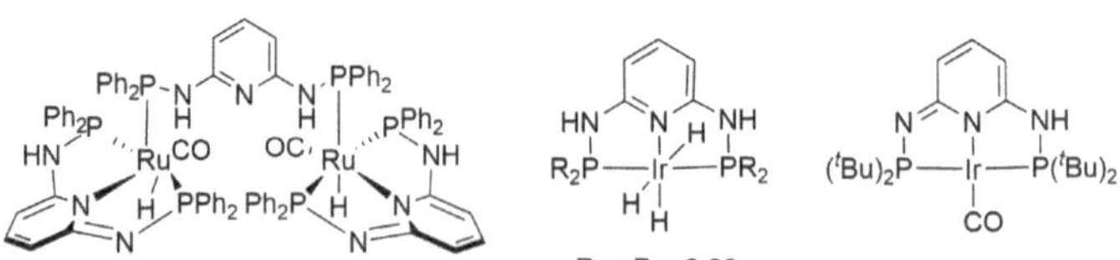

8.29

R= cPe 8.30
tBu 8.31

8.32

Figure 8.5 Structures of PN^3P-pincer complexes **8.29**−**8.32**.

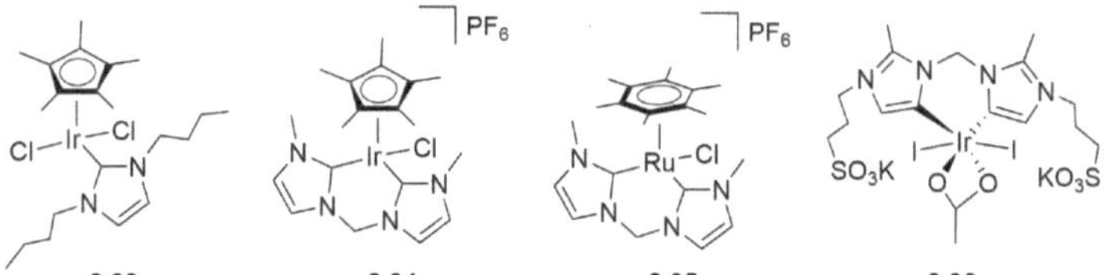

Figure 8.6 Carbene-based transfer hydrogenation catalysts **8.33–8.36**.

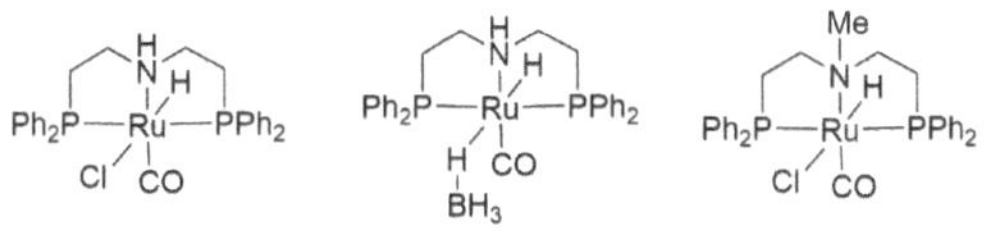

Figure 8.7 Takasago-type ruthenium catalysts.

solution at 80°C. When the analogous Ru complex **8.35** was used, a significant improvement of TON (23,000) was observed under a lower pressure (40 bar equimolar pressure) but at an elevated temperature of 200°C. The importance of NHC ligands was realized as its strong σ donating ability may enhance the thermal stability. Ir complex **8.36** afforded a TON of 190,000 under the equimolar pressure of 60 bar in a 1 M KOH solution at 200°C. However, under the transfer hydrogenation condition using ^{i}PrOH, the catalyst lifetime decreased significantly (TON = 2700) (Dibenedetto et al., 2011).

The Leitner group demonstrated that combination of a suitable catalyst/ionic liquid (IL) matrices in batch reactions could be utilised to produce pure FA using a continuous-flow hydrogenation of $scCO_2$ (Wesselbaum, Hintermair, and Leitner, 2012). The Prakash and Olah group explored the Takasago $PN^{H}P$-pincer Ru catalysts (**8.37** and **8.38**) and a methylated analogue (**8.39**) for H_2 storage in organic solvent/H_2O under variable conditions such as in the presence of carbonates, bicarbonates and alkali hydroxides and under CO_2 pressure (Fig. 8.7) (Kothandaraman, 2015). Employment of catalyst **8.37** and **8.38** yielded around 90% of formate (TOFs of 1688 and 958 h^{-1} respectively) using $NaHCO_3$ in THF/H_2O under 40 bar H_2 at 80°C. Interestingly, **8.39** also showed a similar catalytic activity with a TOF of 1096 h^{-1} in a 1,4-dioxane/H_2O solution of NaOH under 61 bar H_2/20 bar CO_2 at 70°C. Catalyst **8.38** was extended for integrated CO_2 capture in the presence of several amines containing primary, secondary and tertiary amino groups, forming carbamates and/or bicarbonates/carbonates followed by hydrogenation (Kothandaraman et al., 2016).

8.2.2 Non-noble metal-based homogeneous catalysts

In the early report of CO_2 hydrogenation to FA, the Inoue group achieved low activities (TON = 7) using $[Ni(DPPE)_2]$ (DPPE = 1,2-bis(diphenylphosphino)ethane) (**8.40**) catalyst and NEt_3 at ambient temperature (Inoue et al., 1976). In 2003

Jessop and co-workers studied the reactivity of a series of metal precursors (Cr, Fe, Co, Ni, In, Mo, Nb, W) alone or in combination with the phosphine and nitrogen-based ligands, such as PPh_3, DPPE, bis(dicyclohexylphosphino)ethane (DCPE) ($Cy_2PCH_2CH_2PCy_2$), tetramethylethylenediamine and bipyridine (Tai et al., 2003). Combination of $Fe(BF_4)_2 \cdot 6H_2O$/tris[2-(diphenylphosphino)ethyl]phosphine (**8.41**) afforded the formate formation in methanol at 80°C under 60 bar H_2 pressure with a TON of 610 after 20 h. Analogous $Co(BF_4)_2 \cdot 6H_2O$/tris[2-(diphenylphosphino) ethyl]phosphine (**8.42**) showed a similar activity (TON = 645) under otherwise identical conditions (Federsel et al., 2012). However, increasing the temperature to 120°C resulted in an improved TON of 3877 with an average TOF of 190 h^{-1}.

First iron-based catalytic system comprising pincer backbone (Fe-PNP, **8.43**) was demonstrated by Milstein et al. for both CO_2 and bicarbonate hydrogenation under low hydrogen pressures (Fig. 8.8) (Langer et al., 2011). A maximum TON of 788 (TOF of up to 160 h^{-1}) was observed under 10 bar of $H_2/CO_2{:}H_2$ (2:1) in an aqueous NaOH/THF solution at 80°C. Subsequently, the Kirchner group reported a set of iron PN^3P-pincer complexes having either N − H (**8.44**) or N − Me (**8.45**) arms to catalyse the CO_2/bicarbonate hydrogenation under comparable reaction conditions (Fig. 8.8) (Bertini, 2016). The same group subsequently reported Mn(I) analogues (**8.46–8.47**, Fig. 8.8) based on same pincer backbone for CO_2 hydrogenation (Bertini, 2017). It was found that these catalysts are superior compared to the iron counterparts and a TON of 5520 along with quantitative yields could be achieved under 80 bar of $CO_2{:}H_2$ (1:1) in THF/H_2O at 80°C using DBU. Lowering the catalyst loading (0.002 mol.%) and in the presence of LiOTf as an additive, a TON of 30,000 was obtained. Hazari and Bernskoetter examined a number of Iron (II) saturated pincer carbonyl hydrides (**8.48–8.54**) under 69 bar H_2/CO_2 (1:1) in a DBU/THF solution at 80°C and the desired formate was obtained in decent yields (Fig. 8.9) (Zhang et al., 2015). However, addition of a Lewis acid such as LiOTf

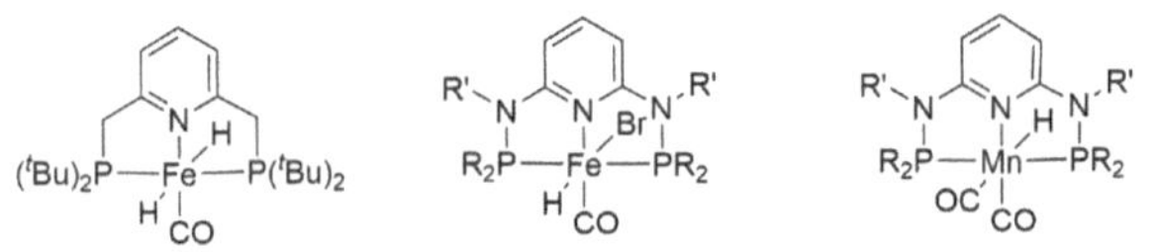

Figure 8.8 Iron- and manganese-based catalysts, having a pyridine-based pincer backbone, for CO_2 hydrogenation.

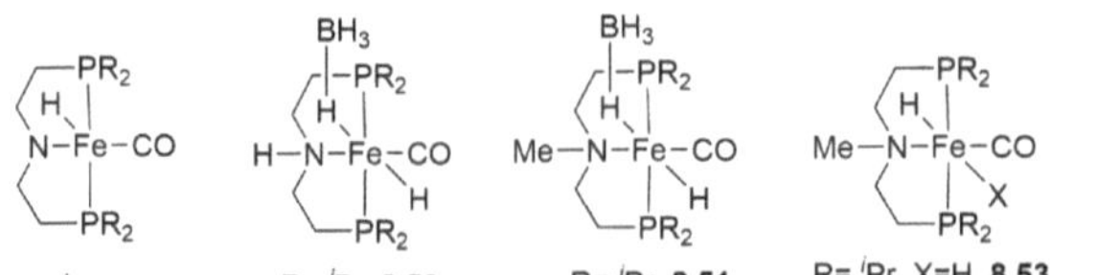

Figure 8.9 Structures of selected Fe catalysts containing saturated pincer carbonyl hydride (**8.48–8.54**).

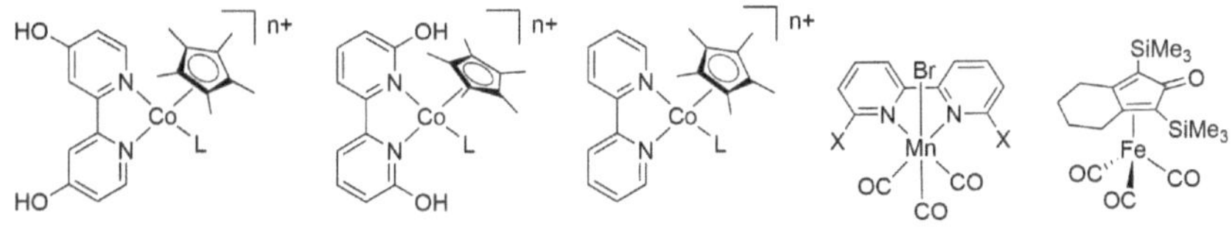

L= Cl, n = 1 **8.55**
L= H_2O, n = 2 **8.56**

L= Cl, n = 1 **8.57**
L= H_2O, n = 2 **8.58**

L= Cl, n = 1 **8.59**
L= H_2O, n = 2 **8.60**

X = OH **8.61**
X = H **8.62**

8.63

Figure 8.10 Structures of selected cobalt and manganese catalysts.

led to enhanced reaction rates. The best results were obtained with TONs of 58,990 and 46,130 for **8.51** and **8.54** respectively after 24 h of reaction time.

The Fujita group synthesised a series of cobalt(III) complexes (**8.55–8.60**) containing proton-responsive dihydroxybipyridine ligand backbone (Fig. 8.10) (Badiei, 2013). Unfortunately, poor catalytic aptitudes were observed due to their poor thermal stability. Based on same ligand architecture, the Khusnutdinova group later reported that Mn complex **8.61** (Fig. 8.10) can serve as an efficient catalyst (TON of 6250) (Dubey et al., 2017). When Mn catalyst **8.62** (Fig. 8.10) with an unsubstituted bipyridine ligand was employed, only a trace amount of formate was detected, suggesting the importance of adjacent −OH group in promoting the reaction efficacy. Zhou and co-workers developed an air-stable iron catalyst **8.63** (Fig. 8.10) for hydrogenation of both bicarbonate and CO_2 in basic media (Zhu et al., 2015). The Knölker complex was proposed as the active catalytic intermediate in this process. Using $NaHCO_3$, optimised activity with a TON of 447 after 24 h was achieved in EtOH/H_2O under 30 bar H_2 pressure at 120°C.

8.2.3 Homogeneous catalysts under base-free conditions

The discussions in the earlier two sections clearly suggest that the majority of homogeneous catalysts for hydrogenation of CO_2 or bicarbonate to FA or formate require the use of basic additives to achieve acceptable efficiencies, as the reaction between FA and the base can generate formate adducts to shift the thermodynamic equilibrium. However, to achieve the production of FA, the resulting formate salts need to be neutralised and purified, generating additional energy demands. Thus developing a catalytic system under base-free conditions is of utmost importance to overcome these issues. In 1989 Khan et al. first reported that base-free CO_2 hydrogenation could be achieved under 3 bar of H_2 and 17 bar of CO_2 in an aqueous solution containing 0.01 M K[Ru(EDTA-H)Cl] · $2H_2O$ (**8.64**) (EDTA = ethylenediaminetetraacetic acid) to generate FA along with a small amount of formaldehyde at 40°C (Khan, Halligudi, and Shukla, 1989). Tsai and Nicholas demonstrated the catalytic activity of [Rh(nbd)($PMe_2Ph)_3$]BF_4 (**8.65**) (nbd = norbornadiene) under base/additive-free conditions for CO_2 hydrogenation in wet THF (Tsai and Nicholas, 1992). A maximum FA concentration of 0.18 M was obtained by performing the reaction in a THF solution containing a trace amount of water under equimolar 96 bar of H_2/CO_2 at 40°C for 48 h. In 2004 the Fukuzumi group synthesised a water-soluble ruthenium catalyst,

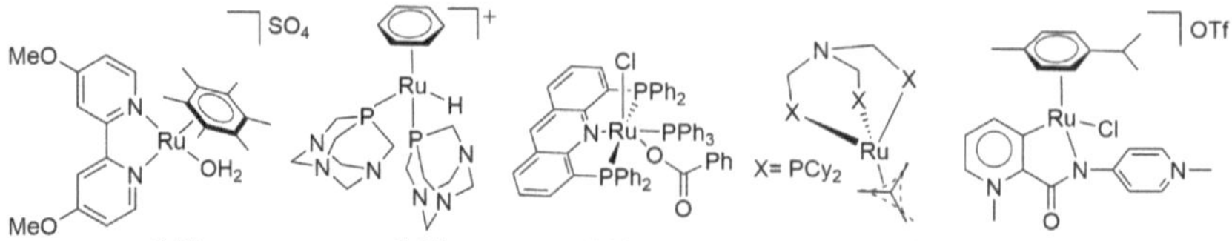

Figure 8.11 Selected Ru catalysts for base-free CO_2 hydrogenation.

[(C_6Me_6)Ru(4,4′-dmbpy)(OH_2)]SO_4 (dmbpy = dimethoxy-2,2′-bipyridine) (**8.66**, Fig. 8.11) that can generate FA with a TON of approximately 55 at 40°C using 25 bar CO_2/55 bar H_2 in H_2O (Hayashi, Ogo, and Fukuzumi, 2004). Laurenczy et al. achieved slightly better yield of FA concentration of 0.2 M using $[(C_6H_6)RuH(pta)_2]^+$ (pta = 1,3,5-triaza-7-phosphaadamantane) (**8.67**, Fig. 8.11) in H_2O under 200 bar H_2/CO_2 (3:1) at 60°C with a TON of 74 (Moret, Dyson, and Laurenczy, 2014). The catalyst was recycled without any loss of catalytic aptitude.

Leithner group further prepared [Ru(acriphos)(PPh_3)(Cl)($PhCO_2$)] (acriphos: 4,5-bis(diphenylphosphino)acridine) (**8.68**, Fig. 8.11) to catalyse CO_2 hydrogenation to FA (Rohmann et al., 2016). The reaction was conducted in DMSO under 80 bar of H_2/CO_2 (1:1) to afford a TON of 1094 (0.09 M FA) after 16 h at 60°C. Changing the solvent to 95% DMSO/5% H_2O (v/v) gave rise to a maximum TON of 4200 (0.33 M FA) under otherwise identical reaction conditions. A recent report from the Klankermayer group discussed a novel [Ru(N-triphosCy)(tmm)] (tmm = trimethylmethane) complex (**8.69**, Fig. 8.11) bearing sterically demanding cyclohexyl groups towards CO_2 hydrogenation in presence of $Al(OTf)_3$ as the Lewis acid additive (Westhues, Belleflamme, and Klankermayer, 2019). Under 120 bar of H_2/CO_2 (3:1) in MeOH/dioxane at 60°C, FA was formed and subsequently converted to methyl formate. Albrecht and co-workers have recently demonstrated Ru catalyst **8.70** (Fig. 8.11) for CO_2 hydrogenation employing an imidazolium-based IL (BMMI · OAc) as the buffering media (Weilhard et al., 2020). The IL may help maintaining the pH value and thus prevent the catalyst deactivation. Under 60 bar equimolar pressure in DMSO/water (5 v/v% water) at 70°C, a maximum TON of 4520 was achieved after 72 h.

8.2.4 Bulk metal catalysts for heterogeneous catalysis

Despite the fact that several homogeneous catalysts achieved impressive catalytic efficiencies in terms of TON and TOF, the major challenge for a large-scale industrial production is the difficulty in catalyst separation from the final reaction mixture (Preti, Squarcialupi, and Fachinetti, 2010). Furthermore, during product separation step(s), homogeneous catalysts may promote the reverse reaction, that is decomposition of the generated formate back into CO_2 and H_2 (Moret et al., 2014; Schaub and Paciello, 2011). In this regard, heterogeneous catalysts may offer merits for convenient product separation and continuous operation.

Hydrogenation of CO_2 to formate using bulk metal catalysts was examined over a century ago. In the presence of 1 M aqueous $NaHCO_3$ solution, a TON of 2.1 was observed by the Wrighton group after carrying out the reaction for 53 h using Pd black (Stalder et al., 1983). Preti et al. proposed that an active catalyst can promote the forward reaction as well as the reverse one (FA decomposition) and screened the gas evolution from HCOOH/NEt_3 adducts in the presence of groups 8–11 metal black as catalysts, such as Raney Ni, Co, Cu, Ru, Rh, Pd, Ag, Ir, Pt and Au, at 40°C (Preti et al., 2011). It was found that only Au metal blacks showed gas evolution and hence it was applied for CO_2 hydrogenation. Srivastava et al. examined the activity of Ru nanoparticles (NPs) as a pure metal catalyst. Ru NPs were synthesised in situ in an IL ([DAMI][NTf_2]) (DAMI, 1,3-di(*N*,*N*-dimethylamino ethyl)-2-methyl imidazolium which was used as the solvent together with H_2O, and a TON of 940 was obtained at 100°C using equimolar 40 MPa pressure (Srivastava, 2014). In 2017 Huo et al. demonstrated nanoporous nickel (NiNPore) for $NaHCO_3$ hydrogenation to FA in aqueous phase (Wang et al., 2017). Under 6 MPa pressure, a TON of 3476 was obtained (86.6% yield) at 200°C for 2 h. Recently, Liu and co-workers have described a protocol of FA synthesis over Pd/C in IL ([Bmim][OAc]) under base-free condition yielding a maximum TON of 594 (Wu et al., 2019). The multiple roles of the IL were responsible for activating CO_2 (*vide infra*), modifying the Pd NPs and stabilizing FA.

8.2.5 Supported metal catalysts for heterogeneous catalysis

As the reaction takes place on surface of heterogeneous catalysts, the bulk metal-based catalysts are economically unproductive. These bulk metals often suffer from the decrease of activity by aggregation. Thus support materials are often employed to increase the number of active sites by dispersing active metals to circumvent such issues. In the earlier report by Preti and co-workers (*vide supra*), as the Au blacks got aggregated, a dispersed Au catalyst on TiO_2 (1 wt.% Au/TiO_2) was tested and showed superior catalytic efficacy yielding a TON of 855 under a total pressure of 18 MPa at 40°C in a NEt_3 solution (Preti et al., 2011). Pidko et al. examined catalytic properties of several supported and unsupported Au NPs towards CO_2 hydrogenation under 40 bar equimolar pressure in EtOH using NEt_3 as the base at 40°C and found higher efficiency per unit mass of Au in case of supported systems (Filonenko et al., 2016).

Among all the supports, viz, Al_2O_3, TiO_2, CeO_2, ZnO, MgCr-HT, MgAl-HT (hydrotalcite) and $CuCr_2O_4$, the Au NP/Al_2O_3 (TON = 215) have shown the best reactivity and this observation was attributed to the synergistic effect between the basic Al_2O_3 and Au^0 NPs. Based on the spectroscopic studies, a plausible mechanism was proposed where heterolytic dissociation of H_2 forms surface hydroxyl and Au–H species at the Au/support interface followed by the formation of an Au-formate intermediate upon CO_2 adsorption. The catalytic cycle closes when this species finally forms stable alumina surface eliminating formate (Fig. 8.12).

The Ma group hypothesised that the presence of surface hydroxyl groups on the support may improve the CO_2 adsorption and in turn enhance the catalytic

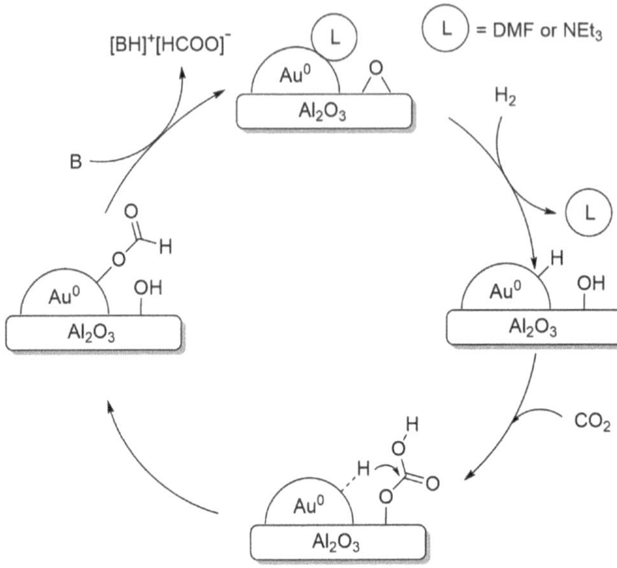

Figure 8.12 Proposed mechanism of CO_2 hydrogenation over supported Au nanoparticles.

efficiency (Hao et al., 2011). To validate this hypothesis, ruthenium catalysts on various supports were prepared, including (1) γ-alumina, having abundant hydroxyl groups on the surface (Ru/γ-Al_2O_3), (2) activated carbon (AC) having a limited number of hydroxyl groups on the surface (Ru/AC), and (3) MgO devoid of any hydroxyl groups (Ru/MgO). Under 13 MPa of H_2/CO_2 (5/8) in EtOH/NEt_3 at 80°C, the catalytic performance followed the expected order Ru/MgO (no activity) < Ru/AC (TON of 10) < Ru/γ-Al_2O_3 (TON of 91). Later, the Li group employed γ-Al_2O_3 nanorods [γ-Al_2O_3 (*n*)] as a support for ruthenium catalyst [Ru/γ-Al_2O_3(*n*)] (Liu et al., 2014; Liu, Du, and Li, 2013). Under otherwise identical reaction conditions, [Ru/γ-Al_2O_3(*n*)] exhibited superior reactivity (TON of 731) compared to that of Ru/γ-Al_2O_3 (TON of 91), likely attributed to the abundant hydroxyl groups, high surface area and increased interaction with ruthenium species. Mori and co-workers anchored a single-atom Ru catalyst on a layered double hydroxide (LDH) and obtained maximum TON of 698 under mild reaction conditions (2.0 MPa, 100°C) (Mori, Taga, and Yamashita, 2017). CO_2 adsorption capacity in the vicinity of the Ru centre had a significant influence on catalytic activity. The CO_2 adsorption capacity near Ru centre was tuned by varying the components and ratio in the LDH and the optimised result was found with $Mg^{2+}/Al^{3+} = 5$.

In 2015 Lin and co-workers observed that hydrogenation of bicarbonates ($NaHCO_3$, $KHCO_3$ and NH_4HCO_3) to formates was more facile than that of carbonates [Na_2CO_3, K_2CO_3 and $(NH_4)_2CO_3$] (Su et al., 2015; Su, Lu, and Lin, 2015). Conducting the reaction under 2.75 MPa pressure at ambient temperature, a porous carbon material supported Pd/AC gave a higher yield for NH_4HCO_3 (42.4%, TON 782) than $NaHCO_3$ (28.6%, TON 527) and $KHCO_3$ (30.8%, TON 567). Nguyen et al. reported bimetallic catalytic systems comprised PdNi alloys supported on carbon nanotube-graphene (PdNi/CNT−GR) and evaluated CO_2 hydrogenation to FA in the absence of base (Nguyen et al., 2015). The composite support was chosen to prevent the stacking of GR and bundling of CNTs. Both can act as a spacers to help to expose their entire surface areas during catalysis. Under 1:1 equimolar 5 MPa pressure, employing the bimetallic PdNi/CNT−GR (Pd_3Ni_7/CNT−GR; Pd-30%, Ni-70%) catalyst afforded a considerable amount of FA (1.92 mmol; TON 6.4) at 40°C along with a trace amount of acetic acid. The poorer efficiency of mono-metallic Pd supported on CNT−GR (Pd/CNT−GR) and mono-metallic Ni supported on CNT−GR (Ni/CNT−GR) compared to that of PdNi/CNT−GR, suggested the involvement of bimetallic cooperativity in the substrate activation. Mu and co-workers demonstrated nano-Ni as an efficient catalyst for FA synthesis (99.8% yield) using 1 mM $NaHCO_3$ (He et al., 2017). The Gai group introduced chitin as a support to load Pd NPs (Song et al., 2017). FA was produced in high efficiency with a TOF of 239 h^{-1} in aqueous medium in the presence of Na_2CO_3 under 40 bar H_2/CO_2 (1:1) at 60°C. The Zhang group devised a Schiff-base-modified Au nanocatalyst supported on SiO_2 and obtained a very high activity towards CO_2 hydrogenation with a maximum TON of 14,470 over 12 h at 90°C in H_2O/MeOH (Liu et al., 2017).

8.2.6 Heterogenised catalysts

The importance of heterogeneous metal catalysts was realized considering easy product separation from the reaction mixtures and recyclability when compared with homogeneous systems. However, the catalytic efficacy of heterogeneous catalysts discussed so far was moderate compared to those of existing homogeneous systems. In this context, strategies involving the heterogenisation of homogeneous complexes have been developed. Materials having advantages of homogeneous metal catalysts, that is activity and selectivity were combined with heterogeneous catalysts having benefits of recyclability and easy product isolation. Such approaches of heterogenisation or immobilisation have been explored for continuous evolution of CO_2 hydrogenation catalysts.

8.2.6.1 Grafted molecular catalysts

Zheng et al. demonstrated the immobilisation of Ru complexes on the surface of support (silica, mesoporous MCM-41, polystyrene) having various functional donor groups such as nitrile (CN), thiol (SH) and amine (NR_2) through the coordination bond between the donor ligand of the support and the metal ion (Fig. 8.13) (Yu et al., 2005, 2006a, 2006b; Zhang et al., 2004, 2006). Among the inorganic supports, the MCM-41 supported catalyst (**8.71**) having amine functionality exhibited a better reactivity with a TON of 1022 than that of the silica-supported one (**8.72**) (TON of 656) under 14.7 MPa $H_2/CO_2/H_2$ (5.4/9.3) at 80°C in EtOH using NEt_3 as the base and PPh_3 as an additive after 1 h (Fig. 8.13). Uniform pores (3.5 nm) and high surface area (852 m^2 g^{-1}) of MCM-41 were accounted for this activity. Under

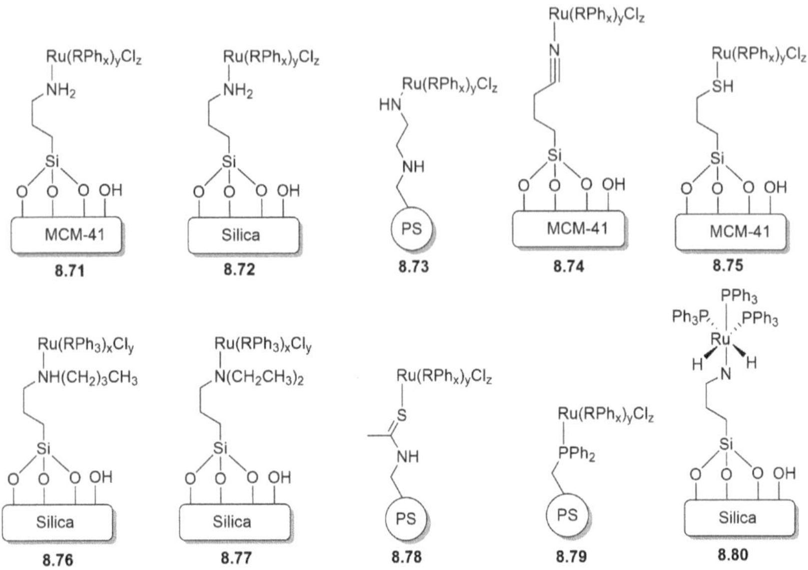

Figure 8.13 Representative grafted Ru catalysts.

otherwise similar reaction conditions, however, the organic support polystyrene (**8.73**, Fig. 8.13) displayed poorer activity (TON of 151). Screening of several functional groups anchored to MCM-41 support suggested that amine group exhibited better activity than nitrile (**8.74**, TON of 723) and thiol (**8.75**, TON of 537) functionalities owing to the stronger electron-donating ability of amine groups to metal ions (Fig. 8.13). Using silica support, the secondary amine-functionalized catalyst (**8.76**, TON of 1384) was superior compared to the primary amine-functionalized (**8.72**, TON of 656) and tertiary amine-functionalized (**8.77**, TON of 868) catalysts (Fig. 8.13). A similar trend was found employing catalysts **8.73**, **8.78** and **8.79**, suggesting that electron-donating ligands increase the efficiency of the catalyst (Fig. 8.13). Several additives were screened, and it was found that the bidentate ligand (DPPE) having a smaller bite angel performed better than the monodentate ligands with relatively larger bite angels. During catalyst recycling, no further addition of ligand was necessary, indicating that an active catalytic species, an octahedral dihydrido Ru complex **8.80** (Fig. 8.13), was formed in the first run.

Amine functionalized ILs were employed as a reusable base to form formate with FA, leading to an easy recovery of catalyst by simple filtration. In the follow-up work, diamine-functionalized ILs were prepared to enhance the catalyst efficiency to give a maximum TOF of 920 h^{-1} under 18 MPa equimolar pressure at 80°C (Zhang et al., 2009). No catalyst leaching was detected up to five runs.

The Hicks group reported the first examples of immobilised Ir complexes for CO_2 hydrogenation through the Schiff base reaction of −*o*-(diphenylphosphino) benzaldehyde and amine groups on the mesoporous silica (**8.81**) (Fig. 8.14) (Xu et al., 2013). Screening different grafted complexes (**8.81–8.84**) and the

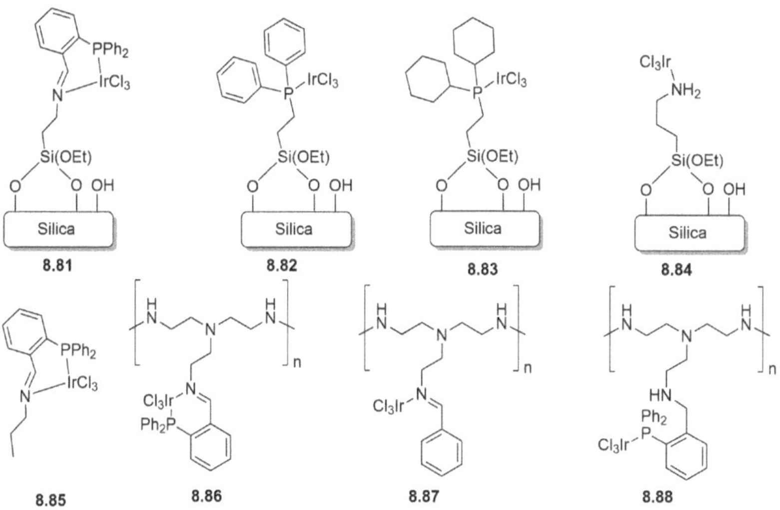

Figure 8.14 Representative grafted Ir catalysts.

homogeneous analogue (**8.85**) revealed that only phosphine containing catalysts (**8.81–8.83** and **8.85**) were active towards CO_2 hydrogenation to formate (Fig. 8.14). Under 4 MPa H_2/CO_2 (1:1) in the presence of NEt_3 base in H_2O, catalyst **8.81** exhibited a superior activity with a TON of 1300 compared to **8.82** (TON of 110), **8.83** (TON of 400) and **8.85** (TON of 70) at 60°C after 2 h. As X-ray photoelectron spectroscopy (XPS) indicated the similar environments around Ir metal in **8.81** (61.6 eV) and **8.85** (61.8 eV), the large difference in catalytic efficacy showed that the stability and activity of the homogeneous system were improved upon heterogenisation. The TON was improved to 2300 at an elevated temperature of 120°C and could reach a maximum value of 2700 for a longer reaction time (20 h). Polyethyleneimine (PEI), an aliphatic amine-based organic polymer containing primary, secondary and tertiary amine groups, was further explored as the support (Fig. 8.14) (McNamara and Hicks, 2014). Such supports could act as CO_2 capturing materials and a formate stabiliser to improve the catalytic activity. Comparison of the catalytic performance of catalyst **8.86**, generated by tethering complex **8.85** on PEI; imine containing catalyst **8.87**, and phosphine containing catalyst **8.88** revealed that **8.87** exhibited a better activity. Nevertheless, in the absence of basic additives, all of them were inactive in spite of the presence of basic amine moieties (Fig. 8.14). The poorer activity of **8.86** (TON of 248) compared to that of **8.81** suggested the weakness of PEI supports.

8.2.6.2 Heterogenised porous polymers

Since conventional silica, polystyrene and aliphatic polymer supports gave only moderate reactivities (*vide supra*), the importance of porous organic framework was investigated as catalyst supports because of their easy synthesis, high surface area and well-defined porosity. The Yoon group prepared Pd NPs supported on mesoporous graphitic carbon nitrides g-C_3N_4 (**8.89**) (Fig. 8.15) (Lee et al., 2014). Employment of this catalyst yielded a TON for 81 under 4 MPa of equimolar pressure in a 20% NEt_3/D_2O solution at 150°C after 24 h. Increasing $p(H_2)/p(CO_2)$ to two under otherwise identical conditions, resulted in an improved TON of 106.

Inspired by the pioneering work of the Himeda group on the development of homogeneous Ir catalysts, Park and co-workers have established a new protocol to immobilise the homogeneous complex (**8.18**) (Fig. 8.15) (Park et al., 2015). A bpy incorporated covalent triazine framework CTF (bpy-CTF) (where CTFs are porous aromatic frameworks made upon trimerization of aromatic nitriles) was synthesised as a support because of its high surface areas, large pore volumes and high thermal stability. Treatment of $[IrCp^*Cl_2]_2$ with bpy-CTF resulted a heterogenised complex, bpy-CTF-[IrCp*Cl]Cl (**8.90**, Fig. 8.15) having a similar coordination environment as that of its homogeneous analogue, which was further confirmed by the XPS measurements with the identical binding energy for Ir $4f_{7/2}$ (62.1 eV). Employment of **8.90** yielded a TON of 500 under equimolar 4 MPa pressure in H_2O at 80°C after 2 h in the presence of NEt_3 as a base. Because of the exothermic nature of the reaction, temperature-dependent study indicated an increase in the activity with increasing temperatures up to 120°C (TON = 3320) and afterwards a decreasing trend was observed (TON = 2720 at 160°C). Catalyst **8.90** was also successfully employed for the recycling experiment up to five runs without any substantial loss of reactivity (~92% catalytic activity was retained after each run). Gunasekar et al. synthesised analogous $[bpy\text{-}CTF\text{-}Ru(acac)_2]Cl$ (**8.91**) by grafting $Ru(acac)_2$ complex onto CTF and obtained an initial TOF of 22,700 h^{-1} (Gunasekar et al., 2018). An excellent recyclability over consecutive runs was also achieved. $RuCl_3 \cdot xH_2O$ was used with the same support to give $[bpy\text{-}CTFRuCl_3]$ (**8.92**) with an unprecedented initial TOF of 38,800 h^{-1}. This is one of the highest value observed using a heterogeneous catalyst (Gunasekar, Jung, and Yoon, 2019).

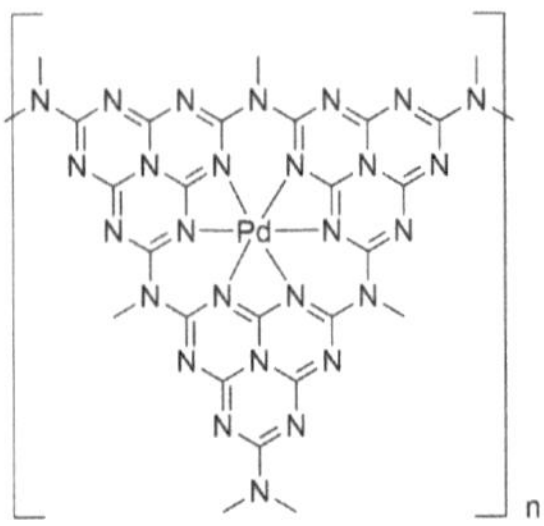

8.89

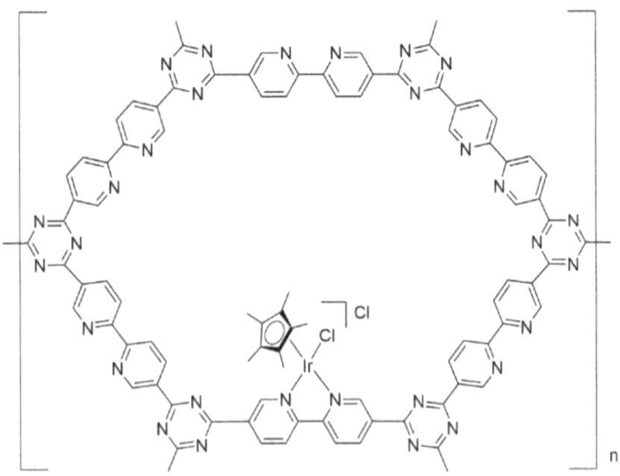

8.90

Figure 8.15 Structural representation of catalysts **8.89** and **8.90**.

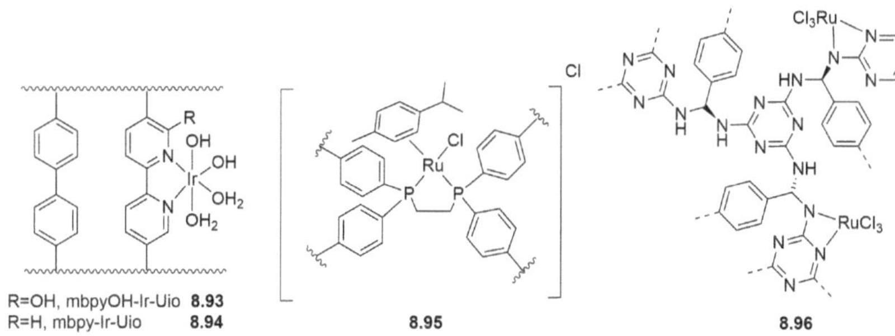

Figure 8.16 Structural representation of catalysts **8.93–8.96**.

In 2017 the Wang group utilised metal-organic framework as the support for Ir-based catalysts and the resultant mbpyOH-Ir-UiO (**8.93**) and mbpy-Ir-UiO (**8.94**) catalysts were employed for CO_2 hydrogenation (Fig. 8.16) (An, 2017). Catalyst **8.93** (TON of 6149) outperformed **8.94** (TON of 417) at 85°C under the atmospheric pressure. Very high kinetic isotope effects and DFT (density functional theory) calculations suggested a concerted proton-hydride transfer in the RDS. The Palkovits group reported a different strategy by cross-linking polyphosphines (pdppe) as a support for [(Ru(*p*-cymene)Cl_2)$_2$] [Ru@pdppe (**8.95**)] and applied in the catalytic hydrogenation of CO_2 to formate (Fig. 8.16) (Kann et al., 2018). Highest activity (TON of 11,250) was achieved at 120°C under 250 bar H_2/CO_2 (2:3) in the presence of K_2CO_3 in water. Recently, Ashan et al. synthesised a novel catalyst Ru@MPN (**8.96**) by grafting $RuCl_3 \cdot xH_2O$ onto melamine polymer-network (MPN) and achieved desired reactivities (Fig. 8.16) (Jaleel, 2020). Under optimised reaction conditions, 0.42 M of formate was produced with initial TOF of 4280 h^{-1}.

8.3 Preparation of formic acid by electrochemical reduction of carbon dioxide

Electrocatalysis for CO_2 conversion offers a direct pathway for power to formic acid. Although it has several advantages, the product selectivity in electrocatalytic CO_2 reduction reaction (eCO_2RR) is challenging because the transfer of different proton and electron equivalents will lead to divergent products. Moreover, the catalytic intermediates are also competent for hydrogen evolution reaction (HER) (Ceballos and Yang, 2018). In the presence of protons, formation of formate ($CO_2 + 2e^- + H^+ \rightarrow HCOO^-$) or FA ($CO_2 + 2e^- + 2H^+ \rightarrow HCOOH$) is one of the most promising hydrogenated products among others (Fig. 8.17).

While the thermodynamic potential for one-electron reduction of CO_2 to $CO_2^{-}\cdot$ is -2.14 V (vs SCE) in aqueous solution, for $2e^-/2H^+$ reduction to HCOOH the value is -0.85 V only because of the involvement of proton-coupled electron transfer and multi-electron transfer pathways, leading to routes that involve low

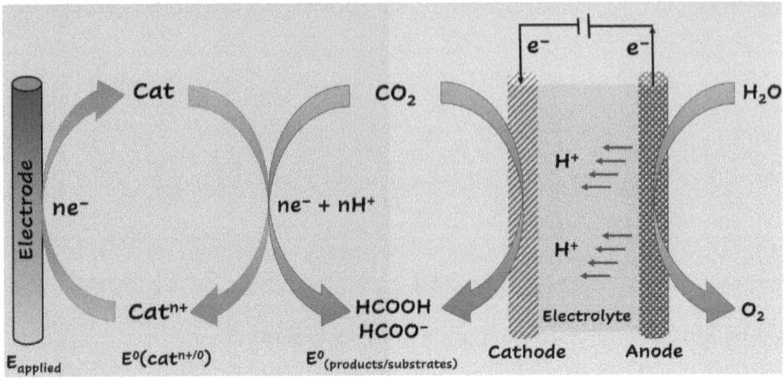

Figure 8.17 Electrocatalysis of CO_2 reduction with the catalyst loaded on an electrode (left); basic representation of CO_2 electroreduction electrolytic cell.

Table 8.2 Multielectron reduction of CO_2 (vs saturated calomel electrode in water).

Reaction	E^0 (V) pH 7	E^0 (V) pH 0
$CO_2 + 2e^- + H^+ \rightarrow HCOO^-$	−0.73	−0.52
$CO_2 + 2e^- + 2H^+ \rightarrow HCOOH$	−0.85	−0.35
$CO_2 + 6e^- + 6H^+ \rightarrow CH_3OH + H_2O$	−0.62	−0.44
$CO_2 + 8e^- + 8H^+ \rightarrow CH_4 + H_2O$	−0.48	−0.07

overpotential processes (Table 8.2) (Taheri and Berben, 2016). Direct eCO_2RR on a bare electrode surface normally requires large overpotentials (η), resulting from disparity between the applied electrode potential ($E_{applied}$) and $E^0_{(products/substrates)}$ at a given current density (CD) (j). Low η and high product selectivity with considerable reaction rate constant (K_{cat}) are highly desired for a catalyst having $E^0(cat^{n+/0})$ well matched to $E^0_{(products/substrates)}$ for an ideal outcome (Zhao et al., 2017). In general, eCO_2RR is accomplished in an electrolytic cell which consists of a pair of electrodes (cathode and anode), catalysts coated on electrode and solvent/electrolyte for transferring ions (Fig. 8.17). General cathodic and anodic reactions for eCO_2RR are

$$\text{Cathode: } 2CO_2 + 4H^+ + 4e^- \Rightarrow 2HCOOH \tag{8.3}$$

$$\text{Anode: } 2H_2O \Rightarrow O_2 + 4H^+ + 4e^- \tag{8.4}$$

The efficient and selective electrocatalytic production of FA/formate is of particular importance for practical applications which is strongly influenced by the common factors and reaction parameters such as solvents/electrolytes, pH of the solution, temperature and pressure, electrode materials, catalysts etc. and will be discussed here.

8.3.1 Solvent and electrolyte

Different solvents and supporting electrolytes that provide a delivery medium for protons and electrons play an essential role in the aqueous eCO_2RR. Thermodynamic potentials as well as pK_a or hydricity of the metal hydride intermediates can vary with solvents. In the case of aqueous electrolytes, HCO_3^- and CO_3^{2-} could act as the source of reductant and there were evidences to show that these ions indeed enhanced the eCO_2RR by effectively increasing the local CO_2 concentration (Köleli, 2003). In addition to these anions, Li^+, Na^+, K^+, H^+ and NH_4^+ were found to favour eCO_2RR (Paik, Andersen, and Eyring, 1969). Since eCO_2RR is often carried out in the aqueous solution to facilitate ion conduction between electrodes, it is inconvenient and energy-intensive to separate liquid products like FA from dissolved salts. To overcome this difficulty, very recently, Wang and co-workers sought to employ a solid-state electrolyte (SSE) in an eCO_2RR cell to produce neat HCOOH (Xia et al., 2019). The continuous production of pure FA solutions [up to (12 M)] was successfully demonstrated with negligible degradation and high selectivity using the 2D-Bi//SSE-50//IrO_2-C (average domain size of SSE = 50 m) cell.

Non-aqueous electrolytes are salts dissolved in organic solvents such as CH_3OH, MeCN, DMF etc. KOH/CH_3OH has been shown to form K^+ and HCO_3^- in MeOH and could produce FA from eCO_2RR over Cu and Pb electrodes, suppressing the HER ($FE_{H2} < 3.5\%$) (Kaneco et al., 1998). ILs are salts in a liquid phase and highly electrically conductive materials that can be employed for eCO_2RR to FA. In 2014 Watkins et al. used 1-ethyl-3-methylimidazolium trifluoroacetate ([EMIM] [TFA]) along with various metals (In, Sn, Pb) as co-catalysts in the electrochemical conversion of CO_2 to HCOOH and found that indium outperformed Sn and Pb (Watkins and Bocarsly, 2014). The Zhang group introduced a novel [EMIM][N $(CN)_2$]/water system for electrochemical reduction of CO_2 to FA on a Sn powder electrode with a maximum faradaic efficiency (FE) of 81.9% during electrolysis ($\eta = 1.08$ V) in 0.5 M of the IL (Zhang et al., 2017). An efficient eCO_2RR into FA with suppression of CO could be obtained using an IL catholyte mixture containing [BMIM]X/H_2O/MeCN (BMIM = 1-butyl-3-methylimidazolium; X = PF_6^- or BF_4^-) for Pb or Sn electrodes with partial current density (PCD) of >35 mA cm^{-2} at a FE of >90% (Wu et al., 2018; Zhu et al., 2016).

8.3.2 pH and pressure

The pH of the eCO_2RR system is an important parameter that directly influences the reaction rate of CO_2 reduction. 0.1–0.5 M $NaHCO_3$ or $KHCO_3$ has always been the best choice for the aqueous eCO_2RR because the bicarbonate ion can act as a buffer to maintain the stability of the pH (pH ~7.3). In general, the sp group metal cathodes, such as Hg, In, Sn and Pb, favour the production of FA in the neutral solution and formate ions in the weak alkaline solution in the presence of aqueous electrolytes (Jitaru et al., 1997). Bumroongsakulsawat et al. have shown that the molar ratio of CO:formate varied from 1 to 0.15 when the pH of the aqueous solution changed from 2.9 to 7.8 during eCO_2RR at Sn electrodes (Bumroongsakulsawat and Kelsall, 2014). A reaction scheme was proposed invoking three different reaction intermediates, $CO_2^{-\cdot}$, $COOH^{\cdot}$ and $CO_2H_2^{+\cdot}$, whose distribution upon protonation was found to be perturbed significantly by the solution pH.

In addition to an optimal solution pH, the influences of elevating pressure in the reactor have also been investigated in many scenarios. Proietto et al. used a high pressure (up to 30 bar) undivided filter-press cell with a continuous recirculation of the electrolytic solution at a Sn cathode to convert CO_2 to FA/formate (Proietto et al., 2018). The increase of CO_2 pressure from 1 to 5 bar and 5 to 15 bar enhanced the final FA concentration from 2 to 5.8 mM and from 5.8 to 9.6 mM respectively, with an increment in the FE_{FA} from 15% →39% →64%. Morrison and co-workers have proposed a model to validate the impacts of the pressure in the CO_2 electrolyser on the CD for CO_2-to-FA/formate conversion. The FE and specific energy of formation were found to reach the highest at pressures above 10–20 bar, at 95% and of 3.7 kWh kg^{-1}, respectively (Morrison et al., 2019). More recently, Dai and co-workers synthesised a novel cathode material comprised of Cu_2O NP film which can reduce CO_2 electrochemically to nearly formate under ≥45 atm CO_2 pressure in bicarbonate catholytes with FE ~98% [at −0.64 V vs RHE (reversible hydrogen electrode)] (Li et al., 2020).

8.3.3 Electrode materials

Metal electrodes, acting both as support and catalysts, have been used as traditional electrodes over many years due to their excellent performance for eCO_2RR. Nevertheless, these electrodes cum catalysts still fail to meet strict criteria such as high selectivity at low η and prolonged durability. In general, an electrode of an electrocatalytic system is comprised of a substrate and a catalyst layer (CL) which usually provides active catalytic sites by adhering the catalysts on its surface and conducts electron with low resistance. However, in the modern fuel cell technology, these electrodes have gradually been replaced with gas diffusion electrodes (GDEs) owing to various advantages (Song et al., 2019). GDEs consists of a conventional CL and a gas diffusion layer. Furuya and co-workers have examined the reactivity and selectivity of eCO_2RR at GDEs made of Ru, Pd and Ru–Pd (1:1) alloy in 0.5 M $KHCO_3$ aqueous solution (Jitaru et al., 1997). It was found that the reactivity reached the highest for the Ru–Pd GDE producing only FA with no CO formation with a current efficiency reaching 90% at −1.1 V versus NHE (normal hydrogen electrode) ($j = 80$ mA cm^{-2}). Although the CO_2-to-FA conversion is not promising using GDEs with Sn NPs, CO_2-to-formate conversion showed encouraging results, particularly in the past few years where FEs could reach up to 70%–90% with a PCD of 150–250 mA cm^{-2} in the presence of a carbon support (Del Castillo et al., 2015; Kopljar et al., 2014).

Cu, as a unique electrocatalyst, reduces CO_2 to various products with low selectivity towards FA but it possesses a good conductivity and a high specific surface area as observed in a recent report from Huang's group where an optimised annealed Cu foam electrode was shown to possess high electroactive surface area and lower overpotential for eCO_2RR. However, the formate selectivity (FE_{FA} ~23% at −0.45 V vs RHE) was inadequate (Min et al., 2016). On the other hand, Sn-based catalysts were found to be highly selective towards formate with FE ~80%−90% but not at high cathodic potentials. Therefore it is desirable to develop an electrode that can integrate the properties of both metals through synergistic effects to achieve high selectivity and high production rate of formate at the same time. Wang et al. deposited Sn on Cu foam substrate (Sn/f-Cu) by electrodeposition in aqueous plating solution, which exhibited excellent catalytic activity with an FE reaching a maximum value of 83.5% (at −1.8 V vs Ag/AgCl and PCD = 8 mA cm^{-2}) when compared to Cu foam and Sn plate (Wang et al., 2016). Guan and co-workers fabricated Bi-doped SnO nanosheets on Cu foam (Bi-SnO/f-Cu) which exhibited a superior FE of 93% at −1.7 V (vs Ag/AgCl) together with a PCD of 12 mA cm^{-2} and stability in 30 h of operation (An, 2019). The experimental results along with DFT calculations indicated that the Bi doping stabilized the Sn^{+2} existing on the surface of the electrocatalyst, making it hard to be reduced to Sn^0 during the electrochemical reduction process. Although metallic Bi promotes eCO_2RR to formate, it requires substantial η and relatively high j to achieve a high FE. Therefore the Zhang group have strategically developed lattice-dislocated Bi nanowires (NWs) on the Cu foam (BiNW/f-Cu) that can convert CO_2 to formate with a FE of 95% at low η and PCD of ~15 mA cm^{-2} at −0.69 V versus RHE (Zhang et al., 2019). The high activity of BiNW/f-Cu was attributed to the high intrinsic activity due to crystal lattice dislocations on the BiNWs and the large catalytic surface area associated with the porous structure.

8.3.4 Catalysts

Electrochemical CO_2-to-FA/formate conversion can only be achieved with certain catalysts that can decrease the overpotential by eliminating some high energy intermediate and suppress the competing H_2 evolution pathways. The cell performance should also be improved by increasing the selectivity, stability and current efficiency. In this section, catalysts that selectively catalyse CO_2-to-FA/formate conversion will be classified into four categories for discussion.

8.3.4.1 Metals and metal oxides

Metallic materials possess simple structures and good electrical conductivity and can be prepared easily. Notably, formate was one of the first products identified from eCO_2RR on metal electrodes more than half a century ago (Teeter and Van Rysselberghe, 1954). Since then, formate has been the primary product observed for various metals such as Pb, Hg, In, Sn, Cd and Tl in the aqueous solution because these metals have high overpotentials for HER and extremely weak CO adsorption capacities (Hori, 2008). However, The Kanan group reported a special nanocrystalline Pb film, prepared by reducing PbO_2 precursors, that showed up to 700-fold lower H^+ reduction activity than those of polycrystalline Pb foil electrodes but maintained the ability to reduce CO_2, that is highly selective for CO_2 reduction to formate in aqueous solutions with almost no competitive H^+ reduction (Lee and Kanan, 2015). Other sp metals, such as Zn, Cu and Cd can also produce FA, but the FEs and PCDs are not satisfactory. It has been previously discussed that slight adjustments in the reaction conditions, such as elevation of pressures, utilisation of GDEs, use of different electrolytes etc., can foster high FEs and PCDs with better cell performances.

Among d-group metals, Pd is the most commonly used metal that can achieve FEs up to 100% at high j (Podlovchenko, Kolyadko, and Lu, 1994). Rh and Ir, on the other hand, only give poor FA selectivity where the products are dominated by H_2 and CO (Hara, Kudo, and Sakata, 1995). While bulk Pd has been considered to be an inefficient eCO_2RR catalyst for FA production, Pd NPs, either dispersed on carbon or embedded in conductive polymers, can exhibit high efficiency and selectivity for the production of FA at minimal overpotentials but only at low j. Min and Kanan showed that modest overpotentials can drive this conversion to high rates with dispersed Pd NPs on a carbon support (Min and Kanan, 2015). Pd/C electrodes could reach geometric CD as high as 10 mA cm^{-2} of formate production at -0.2 V versus RHE with an average FE around 95%. To evaluate the size-dependent activity of Pd NPs towards the production of formate at low η region, three types of Pd NPs (3.8–10.7 nm), capped with polyvinylpyrrolidone (PVP), have been synthesised (Rahaman, Dutta, and Broekmann, 2017). While the FE for 3.8 nm Pd NP exceeded 86% at -0.1 V versus RHE, FE for 6.5 nm was even higher c.98%. However, the FE decreased for larger NPs beyond 6.5 nm. Sargent and co-workers used a field-induced reagent concentration technique to produce formate with a FE $> 90\%$ and an unprecedented geometric CD of 10 mA cm^{-2} at -0.2 V (vs RHE) over the course of 20 h by novel Pd nanoneedle electrocatalysts (Liu et al., 2016). High electric fields were applied locally to concentrate electrolyte cations, which in turn led to a high local CO_2 concentration close to the active eCO_2RR surface.

With substantial research efforts taken in the last decade or so, it has been established that alloys may also be effective catalysts to increase the activity and selectivity of eCO_2RR towards FA production. The Koper group synthesised $Pd_xPt_{(100-x)}$/C NPs which can reduce CO_2 to FA even at a lower onset potential (0 V vs RHE) (Kortlever et al., 2015). The possible influence of varying the Pd to Pt ratio in PdPt/C on catalytic activity was further evaluated. After screening, $Pd_{70}Pt_{30}$/C catalyst showed the maximum catalytic activity toward FA with a FE of 88% at -0.4 V versus RHE, reaching an average CD of $\sim$5 mA cm^{-2}. Sargent and co-workers synthesised sulphur-modulated Sn [Sn(S)] catalyst on Au electrode using atomic layer deposition of SnS_x followed by reduction (Zheng et al., 2017). Sn(S)/Au accelerated eCO_2RR at geometric CD of 55 mA cm^{-2} at -0.75 V (vs RHE) with an FE of 93% and was found to be much superior in formate generation reactivity when compared to Sn NPs/Au (CD = 42 mA cm^{-2} at same potential with a FE of 30.2%). Furthermore, Sn(S)/Au showed excellent stability without deactivation for a period of more than 40 h. The presence of undercoordinated Sn sites due to the incorporation of S-atoms may be responsible for the enhanced activity. The Jiao group reported the synthesis of an Ag–Sn/SnO_x core–shell nanostructure electrocatalyst by first initiating the formation of Sn seeds in the presence of PVP, followed by galvanic displacement of Sn with Ag (Luc et al., 2017). The catalyst with an optimal SnO_x shell thickness of ~1.7 nm, exhibited good selectivity (FE ~80%) and efficiency (PCD ~16 mA cm^{-2} at −0.8 V vs RHE) towards formate production. The bimetallic core with high electronic conductivity and oxygen vacancies on ultrathin partially oxidised shell were found crucial for CO_2 activation and catalytic conversion. Concurrently, Bai et al. developed AC-supported Pd−Sn alloy Pd_xSn/C with a controlled molar ratio of the precursors ($PdCl_2$ and $SnCl_2$) via a modified wetting chemistry reduction method (Bai, 2017). This particular catalyst showed a nearly perfect FE at the lowest overpotential of 0.26 V which outperformed the activity of Ag−Sn core−shell bimetallic catalyst reported by Jiao and co-workers. Ye et al. reconstructed a $Sn_{2.7}Cu$ catalyst with a hierarchical Sn−Cu alloy/Sn core and amorphous SnO_x shell by a facile one-step electrodeposition to achieve a high CD of 406.7 ± 14.4 mA cm^{-2} with C1 FE of 98.0% ± 0.9% at −0.70 V versus RHE, and substantial stability for 40 h at −0.55 V versus RHE (Ye et al., 2020). Formate was the major C1 product while the competitive HER was subdued only because the in situ reconstructed Sn/SnO_x interface facilitated the formate production by optimising the binding of the reaction intermediate HCOO*.

Many studies on Sn-based oxides have been carried out by various research groups in the recent past where tailor-made synthesis with controlled morphologies, structures and chemical compositions have been regulated to further improve the eCO_2RR performance to produce formate/FA exclusively as the only liquid products (Zhao et al., 2019). For example, Chen et al. prepared a thin-film catalyst by simultaneous electrodeposition of Sn^0 and SnO_x on a Ti electrode which exhibited up to seven- to eightfold higher PCD and two- to threefold higher FE for CO_2 conversion to FA than those of an untreated Sn foil electrode (Chen and Kanan, 2012). The Meyer group prepared SnO_2 nanocrystals with a high surface area by a facile hydrothermal method that can selectively reduce CO_2 to formate at η as low as ~340 mV in the aqueous $NaHCO_3$ solution (Zhang, Kang, and Meyer, 2014). It is of particular interest that by controlling the size of SnO_2 NPs (c.5 nm), significant enhancements in both CD and FE could be achieved. The highest activity (FE >93% with 10.2 mA cm^{-2} at −1.8 V vs SCE) and stability (18 h) were observed when the carbon black was replaced by graphene supports with higher surface areas. It is clearly evidenced that the role of SnO_x is paramount in the high-yield production of FA. However, freestanding 2D SnO_x with amorphous structure has never been explored until the Hu group reported the mass-production of amorphous 2D SnO_x nanoflakes with Bi decoration from liquid Sn−Bi alloy by the injection of oxygen and the use of well-selected dispersion non-aqueous solvents (Fig. 8.18A) (Yuan et al., 2020). The as-obtained 2D SnO_x has been evaluated in eCO_2RR with a high FE (>90% at −1.5 V vs RHE) of FA and stable performance over 10 h. To examine the roles of both metal and metal-oxide other than Sn, Gao et al. fabricated two different catalytic sites of four-atom-thick layers: pure Co metal and co-existing domains of Co and CoO (Fig. 8.18B) (Gao, 2016). It was found that the surface Co atoms of the atomically thin layers produced formate during eCO_2RR with a higher intrinsic activity and selectivity at low η than those of surface Co atoms on bulk samples. Partial oxidation of the atomic layers increased their

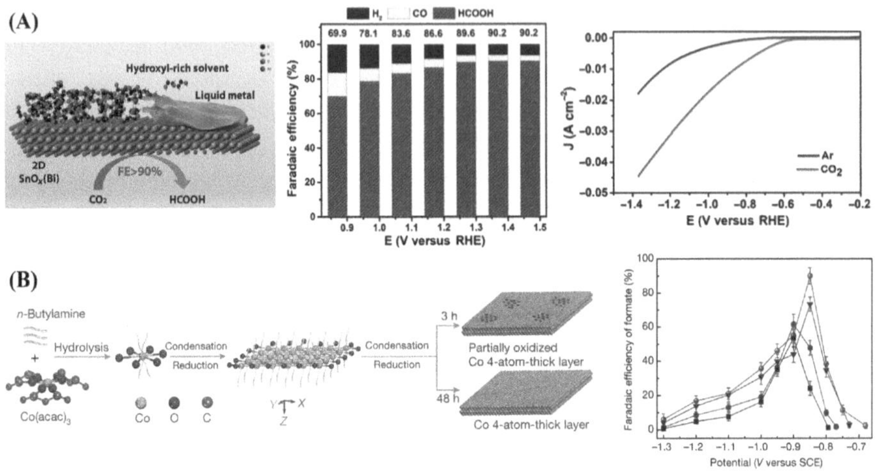

Figure 8.18 (A) Microscopic scheme of the synthetic principle of 2D SnO_x from liquid metals (left), potential dependent FE of products (middle) and the corresponding polarisation curves in CO_2 and Ar atmosphere (right) (Yuan et al., 2020). (B) Schematic formation process of the partially oxidised and pure-Co four-atomic-layer, respectively (left) and FEs of formate at each given potential for 4 h (right) (Gao, 2016). *FE*, Faradaic efficiency.

intrinsic activity by 10 times compared to pure Co four-atom-thick layer, reaching the CD of 10.59 mA cm^{-2} over 40 h, with 90.1% formate selectivity at $\eta = 240$ mV which outperformed many previously reported metal or metal oxide electrodes under comparable conditions. These results strongly suggested that the correct morphology and oxidation state can transform a non-catalytic material into an active catalyst.

8.3.4.2 Molecular complexes

Although previous studies with metals including their alloys and oxides have demonstrated some potentials to electrochemical CO_2-to-FA conversion, their large negative potentials may force them to take a back seat because of the formation of intermediate species such as $CO_2^{\cdot -}$. Particularly in the last decade or so, various highly active transition metal complexes, spanning over first to third row, have been developed to alleviate this barrier. The Yang group has demonstrated that Pt-complexes, $[Pt(dmpe)_2^{+2}]$ [**8.97**, dmpe: 1,2-bis(dimethylphosphino)ethane] and $[Pt(depe)_2^{+2}]$ [**8.98**, depe: 1,2-bis(diethylphosphino)ethane] mediated highly selective electrocatalytic CO_2 reduction to formate through thermodynamic and kinetic control over the metal hydrides (Ceballos and Yang, 2018; Cunningham et al., 2020). These catalysts could reduce CO_2 selectively to formate in acetonitrile with FE > 90% at η < 100 mV with no evidence of catalyst degradation. On the basis of homogeneous CO_2 hydrogenation studies, two different Ir(III)-dihydride pincer complexes (**8.99** and **8.100**) with modified PCP ligand scaffold were reported by

Meyer and co-workers that could be used as electrocatalysts for the selective reduction of CO_2 to FA/formate (Fig. 8.19) (Kang, 2012). (POCOP)IrH_2(MeCN) complex (**8.99**) was shown to be selective in producing formate from MeCN/H_2O (95:5) mixture with no CO formed at -1.45 V versus NHE with a $k_{cat} = 20\ s^{-1}$ and FE = 85%. This active complex was stable but unreactive towards water which was supposed to solvate formate and assist in dissociation from the Ir centre. Another (POC′OP)IrH_2(MeCN) pincer complex (**8.100**) was designed to contain a water solubilizing functional group (Fig. 8.19). Complex **8.100** was shown to selectively reduce CO_2 to formate in the aqueous buffer (pH = 7). Under the neutral pH, complex **8.100** did not significantly catalyse HER (c.7%), making water a solvent for use. Interestingly, the addition of 1% MeCN into the aqueous buffer reduced the overpotential and rendered the catalysis sustainable. The catalyst was 93% efficient for formate production over 24 h during the controlled potential electrolysis (CPE) experiment at -1.41 V versus NHE with CD of 0.6 mA cm^{-2}. Ahn et al. used the Ir(III)-H_3 complex **8.23** for CO_2 reduction to formate in 12% H_2O-MeCN solvent under 1 atm CO_2, with FE >99% and an onset potential at -1.5 V versus Fc/Fc^+ (Ahn, 2015). The observed k_{cat} ($0.56\ s^{-1}$) was markedly lower than that of **8.99** under their respective optimum conditions. Preliminary mechanistic studies indicated that while the formate formation via CO_2 insertion into Ir − H bond was facile, the extrusion of formate was a kinetically difficult step, in turn limiting the performance of the catalyst.

There was a continuous thrust for the development of highly stable first row earth-abundant molecular catalysts for the FA/formate production. One approach put forwarded by the Berben group was to use a metal−metal clusters with delocalised bonding within the metal−core and electron delocalisation extending out to electron-withdrawing CO ligands (Loewen, Neelakantan, and Berben, 2017). The Fe-carbonyl cluster containing interstitial nitride, $[Fe_4N(CO)_{12}]^-$ (**8.101**, Fig. 8.19), has been studied in both MeCN and aqueous solutions (Rail and Berben, 2011; Taheri et al., 2015). This Fe − CO cluster was found to be non-selective in producing formate in the presence of weak acid in MeCN. However, it produced formate selectively over H_2 over the entire pH range of 5−13 in aqueous buffer and also in MeCN with 5% water added. In both aqueous and mixed solvent conditions, **8.101** formed a reduced hydride intermediate through stepwise electron and proton

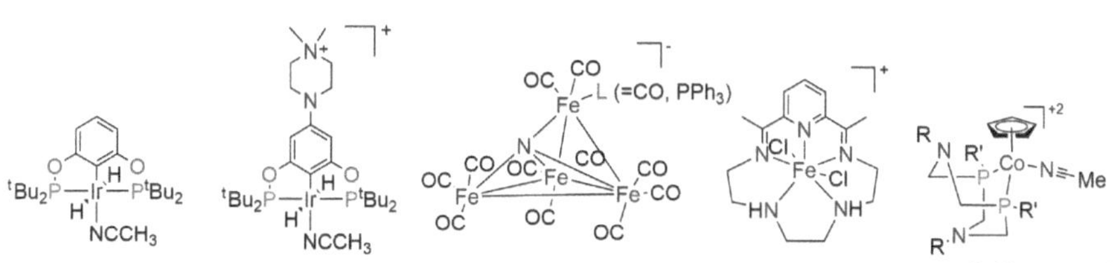

(POCOP)IrH_2(MeCN) 8.99 | (POC'OP)IrH_2(MeCN) 8.100 | $[Fe_4N(CO)_{11}L]^-$ 8.101 | $[FeN_5Cl_2]^+$ 8.103 | $[CpCo(N^R{}_2P^{R'}{}_2)(MeCN)]^{+2}$ 8.104-8.107

Figure 8.19 Transition metal-based molecular electrocatalysts for selective reduction of CO_2 to formate.

transfers. Maximum efficiency was observed in the pH 7 buffer where a CD reached up to 4 mA cm^{-2} with a TOF of 106 ± 6 h^{-1}. Another derivative of this carbonyl cluster was synthesised by replacing one CO-ligand with weaker π-accepting PPh_3 ligand which yielded formate selectively from a CO_2-saturated solution of $MeCN/H_2O$ (95:5) with reduced FE of 61% (Loewen et al., 2017). Robert co-workers synthesised a Co^{II} complex $[CoN_5]^{+2}$ (N_5 = 2,13-dimethyl-3,6,9,12,18-pentaazabicyclo-[12.3.1]octadeca-1(18),2,12,14,16-pentaene) (**8.102**) which when reacted with CO_2, produced CO upon electrochemical reduction in DMF (Chen, 2015). However, replacing the metal centre with Fe (**8.103**) changed the selectivity completely toward FA with no detection of CO (Fig. 8.19). Roy et al. reported a new family of cobalt-based CO_2-reduction catalysts, $[CpCo(N^R{}_2P^{R'}{}_2)L]^{+n}$ (L = I, MeCN; $N^R{}_2P^{R'}{}_2$: 1,5-diaza-3,7-diphosphacyclooctane) bearing four different $N^R{}_2P^{R'}{}_2$ ligands with cyclohexyl or phenyl substituents on phosphorus (P^{Cy} or P^{Ph}) and benzyl or phenyl substituents on nitrogen (N^{Bn} or N^{Ph}) (**8.104**: R = Ph, R' = Cy; **8.105**: R = Ph, R' = Ph; **8.106**: R = Bn, R' = Cy; **8.107**: R = Bn, R' = Ph) (Fig. 8.19) (Roy et al., 2017). These compounds exhibited excellent catalytic activities towards eCO_2RR to FA with remarkably high FEs (90% ± 10%) and good TONs at moderate overpotentials (400 − 600 mV in DMF). Complex **8.106** containing the most electron-donating phosphine and the most basic amine was identified to perform best among the series with a remarkable TOF of >1000 s^{-1}. The pendant amines were shown not to be involved in the direct proton transfer to CO_2 but rather in the stabilisation of key intermediates through hydrogen bonding with water molecules during the hydride transfer.

8.3.4.3 Immobilised complexes/composites/hybrid materials

Enthused by the high activities of molecular catalysts in the homogeneous phase, immobilising such molecular catalysts onto electrode or conductive surfaces either directly via modified linker present in the ligand arms or with the help of any assembly or co-catalysts have been considered as promising approaches to enhance the selectivity and catalyst stability for heterogeneous eCO_2RR in aqueous media. Isaacs et al. electro-polymerised cobalt tetra-amino phthalocyanine (CoTAPc) on a glassy carbon electrode to selectively catalyse the electroreduction of CO_2 to FA at an onset potential of −1.0 V (vs Ag/AgCl) (Isaacs, 2005). Later, Zhao et al. introduced multi-walled carbon nanotubes (MWCNT) into this system to develop a composite of MWCNT and CoTAPc that can form single and/or multiple layers with layer-by-layer assembly technique onto positively charged (with c. PEI) ITO or graphite electrode (GE), marked as ITO-$(MWCNT/CoTAPc)_n$ or GE-$(MWCNT/CoTAPc)_n$ (n = 1−5) (Zhao et al., 2012). This study was aimed to produce FA in a microbial electrolysis cell driven by a microbial fuel cell. The electrochemical results showed that the best performance was obtained for GE-$(MWCNT/CoTAPc)_5$ composite modified electrode where peak potential for CO_2R was observed at −0.5 V (vs Ag/AgCl) with increased CD. After 4 h of electrolysing CO_2 under the peak voltage, only FA was detected in the catholyte with a production rate of 21.0 ± 0.2 $mgL^{-1}h^{-1}$ and a FE of 73.5% ± 0.6% (Fig. 8.20, left).

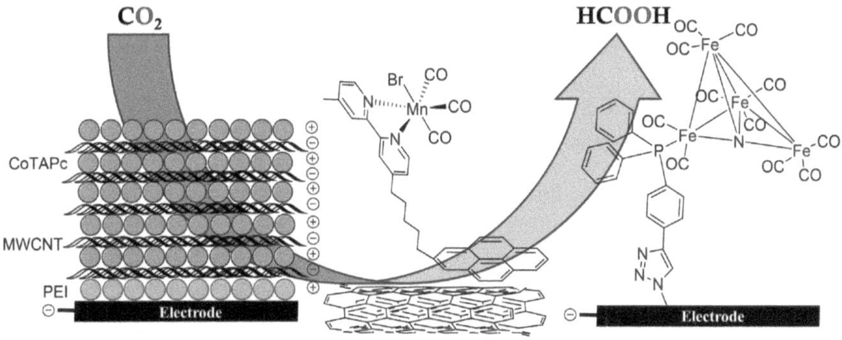

Figure 8.20 Schematics for electrochemical reduction of CO_2 to FA by various catalysts immobilised on electrode surfaces. *FA*, Formic acid.

Compared with the electrode modified by CoTAPc alone, the MWCNT/CoTAPc composite modified electrode could increase the current and FA production rate by approximately 20% and 100% respectively. The Reisner group immobilised [MnBr (2,2′-bpy)$(CO)_3$] complex (**8.62**) to a CNT electrode via an anchoring pyrene unit, protruding from the bpy arm (Reuillard et al., 2017). Unlike previous report by Zhao et al., only single layers were formed where the catalyst concentration could be varied. This Mn-modified CNT electrode allowed eCO_2RR to formate under fully aqueous conditions only when the Mn-catalyst loading was low, whereas the high loading resulted in a higher selectivity of gaseous CO (Fig. 8.20, middle). Despite the low FE, this hybrid electrode displayed high activity with TONs up to 3920 ± 230 for formate at an applied potential of −1.1 V versus SHE ($\eta = 590$ mV). This strategy enabled controlled surface loading of the catalysts that can eventually tune the selectivity of the products. To further increase the stability of catalyst **8.101**, the Berben group has further immobilised alkyne-modified **8.101** onto the azide-terminated GC electrode via covalent attachment using the 'Click chemistry' (Fig. 8.20, right) (Cluff et al., 2019). CPE experiments at −1.25 V versus SCE in pH 9 buffer yielded formate with a FE of 75% ± 20% and a TON of 52,500. Interestingly, the modified electrode was stable for at least 4 days with almost no degradation.

Inspired by the seminal works of molecular catalysts immobilisation on PEI and MWCNT modified electrode surfaces, the Meyer group used PEI as a co-catalyst to enhance the eCO_2RR to formate at N-doped CNT (PEI-NCNT) without using any metal (Zhang et al., 2014). For the electrode fabrication, the homogeneous CNT suspension, obtained by sonication, was first spin-coated on a GC electrode followed by the ammonia plasma treatment (for N-doping) to afford NCNT/GC. A PEI overlayer was further applied to NCNT/GC by dip-coating to finally obtain the desired PEI-NCNT/GC electrode. Although NCNT/GC can act as a selective and robust electrocatalyst for CO_2-to-formate conversion, PEI functionalisation reduced the catalytic overpotential by 200 mV and enhanced both FEs and CDs for the formate production. CPE experiments at −1.8 V versus SCE for 24 h in 0.1 M

$KHCO_3/CO_2$-saturated water resulted in a steady-state CD of 7.2 mA cm^{-2}, comparable to those of Sn catalysts, with a maximum FE value of 85%. This report opens a new avenue for eCO_2RR employing a metal-free co-catalysis strategy.

8.3.4.4 Bio-inspired electrocatalysts

Idealised catalysts can be often found as metalloenzymes in biological systems and CO_2 reduction is not an exception (Amao, 2018; Oliveira et al., 2020). One typical example is the tungsten-containing formate dehydrogenase 1 (*Sf*FDH1) enzyme from *Syntrophobacter fumaroxidans (Sf)*. When adsorbed to a freshly polished pyrolytic graphite edge electrode surface, *Sf*FDH1 catalysed the efficient electrochemical reduction of CO_2 to formate (Reda et al., 2008). Due to several major drawbacks related to the handling and mechanistic studies of W-containing *Sf*FDH1 enzyme, Hirst's and Reisner's groups further investigated the catalytic properties of structurally defined Mo-containing formate dehydrogenase H (*Ec*FDH-H) from the model organism *Escherichia coli* (*Ec*) by adsorbing on a graphite-epoxy electrode (Bassegoda et al., 2014). The protein film voltammetry under mild conditions (in pH 6 at 23°C) was used to demonstrate that *Ec*FDH-H is a highly active and reversible electrocatalyst for producing formate specifically with a FE of close to unity.

Inspired by the metal active sites of FDH (Fig. 8.21A), scientists were intrigued to explore bio-inspired catalysts, with a specific interest for dithiolate-based ligands which are believed to play a pivotal role in enzymatic catalysis. The W centre in the FDH is stabilized by two redox non-innocent pyranopterin dithiolate ligands or molybdopterin cofactor (MPT) (Fig. 8.21B). In the past, several dithiolate-containing W-complexes have been synthesised but none of them showed enzyme-like activity until Kim and co-workers prepared a novel biomimetic $W^{IV} = O$ bis-dithiolene complex (**8.108**, Fig. 8.21C) to reduce gaseous CO_2 to formate at 90°C in MeCN (Seo et al., 2019). While the reported data did not provide any proof for the catalytic formation of formate, this reaction seemed to be stoichiometric. These results provide insights into the importance of the secondary coordination environment as well as synthetic strategies. Recently, the Mougel and the Fontecave groups, in pursuit of mimicking the active site of CO-dehydrogenase (CODH2) (Fig. 8.21D) that commonly catalyses CO oxidation to CO_2, synthesised a heterobimetallic Mo–Cu active site, $[(bdt)Mo^{VI}(O)S_2Cu^{I}CN]^{2-}$ (**8.109**, Fig. 8.21E) where the two ions are bridged by two μ_2-sulphide ligands and the MPT fragment is mimicked by the benzenedithiolate (bdt) ligand chelating the Mo ion (Mouchfiq et al., 2020).

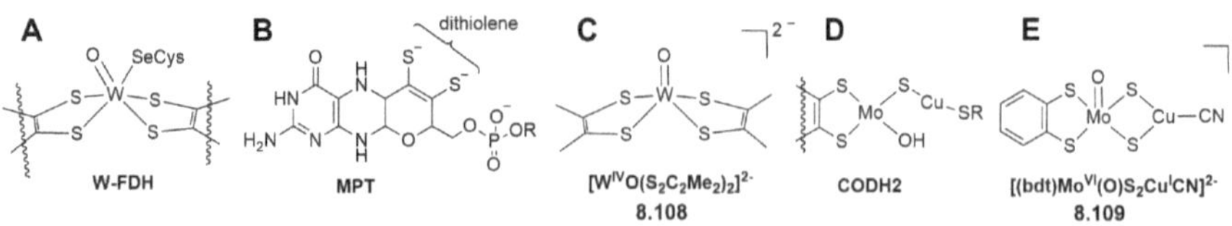

Figure 8.21 Structural representations of bio-inspired electrocatalysts. (A) Active site structure and (B) molybdopterin cofactor of FDH enzymes. (C–E) Structural representations of some bio-inspired synthetic electrocatalysts.

Complex **8.109** was able to reproduce the function of FDH to catalyse CO_2 reduction with a high selectivity for the formate production (FE_{FA} ~75%). Infrared spectroelectrochemical (IR-SEC) studies coupled with DFT computations revealed that the complex is only a pre-catalyst, the active catalyst is generated upon reduction in the presence of CO_2. It was found that the two-electron reduction of complex **8.109** triggers the transfer of the oxo moiety to CO_2 to give CO_3^{2-} and a further one-electron reduction is needed to generate the active catalyst. This study also suggests that the presence of the Cu^I site is essential for the catalytic function, although more efforts are needed to understand the interplay between the two metal centres.

8.4 Cost of formic acid production

FA is currently a high-priced product, with a concentrated, small and mature market along with low risk of substitution. Apart from CO, FA is likely the only profitable product obtained from CO_2 electrolysis at the base case conditions (Jouny, Luc, and Jiao, 2018). FA is usually found in the market at concentrations ranging from 85% to 99%, with 85% being the most common form. In 2014 the price for 85% FA was in the range of \$570−670/ton. Conventional production of FA in industry mainly utilises fossil fuels. Out of several commercial production processes, methyl formate hydrolysis is the dominant, whose capacity is estimated to be ~\$770 kton/year in 2014, covering approximately 90% of the overall installed capacity (Hietala et al., 2000). Although electrocatalytic CO_2 reduction process is in its research phase and farther from its industrial applications, it can become useful and interesting process from the environmental point of view. Unfortunately, at this point it is unclear whether eCO_2RR can make up for the higher energy consumption because of the inherent final low FA concentrations and high η. Researchers performed life cycle assessment to demonstrate the potential of electrocatalytic process over conventional commercial process as well as catalytic conversion of CO_2 (captured directly from the air in a carbon capture unit) and H_2 to FA by evaluating energy cost of electricity and heat (Aldaco, 2019; Rumayor, Dominguez-Ramos, and Irabien, 2018). The electricity consumption was evaluated to be 4.59−11.93 kWh_e/kg_{FA} for eCO_2RR compared to 3.1−4.07 kWh_e/kg_{FA} and 0.29−1.55 kWh_e/kg_{FA} for catalytic hydrogenation and conventional methods respectively. The use of renewable energy could be promising for electrocatalytic process, but the wide range consumption of steam (from 401−25.7 MJ/kg_{FA}), produced by natural gas, in the purification step to increase the final FA concentration poses serious restriction for practical use at this moment. Pérez-Fortes et al. in their policy report provided a techno-economic and financial analyses on the feasibility of FA production from a carbon capture and utilisation (CCU) plant to estimate the production cost and economic viability and compared to conventional methyl formate hydrolysis process (Pérez-Fortes and Tzimas, 2016). Their analyses proposed the costs of \$1700/ton_{FA}, and ~\$530/ton_{FA} for FA production from CCU and from the conventional plant, respectively. FA production from CCU can become profitable if further development in the CCU and energy generation sectors including heavy and chemical industries are undertaken.

According to a recent analysis by Haperen, ~4 MW_e is required to synthesise 1 ton FA/h via electrocatalytic reduction of CO_2 of FA (24 tons FA/day and 8600 tons FA/year) (Harepen, 2016). A possible configuration would involve 10 electrocatalytic units each of 0.4 MW with a capacity of 0.1 ton FA/h. For a well-established H_2 production electrolyser of 1 MW, the price is ~\$1.2 million. Owing to a more complex reactor for CO_2-to-FA conversion than H_2O electrolysis to produce H_2, 25% higher price has been estimated. Hence, for 10 electrolysers with 0.4 MW power, the total estimated cost will be approx. \$6 million. This report estimated a total investment or capital expenditure (CAPEX) of ~\$7 million if proper handling, storage, and separation unit are being considered. Evidently, for eCO_2RR to FA, the prices of CO_2 and electricity are one of the important factors that also depend on a range of parameters, including location, energy source, CCU capacity, consumption rate and scale etc. Current market prices of electricity and CO_2 fall in the range of \$30–40 MWh and \$40–70/ton_{CO2} respectively (Jouny et al., 2018). Different approaches were adopted to predict a more accurate estimate for the FA production cost, including discounted cash flow, internal rate of return and net present value. While there is some uncertainty in the CAPEX due to unavailability of commercial electrolysers for CO_2 reduction to FA, operational cost (OPEX) can be well-defined (\$170–330/ton) considering the electricity and CO_2 costs. A simplified relation between the CAPEX and OPEX [Eq. (8.5)] is commonly used to estimate the total cost for FA (Douglas, 1988):

$$\mathrm{OPEX} = 0.15 \times \mathrm{CAPEX} \tag{8.5}$$

Assuming an average lifetime of around 15 years for the major equipment, the CAPEX/year can be related to the OPEX, according to Eq. (8.6):

$$\mathrm{CAPEX/year} = \mathrm{OPEX}/(0.15 \times 15) \tag{8.6}$$

Depending on the different ranges of electricity and CO_2 prices, the production costs of FA through electrochemical conversion of CO_2 are estimated under numerous scenarios, including a future scenario where the CO_2 price/ton is significantly below the current price and seemed to be competitive with existing technology (Table 8.3). In the coming few years, substantial changes in policies related to anthropogenic CO_2 emission are to be expected with the implementation of more stringent regulations that can reduce the CO_2 price significantly.

8.5 Concluding remarks

FA is in principle easy to decompose back to H_2 and CO_2 on demand. Thus FA may serve as a promising candidate as a H_2 energy carrier with a short path to

Table 8.3 Effect of different price scenarios on the production cost of 1 ton formic acid (FA).

Scenario	Electricity price (US$/MWh)	CO_2 price (US$/ton-$CO_2$)	FA price in US$/ton OPEX[a]/total
High electricity, high CO_2	70	80	358/464
High electricity, low CO_2	70	50	330/436
Low electricity, high CO_2	30	80	198/304
Low electricity, low CO_2	30	50	170/276
Low electricity, low CO_2[b]	30	25	146/252
Low electricity, low CO_2[b]	30	0	122/228

[a]Future scenarios where CO_2 capture is regulated and possibly even taxed.
[b]OPEX includes only the production cost, which is given by the electricity price, and no handling costs are included.

market. On the basis of its physical and chemical properties, existing infrastructures for fossil fuel distribution may be adapted by employing some proper anti-corrosiveness design (Eppinger and Huang, 2017). Moreover, FA-based electricity generation systems have been successfully developed (EPFL, 2018; Huang and Zheng, 2019; Van Putten et al., 2019), and 45% overall energy conversion efficiency has been demonstrated. FA will very likely play an important role in the hydrogen energy future once the practical solutions for FA synthesis from low-carbon energy, power to FA, are established. The influences of many factors, including the choice and design of catalyst, solvent, electrolyte, base, pressure, support, reactor etc. on the efficiency of the CO_2 reduction process have been researched. While CO_2 to FA may appear to a relatively simple $2e^-$ reduction process, there are still several challenges to overcome before practical applications can be realized. A remarkable example of pure FA production from eCO_2RR was achieved with an innovative cell design (Xia et al., 2019). However, R&D resources have to be continuously invested to identify a scalable process with high efficiency. With the current worldwide efforts in this direction, FA economy may be realized, hopefully, in a not too distanced future.

References

Ahn, S. T., et al. (2015). Enhanced CO_2 electroreduction efficiency through secondary coordination effects on a pincer iridium catalyst. *Chemical Communications*, *51*(27), 5947–5950. Available from https://doi.org/10.1039/c5cc00458f.

Aldaco, R., et al. (2019). Bringing value to the chemical industry from capture, storage and use of CO_2: A dynamic LCA of formic acid production. *Science of the Total Environment*, *663*, 738–753. Available from https://doi.org/10.1016/j.scitotenv.2019.01.395.

Amao, Y. (2018). Formate dehydrogenase for CO_2 utilization and its application. *Journal of CO_2 Utilization*, *26*, 623–641. Available from https://doi.org/10.1016/j.jcou.2018.06.022.
An, B., et al. (2017). Molecular iridium complexes in metal-organic frameworks catalyze CO_2 hydrogenation via concerted proton and hydride transfer. *Journal of the American Chemical Society*, *139*(49), 17747–17750. Available from https://doi.org/10.1021/jacs.7b10922.
An, X., et al. (2019). Bi-doped SnO nanosheets supported on Cu foam for electrochemical reduction of CO_2 to HCOOH. *ACS Applied Materials and Interfaces*, *11*(45), 42114–42122. Available from https://doi.org/10.1021/acsami.9b13270.
Badiei, Y. M., et al. (2013). Cp*Co(III) catalysts with proton-responsive ligands for carbon dioxide hydrogenation in aqueous media. *Inorganic Chemistry*, *52*(21), 12576–12586. Available from https://doi.org/10.1021/ic401707u.
Bai, X., et al. (2017). Exclusive formation of formic acid from CO_2 electroreduction by a tunable Pd-Sn alloy. *Angewandte Chemie - International Edition*, *56*(40), 12219–12223. Available from https://doi.org/10.1002/anie.201707098.
Bassegoda, A., et al. (2014). Reversible interconversion of CO_2 and formate by a molybdenum-containing formate dehydrogenase. *Journal of the American Chemical Society*, *136*(44), 15473–15476. Available from https://doi.org/10.1021/ja508647u.
Bertini, F., et al. (2016). Efficient and mild carbon dioxide hydrogenation to formate catalyzed by Fe(II) hydrido carbonyl complexes bearing 2,6-(diaminopyridyl)diphosphine pincer ligands. *ACS Catalysis*, *6*(5), 2889–2893. Available from https://doi.org/10.1021/acscatal.6b00416.
Bertini, F., et al. (2017). Carbon dioxide hydrogenation catalysed by well-defined Mn(i) PNP pincer hydride complexes. *Chemical Science*, *8*(7), 5024–5029. Available from https://doi.org/10.1039/c7sc00209b.
Bulgarian Ministry of Environment and Water (2017) *Risk management option analysis conclusion document*, (Bulgarian).
Bulushev, D. A., and Ross, J. R. H. (2018). Heterogeneous catalysts for hydrogenation of CO_2 and bicarbonates to formic acid and formates. *Catalysis Reviews - Science and Engineering*, *60*(4), 566–593. Available from https://doi.org/10.1080/01614940.2018.1476806.
Bumroongsakulsawat, P., and Kelsall, G. H. (2014). Effect of solution pH on CO: Formate formation rates during electrochemical reduction of aqueous CO_2 at Sn cathodes. *Electrochimica Acta*, *141*, 216–225. Available from https://doi.org/10.1016/j.electacta.2014.07.057.
Ceballos, B. M., and Yang, J. Y. (2018). Directing the reactivity of metal hydrides for selective CO_2 reduction. *Proceedings of the National Academy of Sciences of the United States of America*, *115*(50), 12686–12691. Available from https://doi.org/10.1073/pnas.1811396115.
Chatterjee, S., et al. (2021). Enabling storage and utilization of low-carbon electricity: power to formic acid. *Energy Environmental Science*, *14*, 1194–1246. Available from https://doi.org/10.1039/D0EE03011B.
Chen, L., et al. (2015). Molecular catalysis of the electrochemical and photochemical reduction of CO_2 with earth-abundant metal complexes. selective production of CO vs HCOOH by switching of the metal center. *Journal of the American Chemical Society*, *137*(34), 10918–10921. Available from https://doi.org/10.1021/jacs.5b06535.
Chen, Y., and Kanan, M. W. (2012). Tin oxide dependence of the CO_2 reduction efficiency on tin electrodes and enhanced activity for tin/tin oxide thin-film catalysts. *Journal of the American Chemical Society*, *134*(4), 1986–1989. Available from https://doi.org/10.1021/ja2108799.

Cluff, D. B., et al. (2019). Electrocatalytic reduction of CO_2 into formate with glassy carbon modified by $[Fe_4N(CO)_{11}(PPh_2Ph\text{-linker})]^-$. *Organometallics*, *38*(6), 1230–1235. Available from https://doi.org/10.1021/acs.organomet.8b00396.

Cunningham, D. W., et al. (2020). Reversible and selective CO_2 to HCO_2^- electrocatalysis near the thermodynamic potential. *Angewandte Chemie - International Edition*, *59*(11), 4443–4447. Available from https://doi.org/10.1002/anie.201913198.

Del Castillo, A., et al. (2015). Electrocatalytic reduction of CO2 to formate using particulate Sn electrodes: Effect of metal loading and particle size. *Applied Energy*, *157*, 165–173. Available from https://doi.org/10.1016/j.apenergy.2015.08.012.

Dibenedetto, A., et al. (2011). RuII-mediated hydrogen transfer from aqueous glycerol to CO_2: From waste to value-added products. *ChemSusChem*, *4*(9), 1311–1315. Available from https://doi.org/10.1002/cssc.201000434.

Douglas, J. M. (1988). *Conceptual design of chemical processes*. McGraw-Hill.

Dubey, A., et al. (2017). Bio-inspired Mn(I) complexes for the hydrogenation of CO_2 to formate and formamide. *ACS Catalysis*, *7*(6), 3864–3868. Available from https://doi.org/10.1021/acscatal.7b00943.

EPFL (2018) *The world's first formic acid-based fuel cell*, [Online]. Available at https://actu.epfl.ch/news/the-world-s-first-formic-acid-based-fuel-cell/ Accessed 15.08.20.

Eppinger, J., and Huang, K.-W. (2017). Formic acid as a hydrogen energy carrier. *ACS Energy Letters*, *2*(1), 188–195. Available from https://doi.org/10.1021/acsenergylett.6b00574.

Federsel, C., et al. (2012). Catalytic hydrogenation of carbon dioxide and bicarbonates with a well-defined cobalt dihydrogen complex. *Chemistry - A European Journal*, *18*(1), 72–75. Available from https://doi.org/10.1002/chem.201101343.

Filonenko, G. A., Hensen, E. J. M., and Pidko, E. A. (2014). Mechanism of CO_2 hydrogenation to formates by homogeneous Ru-PNP pincer catalyst: from a theoretical description to performance optimization. *Catalysis Science and Technology*, *4*(10), 3474–3485. Available from https://doi.org/10.1039/c4cy00568f.

Filonenko, G. A., et al. (2013). The impact of metal-ligand cooperation in hydrogenation of carbon dioxide catalyzed by ruthenium PNP pincer. *ACS Catalysis*, *3*(11), 2522–2526. Available from https://doi.org/10.1021/cs4006869.

Filonenko, G. A., et al. (2014). Highly efficient reversible hydrogenation of carbon dioxide to formates using a ruthenium PNP-pincer catalyst. *ChemCatChem*, *6*(6), 1526–1530. Available from https://doi.org/10.1002/cctc.201402119.

Filonenko, G. A., et al. (2016). On the activity of supported Au catalysts in the liquid phase hydrogenation of CO_2 to formates. *Journal of Catalysis*, *343*, 97–105. Available from https://doi.org/10.1016/j.jcat.2015.10.002.

Gao, S., et al. (2016). Partially oxidized atomic cobalt layers for carbon dioxide electroreduction to liquid fuel. *Nature*, *529*(7584), 68–71. Available from https://doi.org/10.1038/nature16455.

Gassner, F., and Leitner, W. (1993). Hydrogenation of carbon dioxide to formic acid using water-soluble rhodium catalysts. *Journal of the Chemical Society, Chemical Communications* (19), 1465–1466. Available from https://doi.org/10.1039/C39930001465.

Graf, E., and Leitner, W. (1992). Direct formation of formic acid from carbon dioxide and dihydrogen using the $[\{Rh(cod)Cl\}_2]\text{-}Ph_2P(CH_2)_4PPh_2$ catalyst system. *Journal of the Chemical Society, Chemical Communications* (8), 623–624. Available from https://doi.org/10.1039/C39920000623.

Guan, C., et al. (2018). Conversion of CO_2 from air into formate using amines and phosphorus-nitrogen PN^3P-Ru(ii) pincer complexes. *Green Chemistry*, *20*(18), 4201–4205. Available from https://doi.org/10.1039/c8gc02186d.

Gunasekar, G. H., Jung, K. D., and Yoon, S. (2019). Hydrogenation of CO_2 to formate using a simple, recyclable, and efficient heterogeneous catalyst. *Inorganic Chemistry*, *58*(6), 3717–3723. Available from https://doi.org/10.1021/acs.inorgchem.8b03336.

Gunasekar, G. H., et al. (2018). Design strategy toward recyclable and highly efficient heterogeneous catalysts for the hydrogenation of CO_2 to formate. *ACS Catalysis*, *8*(5), 4346–4353. Available from https://doi.org/10.1021/acscatal.8b00392.

Hao, C., et al. (2011). Hydrogenation of CO_2 to formic acid on supported ruthenium catalysts. *Catalysis Today*, *160*(1), 184–190. Available from https://doi.org/10.1016/j.cattod.2010.05.034.

Hara, K., Kudo, A., and Sakata, T. (1995). Electrochemical reduction of carbon dioxide under high pressure on various electrodes in an aqueous electrolyte. *Journal of Electroanalytical Chemistry*, *391*(1–2), 141–147. Available from https://doi.org/10.1016/0022-0728(95)03935-A.

Harepen, R. (2016) *Formic acid as energy carrier*, [Online]. Available at https://projecten.topsectorenergie.nl/storage/app/uploads/public/5b1/551/e4f/5b1551e4f1a85026449477.pdf Accessed 15.08.20.

Hayashi, H., Ogo, S., and Fukuzumi, S. (2004). Aqueous hydrogenation of carbon dioxide catalysed by water-soluble ruthenium aqua complexes under acidic conditions. *Chemical Communications* (23), 2714–2715. Available from https://doi.org/10.1039/b411633j.

He, C. S., et al. (2017). Highly selective hydrogenation of CO_2 into formic acid on a nano-Ni catalyst at ambient temperature: Process, mechanisms and catalyst stability. *Journal of CO2 Utilization*, *19*, 157–164. Available from https://doi.org/10.1016/j.jcou.2017.03.012.

HietalaVouri, J. A., et al. (2000). *Formic acid. Ullmann's encyclopedia of industrial chemistry* (pp. 1–22). Weinheim, Germany: Wiley-VCH.

Himeda, Y., et al. (2004). Half-sandwich complexes with 4,7-dihydroxy-1,10-phenanthroline: Water-soluble, highly efficient catalysts for hydrogenation of bicarbonate attributable to the generation of an oxyanion on the catalyst ligand. *Organometallics*, *23*(7), 1480–1483. Available from https://doi.org/10.1021/om030382s.

Himeda, Y., et al. (2005). Recyclable catalyst for conversion of carbon dioxide into formate attributable to an oxyanion on the catalyst ligand. *Journal of the American Chemical Society*, *127*(38), 13118–13119. Available from https://doi.org/10.1021/ja054236k.

Hori, Y. (2008). Electrochemical CO_2 reduction on metal electrodes. In C. G. Vayenas, R. E. White, and M. E. Gamboa-Aldeco (Eds.), *Modern aspects of electrochemistry* (pp. 89–189). New York: Springer. Available from https://doi.org/10.1007/978-0-387-49489-0_3.

Huang, K.-W. and Zheng, J. (2019) Electricity generation devices using formic acid. EP3391447B1.

Huang, K.-W., et al. (2013) Phospho-amino pincer-type ligands and catalytic metal complexes thereof. United states patent application US8598351B2.

Huff, C. A., and Sanford, M. S. (2013). Catalytic CO_2 hydrogenation to formate by a ruthenium pincer complex. *ACS Catalysis*, *3*(10), 2412–2416. Available from https://doi.org/10.1021/cs400609u.

Hull, J. F., et al. (2012). Reversible hydrogen storage using CO_2 and a proton-switchable iridium catalyst in aqueous media under mild temperatures and pressures. *Nature Chemistry*, *4*(5), 383–388. Available from https://doi.org/10.1038/nchem.1295.

Inoue, Y., et al. (1976). Catalytic fixation of carbon dioxide to formic acid by transition-metal complexes under mild conditions. *Chemistry Letters*, *5*(8), 863–864. Available from https://doi.org/10.1246/cl.1976.863.

Isaacs, M., et al. (2005). Electrochemical reduction of CO_2 mediated by poly-M-aminophthalocyanines (M = Co, Ni, Fe): Poly-Co-tetraaminophthalocyanine, a selective catalyst. *Journal of Molecular Catalysis A: Chemical*, *229*(1–2), 249–257. Available from https://doi.org/10.1016/j.molcata.2004.11.026.

Jaleel, A., et al. (2020). Hydrogenation of CO_2 to formates on ruthenium(III) coordinated on melamine polymer network. *Journal of CO2 Utilization*, *35*, 245–255. Available from https://doi.org/10.1016/j.jcou.2019.10.003.

Jessop, P. G. (2007). Homogeneous Hydrogenation of carbon dioxide. In J. G. De Vries, and C. J. Elsevier (Eds.), *The handbook of homogeneous hydrogenation* (pp. 489–511). WILEY-VCH Verlag GmbH & Co. Available from https://doi.org/10.1002/9783527619382.ch17.

Jessop, P. G., Ikariya, T., and Noyori, R. (1994). Homogeneous catalytic hydrogenation of supercritical carbon dioxide. *Nature*, *368*(6468), 231–233. Available from https://doi.org/10.1038/368231a0.

Jitaru, M., et al. (1997). Electrochemical reduction of carbon dioxide on flat metallic cathodes. *Journal of Applied Electrochemistry*, *27*(8), 875–889. Available from https://doi.org/10.1023/A:1018441316386.

Joó, F., et al. (1999). Homogeneous hydrogenation of aqueous hydrogen carbonate to formate under exceedingly mild conditions - A novel possibility of carbon dioxide activation. *Chemical Communications* (11), 971–972. Available from https://doi.org/10.1039/a902368b.

Joó, F., et al. (2000). Homogeneous hydrogenation of aqueous hydrogen carbonate to formate under mild conditions with water soluble rhodium(I)- and ruthenium(II)-phosphine catalysts. *Applied Organometallic Chemistry*, *14*(12), 857–859. Available from https://doi.org/10.1002/1099-0739(200012)14:12<857::AID-AOC86>3.0.CO;2-9.

Jouny, M., Luc, W., and Jiao, F. (2018). General techno-economic analysis of CO_2 electrolysis systems. *Industrial and Engineering Chemistry Research*, *57*(6), 2165–2177. Available from https://doi.org/10.1021/acs.iecr.7b03514.

Kaneco, S., et al. (1998). Electrochemical conversion of carbon dioxide to formic acid on Pb in KOH/methanol electrolyte at ambient temperature and pressure. *Energy*, *23*(12), 1107–1112. Available from https://doi.org/10.1016/S0360-5442(98)00054-1.

Kang, P., et al. (2012). Selective electrocatalytic reduction of CO_2 to formate by water-stable iridium dihydride pincer complexes. *Journal of the American Chemical Society*, *134*(12), 5500–5503. Available from https://doi.org/10.1021/ja300543s.

Kann, A., et al. (2018). Hydrogenation of CO_2 to formate over ruthenium immobilized on solid molecular phosphines. *ChemSusChem*, *11*(11), 1857–1865. Available from https://doi.org/10.1002/cssc.201800413.

Khan, M. M. T., Halligudi, S. B., and Shukla, S. (1989). Reduction of CO_2 by molecular hydrogen to formic acid and formaldehyde and their decomposition to CO and H_2O. *Journal of Molecular Catalysis*, *57*(1), 47–60. Available from https://doi.org/10.1016/0304-5102(89)80126-9.

Köleli, F., et al. (2003). Electrochemical reduction of CO_2 at Pb- and Sn-electrodes in a fixed-bed reactor in aqueous K_2CO_3 and $KHCO_3$ media. *Journal of Applied Electrochemistry*, *33*(5), 447–450. Available from https://doi.org/10.1023/A:1024471513136.

Kopljar, D., et al. (2014). Electrochemical reduction of CO_2 to formate at high current density using gas diffusion electrodes. *Journal of Applied Electrochemistry*, *44*(10), 1107–1116. Available from https://doi.org/10.1007/s10800-014-0731-x.

Kortlever, R., et al. (2015). Electrochemical CO_2 reduction to formic acid at low overpotential and with high faradaic efficiency on carbon-supported bimetallic Pd-Pt nanoparticles. *ACS Catalysis*, *5*(7), 3916–3923. Available from https://doi.org/10.1021/acscatal.5b00602.

Kothandaraman, J., et al. (2015). Amine-free reversible hydrogen storage in formate salts catalyzed by ruthenium pincer complex without pH control or solvent change. *ChemSusChem, 8* (8), 1442–1451. Available from https://doi.org/10.1002/cssc.201403458.

Kothandaraman, J., et al. (2016). CO_2 capture by amines in aqueous media and its subsequent conversion to formate with reusable ruthenium and iron catalysts. *Green Chemistry, 18* (21), 5831–5838. Available from https://doi.org/10.1039/c6gc01165a.

Langer, R., et al. (2011). Low-pressure hydrogenation of carbon dioxide catalyzed by an iron pincer complex exhibiting noble metal activity. *Angewandte Chemie - International Edition, 50*(42), 9948–9952. Available from https://doi.org/10.1002/anie.201104542.

Lau, C. P., and Chen, Y. Z. (1995). Hydrogenation of carbon dioxide to formic acid using a 6,6′-dichloro-2,2′-bipyridine complex of ruthenium,cis-$[Ru(6,6'-Cl_2bpy)_2(H_2O)_2](CF_3SO_3)_2$. *Journal of Molecular Catalysis. A, Chemical, 101*(1), 33–36. Available from https://doi.org/10.1016/1381-1169(95)00068-2.

Lee, C. H., and Kanan, M. W. (2015). Controlling H^+ vs CO_2 reduction selectivity on Pb electrodes. *ACS Catalysis, 5*(1), 465–469. Available from https://doi.org/10.1021/cs5017672.

Lee, J. H., et al. (2014). Carbon dioxide mediated, reversible chemical hydrogen storage using a Pd nanocatalyst supported on mesoporous graphitic carbon nitride. *Journal of Materials Chemistry A, 2*(25), 9490–9495. Available from https://doi.org/10.1039/c4ta01133c.

Li, H., Zheng, B., and Huang, K.-W. (2015). A new class of PN^3-pincer ligands for metal-ligand cooperative catalysis. *Coordination Chemistry Reviews, 293–294*, 116–138. Available from https://doi.org/10.1016/j.ccr.2014.11.010.

Li, H., et al. (2019). PN^3(P)-pincer complexes: Cooperative catalysis and beyond. *ACS Catalysis, 9*(3), 1619–1629. Available from https://doi.org/10.1021/acscatal.8b04495.

Li, J., et al. (2020). Electroreduction of CO_2 to formate on a copper-based electrocatalyst at high pressures with high energy conversion efficiency. *Journal of the American Chemical Society, 142*(16), 7276–7282. Available from https://doi.org/10.1021/jacs.0c00122.

Liu, M., et al. (2016). Enhanced electrocatalytic CO_2 reduction via field-induced reagent concentration. *Nature, 537*(7620), 382–386. Available from https://doi.org/10.1038/nature19060.

Liu, N., et al. (2014). The influence of preparation procedures on hydrogenation CO_2 to formic acid over supported Ru catalysts. *Advanced Materials Research, 881–883*, 283–286. Available from https://doi.org/10.4028/www.scientific.net/AMR.881-883.283.

Liu, N., Du, R. J., and Li, W. (2013). Hydrogenation CO_2 to formic acid over Ru supported on γ-Al_2O_3 nanorods. *Advanced Materials Research, 821–822*, 1330–1335. Available from https://doi.org/10.4028/www.scientific.net/AMR.821-822.1330.

Liu, Q., et al. (2017). Direct catalytic hydrogenation of CO_2 to formate over a Schiff-base-mediated gold nanocatalyst. *Nature Communications, 8*(1), 1407. Available from https://doi.org/10.1038/s41467-017-01673-3.

Loewen, N. D., Neelakantan, T. V., and Berben, L. A. (2017). Renewable formate from C-H bond formation with CO_2: Using iron carbonyl clusters as electrocatalysts. *Accounts of Chemical Research, 50*(9), 2362–2370. Available from https://doi.org/10.1021/acs.accounts.7b00302.

Luc, W., et al. (2017). Ag-Sn bimetallic catalyst with a core-shell structure for CO_2 reduction. *Journal of the American Chemical Society, 139*(5), 1885–1893. Available from https://doi.org/10.1021/jacs.6b10435.

McNamara, N. D., and Hicks, J. C. (2014). CO_2 capture and conversion with a multifunctional polyethyleneimine-tethered iminophosphine iridium catalyst/adsorbent. *ChemSusChem, 7* (4), 1114–1124. Available from https://doi.org/10.1002/cssc.201301231.

Min, S., et al. (2016). Low overpotential and high current CO_2 reduction with surface reconstructed Cu foam electrodes. *Nano Energy*, *27*, 121–129. Available from https://doi.org/10.1016/j.nanoen.2016.06.043.

Min, X., and Kanan, M. W. (2015). Pd-catalyzed electrohydrogenation of carbon dioxide to formate: High mass activity at low overpotential and identification of the deactivation pathway. *Journal of the American Chemical Society*, *137*(14), 4701–4708. Available from https://doi.org/10.1021/ja511890h.

Moret, S., Dyson, P. J., and Laurenczy, G. (2014). Direct synthesis of formic acid from carbon dioxide by hydrogenation in acidic media. *Nature Communications*, *5*(1), 4017. Available from https://doi.org/10.1038/ncomms5017.

Mori, K., Taga, T., and Yamashita, H. (2017). Isolated Single-atomic Ru catalyst bound on a layered double hydroxide for hydrogenation of CO_2 to formic acid. *ACS Catalysis*, *7*(5), 3147–3151. Available from https://doi.org/10.1021/acscatal.7b00312.

Morrison, A. R. T., et al. (2019). Modeling the electrochemical conversion of carbon dioxide to formic acid or formate at elevated pressures. *Journal of the Electrochemical Society*, *166*(4), E77–E86. Available from https://doi.org/10.1149/2.0121904jes.

Mouchfiq, A., et al. (2020). A bioinspired molybdenum-copper molecular catalyst for CO_2 electroreduction. *Chemical Science*, *11*(21), 5503–5510. Available from https://doi.org/10.1039/d0sc01045f.

Nguyen, L. T. M., et al. (2015). Catalytic CO_2 hydrogenation to formic acid over carbon nanotube-graphene supported PdNi alloy catalysts. *RSC Advances*, *5*(128), 105560–105566. Available from https://doi.org/10.1039/c5ra21017h.

Oliveira, A. R., et al. (2020). Toward the mechanistic understanding of enzymatic CO_2 reduction. *ACS Catalysis*, *10*(6), 3844–3856. Available from https://doi.org/10.1021/acscatal.0c00086.

Paik, W., Andersen, T. N., and Eyring, H. (1969). Kinetic studies of the electrolytic reduction of carbon dioxide on the mercury electrode. *Electrochimica Acta*, *14*(12), 1217–1232. Available from https://doi.org/10.1016/0013-4686(69)87019-2.

Pan, Y., et al. (2019). CO_2 hydrogenation by phosphorus-nitrogen PN^3P-pincer iridium hydride complexes: Elucidation of the deactivation pathway. *Dalton Transactions*, *48*(34), 12812–12816. Available from https://doi.org/10.1039/c9dt01319a.

Park, K., et al. (2015). A highly efficient heterogenized iridium complex for the catalytic hydrogenation of carbon dioxide to formate. *ChemSusChem*, *8*(20), 3410–3413. Available from https://doi.org/10.1002/cssc.201500436.

Pérez-Fortes, M. and Tzimas, E. (2016) *Techno-economic and environmental evaluation of carbon dioxide utilisation for fuel production. Synthesis of methanol and formic acid. EUR 27629 EN*. Available from https://doi.org/10.2790/981669.

PHE Centre for Radiation (2015) *Chemical and Environmental Hazards: Formic Acid Incident Management*.

Podlovchenko, B. I., Kolyadko, E. A., and Lu, S. (1994). Electroreduction of carbon dioxide on palladium electrodes at potentials higher than the reversible hydrogen potential. *Journal of Electroanalytical Chemistry*, *373*(1–2), 185–187. Available from https://doi.org/10.1016/0022-0728(94)03324-2.

Preti, D., Squarcialupi, S., and Fachinetti, G. (2010). Production of $HCOOH/NEt_3$ adducts by CO_2/H_2 incorporation into Neat NEt_3. *Angewandte Chemie - International Edition*, *49*(14), 2581–2584. Available from https://doi.org/10.1002/anie.200906054.

Preti, D., et al. (2011). Carbon dioxide hydrogenation to formic acid by using a heterogeneous gold catalyst. *Angewandte Chemie - International Edition*, *50*(52), 12551–12554. Available from https://doi.org/10.1002/anie.201105481.

Proietto, F., et al. (2018). Electrochemical conversion of CO_2 to HCOOH at tin cathode in a pressurized undivided filter-press cell. *Electrochimica Acta*, *277*, 30–40. Available from https://doi.org/10.1016/j.electacta.2018.04.159.

Rahaman, M., Dutta, A., and Broekmann, P. (2017). Size-dependent activity of palladium nanoparticles: Efficient conversion of CO_2 into formate at low overpotentials. *ChemSusChem*, *10*(8), 1733–1741. Available from https://doi.org/10.1002/cssc.201601778.

Rail, M. D., and Berben, L. A. (2011). Directing the reactivity of $[HFe_4N(CO)_{12}]^-$ toward H^+ or CO_2 reduction by understanding the electrocatalytic mechanism. *Journal of the American Chemical Society*, *133*(46), 18577–18579. Available from https://doi.org/10.1021/ja208312t.

Reda, T., et al. (2008). Reversible interconversion of carbon dioxide and formate by an electroactive enzyme. *Proceedings of the National Academy of Sciences of the United States of America*, *105*(31), 10654–10658. Available from https://doi.org/10.1073/pnas.0801290105.

Reuillard, B., et al. (2017). Tuning product selectivity for aqueous CO_2 reduction with a Mn (bipyridine)-pyrene catalyst immobilized on a carbon nanotube electrode. *Journal of the American Chemical Society*, *139*(41), 14425–14435. Available from https://doi.org/10.1021/jacs.7b06269.

Rohmann, K., et al. (2016). Hydrogenation of CO_2 to formic acid with a highly active ruthenium acriphos complex in DMSO and DMSO/water. *Angewandte Chemie - International Edition*, *55*(31), 8966–8969. Available from https://doi.org/10.1002/anie.201603878.

Roy, S., et al. (2017). Molecular cobalt complexes with pendant amines for selective electrocatalytic reduction of carbon dioxide to formic acid. *Journal of the American Chemical Society*, *139*(10), 3685–3696. Available from https://doi.org/10.1021/jacs.6b11474.

Rumayor, M., Dominguez-Ramos, A., and Irabien, A. (2018). Formic acid manufacture: Carbon dioxide utilization alternatives. *Applied Sciences*, *8*(6), 914. Available from https://doi.org/10.3390/app8060914.

Sanz, S., Azua, A., and Peris, E. (2010). "(η^6-arene)Ru(bis-NHC)" complexes for the reduction of CO_2 to formate with hydrogen and by transfer hydrogenation with *i*PrOH. *Dalton Transactions*, *39*(27), 6339–6343. Available from https://doi.org/10.1039/c003220d.

Sanz, S., Benítez, M., and Peris, E. (2010). A new approach to the reduction of carbon dioxide: CO_2 reduction to formate by transfer hydrogenation in *i*PrOH. *Organometallics*, *29* (1), 275–277. Available from https://doi.org/10.1021/om900820x.

Schaub, T., and Paciello, R. A. (2011). A process for the synthesis of formic acid by CO_2 hydrogenation: Thermodynamic aspects and the role of CO. *Angewandte Chemie - International Edition*, *50*(32), 7278–7282. Available from https://doi.org/10.1002/anie.201101292.

Schmeier, T. J., et al. (2011). Secondary coordination sphere interactions facilitate the insertion step in an iridium(III) CO_2 reduction catalyst. *Journal of the American Chemical Society*, *133*(24), 9274–9277. Available from https://doi.org/10.1021/ja2035514.

Seo, J., et al. (2019). Reactivity of a biomimetic W(IV) bis-dithiolene complex with CO_2 leading to formate production and structural rearrangement. *Dalton Transactions*, *48* (47), 17441–17444. Available from https://doi.org/10.1039/c9dt03906f.

Song, H., et al. (2017). Hydrogenation of CO_2 into formic acid using a palladium catalyst on chitin. *New Journal of Chemistry*, *41*(17), 9170–9177. Available from https://doi.org/10.1039/c7nj00460e.

Song, J. T., et al. (2019). Towards higher rate electrochemical CO_2 conversion: From liquid-phase to gas-phase systems. *Catalysts*, *9*(3), 224. Available from https://doi.org/10.3390/catal9030224.

Sordakis, K., et al. (2018). Homogeneous catalysis for sustainable hydrogen storage in formic acid and alcohols. *Chemical Reviews*, *118*(2), 372–433. Available from https://doi.org/10.1021/acs.chemrev.7b00182.

Srivastava, V. (2014). In situ generation of ru nanoparticles to catalyze CO_2 hydrogenation to formic acid. *Catalysis Letters*, *144*(10), 1745–1750. Available from https://doi.org/10.1007/s10562-014-1321-6.

Stalder, C. J., et al. (1983). Supported palladium catalysts for the reduction of sodium bicarbonate to sodium formate in aqueous solution at room temperature and one atmosphere of hydrogen. *Journal of the American Chemical Society*, *105*(20), 6318–6320. Available from https://doi.org/10.1021/ja00358a026.

Su, J., Lu, M., and Lin, H. (2015). High yield production of formate by hydrogenating CO_2 derived ammonium carbamate/carbonate at room temperature. *Green Chemistry*, *17*(5), 2769–2773. Available from https://doi.org/10.1039/c5gc00397k.

Su, J., et al. (2015). Highly efficient hydrogen storage system based on ammonium bicarbonate/formate redox equilibrium over palladium nanocatalysts. *ChemSusChem*, *8*(5), 813–816. Available from https://doi.org/10.1002/cssc.201403251.

Taheri, A., and Berben, L. A. (2016). Making C-H bonds with CO_2: production of formate by molecular electrocatalysts. *Chemical Communications*, *52*(9), 1768–1777. Available from https://doi.org/10.1039/c5cc09041e.

Taheri, A., et al. (2015). An iron electrocatalyst for selective reduction of CO_2 to formate in water: Including thermochemical insights. *ACS Catalysis*, *5*(12), 7140–7151. Available from https://doi.org/10.1021/acscatal.5b01708.

Tai, C. C., et al. (2003). High-pressure combinatorial screening of homogeneous catalysts: Hydrogenation of carbon dioxide. *Inorganic Chemistry*, *42*(23), 7340–7341. Available from https://doi.org/10.1021/ic034881x.

Tanaka, R., Yamashita, M., and Nozaki, K. (2009). Catalytic hydrogenation of carbon dioxide using Ir(III)-pincer complexes. *Journal of the American Chemical Society*, *131*(40), 14168–14169. Available from https://doi.org/10.1021/ja903574e.

Teeter, T. E., and Van Rysselberghe, P. (1954). Reduction of carbon dioxide on mercury cathodes. *The Journal of Chemical Physics*, *22*(4), 759–760. Available from https://doi.org/10.1063/1.1740178.

Thomas, C. A., et al. (2001). Hydrogenation of carbon dioxide catalyzed by ruthenium trimethylphosphine complexes—Effect of gas pressure and additives on rate in the liquid phase. *Canadian Journal of Chemistry*, *79*(5–6), 719–724. Available from https://doi.org/10.1139/v00-200.

Tsai, J. C., and Nicholas, K. M. (1992). Rhodium-catalyzed hydrogenation of carbon dioxide to formic acid. *Journal of the American Chemical Society*, *114*(13), 5117–5124. Available from https://doi.org/10.1021/ja00039a024.

Tsang, S. (2005) *A feasibility study into the formic acid economy*. Available at https://gtr.rcuk.ac.uk.

Van Putten, R., et al. (2019). Fuelling the hydrogen economy: Scale-up of an integrated formic acid-to-power system. *International Journal of Hydrogen Energy*, *44*(53), 28533–28541. Available from https://doi.org/10.1016/j.ijhydene.2019.01.153.

Wang, T., et al. (2017). A nanoporous nickel catalyst for selective hydrogenation of carbonates into formic acid in water. *Green Chemistry*, *19*(3), 716–721. Available from https://doi.org/10.1039/c6gc02866g.

Wang, W., et al. (2011). Recent advances in catalytic hydrogenation of carbon dioxide. *Chemical Society Reviews*, *40*(7), 3703–3727. Available from https://doi.org/10.1039/c1cs15008a.

Wang, W. H., et al. (2015). CO_2 hydrogenation to formate and methanol as an alternative to photo- and electrochemical CO_2 reduction. *Chemical Reviews*, *115*(23), 12936–12973. Available from https://doi.org/10.1021/acs.chemrev.5b00197.

Wang, Y., et al. (2016). Electrochemical reduction of CO_2 to formate catalyzed by electroplated tin coating on copper foam. *Applied Surface Science*, *362*, 394–398. Available from https://doi.org/10.1016/j.apsusc.2015.11.255.

Watkins, J. D., and Bocarsly, A. B. (2014). Direct reduction of carbon dioxide to formate in high-gas-capacity ionic liquids at post-transition-metal electrodes. *ChemSusChem*, *7*(1), 284–290. Available from https://doi.org/10.1002/cssc.201300659.

Weilhard, A., et al. (2020). Catalyst design for highly efficient carbon dioxide hydrogenation to formic acid under buffering conditions. *Journal of Catalysis*, *385*, 1–9. Available from https://doi.org/10.1016/j.jcat.2020.02.027.

Wesselbaum, S., Hintermair, U., and Leitner, W. (2012). Continuous-flow hydrogenation of carbon dioxide to pure formic acid using an integrated $scCO_2$ process with immobilized catalyst and base. *Angewandte Chemie - International Edition*, *51*(34), 8585–8588. Available from https://doi.org/10.1002/anie.201203185.

Westhues, N., Belleflamme, M., and Klankermayer, J. (2019). Base-free hydrogenation of carbon dioxide to methyl formate with a molecular ruthenium-phosphine catalyst. *ChemCatChem*, *11*(21), 5269–5274. Available from https://doi.org/10.1002/cctc.201900627.

Wu, H., et al. (2018). Highly efficient electrochemical reduction of CO_2 into formic acid over lead dioxide in an ionic liquid-catholyte mixture. *Green Chemistry*, *20*(8), 1765–1769. Available from https://doi.org/10.1039/c8gc00471d.

Wu, Y., et al. (2019). 110th anniversary: Ionic liquid promoted CO_2 hydrogenation to free formic acid over Pd/C. *Industrial and Engineering Chemistry Research*, *58*(16), 6333–6339. Available from https://doi.org/10.1021/acs.iecr.9b00654.

Xia, C., et al. (2019). Continuous production of pure liquid fuel solutions via electrocatalytic CO_2 reduction using solid-electrolyte devices. *Nature Energy*, *4*(9), 776–785. Available from https://doi.org/10.1038/s41560-019-0451-x.

Xu, S., et al. (2015). Efficient Cp*Ir catalysts with imidazoline ligands for CO_2 hydrogenation. *European Journal of Inorganic Chemistry*, *2015*(34), 5591–5594. Available from https://doi.org/10.1002/ejic.201501030.

Xu, Z., et al. (2013). Catalytic hydrogenation of CO_2 to formic acid with silica-tethered iridium catalysts. *ChemCatChem*, *5*(7), 1769–1771. Available from https://doi.org/10.1002/cctc.201200839.

Ye, K., et al. (2020). In situ reconstruction of a hierarchical Sn-Cu/SnO_x core/shell catalyst for high-performance CO_2 electroreduction. *Angewandte Chemie - International Edition*, *59*(12), 4814–4821. Available from https://doi.org/10.1002/anie.201916538.

Yin, C., et al. (2001). Promoting effect of water in ruthenium-catalyzed hydrogenation of carbon dioxide to formic acid. *Organometallics*, *20*(6), 1216–1222. Available from https://doi.org/10.1021/om000944x.

Yu, X., and Pickup, P. G. (2008). Recent advances in direct formic acid fuel cells (DFAFC). *Journal of Power Sources*, *182*(1), 124–132. Available from https://doi.org/10.1016/j.jpowsour.2008.03.075.

Yu, Y. M., et al. (2005). Silica immobilized ruthenium catalyst for formic acid synthesis from supercritical carbon dioxide hydrogenation II: Effect of reaction conditions on the catalyst performance. *Chinese Journal of Chemistry*, *23*(8), 977–982. Available from https://doi.org/10.1002/cjoc.200590977.

Yu, Y. M., et al. (2006a). Immobilized ruthenium catalyst for carbon dioxide hydrogenation. *Chinese Chemical Letters*, *17*(8), 1097–1100.

Yu, Y. M., et al. (2006b). MCM-41 bound ruthenium complex as heterogeneous catalyst for hydrogenation I: Effect of support, ligand and solvent on the catalyst performance. *Chinese Journal of Chemistry*, *24*(7), 840–844. Available from https://doi.org/10.1002/cjoc.200690160.

Yuan, T., et al. (2020). Two-dimensional amorphous SnO_x from liquid metal: Mass production, phase transfer, and electrocatalytic CO_2 reduction toward formic acid. *Nano Letters*, *20*(4), 2916–2922. Available from https://doi.org/10.1021/acs.nanolett.0c00844.

Zhang, S., Kang, P., and Meyer, T. J. (2014). Nanostructured tin catalysts for selective electrochemical reduction of carbon dioxide to formate. *Journal of the American Chemical Society*, *136*(5), 1734–1737. Available from https://doi.org/10.1021/ja4113885.

Zhang, S., et al. (2014). Polyethylenimine-enhanced electrocatalytic reduction of CO_2 to formate at nitrogen-doped carbon nanomaterials. *Journal of the American Chemical Society*, *136*(22), 7845–7848. Available from https://doi.org/10.1021/ja5031529.

Zhang, X., et al. (2017). Electrochemical reduction of carbon dioxide to formic acid in ionic liquid [Emim][$N(CN)_2$]/water system. *Electrochimica Acta*, *247*, 281–287. Available from https://doi.org/10.1016/j.electacta.2017.06.112.

Zhang, X., et al. (2019). Formation of lattice-dislocated bismuth nanowires on copper foam for enhanced electrocatalytic CO_2 reduction at low overpotential. *Energy and Environmental Science*, *12*(4), 1334–1340. Available from https://doi.org/10.1039/c9ee00018f.

Zhang, Y., et al. (2004). Silica immobilized ruthenium catalyst used for carbon dioxide hydrogenation to formic acid (I): The effect of functionalizing group and additive on the catalyst performance. *Catalysis Communications*, *5*(10), 643–646. Available from https://doi.org/10.1016/j.catcom.2004.08.001.

Zhang, Y., et al. (2015). Iron catalyzed CO_2 hydrogenation to formate enhanced by Lewis acid co-catalysts. *Chemical Science*, *6*(7), 4291–4299. Available from https://doi.org/10.1039/c5sc01467k.

Zhang, Y. P., et al. (2006). Ruthenium bisphosphine catalyst on functionalized silica: Novel efficient catalyst for carbon dioxide hydrogenation to formic acid. *Chinese Chemical Letters*, *17*(2), 261–264.

Zhang, Z., et al. (2009). Hydrogenation of CO_2 to formic acid promoted by a diamine-functionalized ionic liquid. *ChemSusChem*, *2*(3), 234–238. Available from https://doi.org/10.1002/cssc.200800252.

Zhao, G., et al. (2017). Progress in catalyst exploration for heterogeneous CO_2 reduction and utilization: A critical review. *Journal of Materials Chemistry A*, *5*(41), 21625–21649. Available from https://doi.org/10.1039/c7ta07290b.

Zhao, H., et al. (2012). Electrochemical reduction of carbon dioxide in an MFC-MEC system with a layer-by-layer self-assembly carbon nanotube/cobalt phthalocyanine modified electrode. *Environmental Science and Technology*, *46*(9), 5198–5204. Available from https://doi.org/10.1021/es300186f.

Zhao, S., et al. (2019). Advances in Sn-based catalysts for electrochemical CO_2 reduction. *Nano-Micro Letters*, *11*(1), 62. Available from https://doi.org/10.1007/s40820-019-0293-x.

Zheng, X., et al. (2017). Sulfur-modulated tin sites enable highly selective electrochemical reduction of CO_2 to formate. *Joule*, *1*(4), 794–805. Available from https://doi.org/10.1016/j.joule.2017.09.014.

Zhu, F., et al. (2015). Iron-catalyzed hydrogenation of bicarbonates and carbon dioxide to formates. *ChemSusChem*, *8*(4), 609–612. Available from https://doi.org/10.1002/cssc.201403234.

Zhu, Q., et al. (2016). Efficient reduction of CO_2 into formic acid on a lead or tin electrode using an ionic liquid catholyte mixture. *Angewandte Chemie - International Edition*, *55* (31), 9012–9016. Available from https://doi.org/10.1002/anie.201601974.

Power-to-Fuel existing plants and pilot projects

9

Eleonora Bargiacchi
Department of Energy, Systems, Territory and Constructions Engineering, University of Pisa, Pisa, Italy

In recent years, Power-to-X technologies received an increasing interest for their potential of storing renewable electricity into a versatile energy vector and for sector coupling purposes.

This chapter reviews 48 pilot and demonstration worldwide Power-to-X projects. In the first three sections the most relevant Power-to-Methane, Power-to-Methanol and Power-to-Ammonia plants and their specific features are reported and commented. Sections 9.4 and 9.5 present an analysis of the plant data which comprises the number of projects per technology, the average and total installed power based on the electrolyser size and the project geographic distribution. It is worth mentioning that the chapter presents only projects where the hydrogen is partially or totally produced by an electrolyser. This choice allows a fairer size comparison among different Power-to-X plants (if based on the electrolyser size), independently from the final product, and it is considered in line with the book purpose.

9.1 Power-to-Methane projects

This section presents a selection of 37 Power-to-Methane pilot and demonstration plants worldwide (see Table 9.1). The first Power-to-Methane plant was realized in 1996 by Japan who built a prototype plant in Tohoku University (Hashimoto et al., 2002, 2014). Since 2009, the biggest projects have been carried out in Europe, with Germany holding the leadership of Power-to-Methane projects followed by Switzerland, Denmark and Netherlands. Nowadays more than 50 Power-to-Gas projects have been developed, which involved more than 180 public institutions and companies (Bailera et al., 2020). Electrolyser sizes for Power-to-Methane vary from 1 kW to 12 MW, and the carbon dioxide conversion into methane can be performed via biological or chemical pathway.

9.1.1 Chemical methanation

In 2013 Audi built the world's first industrial scale plant (6 MW) to generate Substitute Natural Gas (SNG) from CO_2 and renewable electricity in Werlte (Northern Germany). The facility comprises three alkaline electrolysers of 2 MW

Power to Fuel. DOI: https://doi.org/10.1016/B978-0-12-822813-5.00004-7

Table 9.1 Power-to-Methane projects, listed alphabetically by country code and chronologically within the same country code.

Process type	Project/ plant name	Year	Location	Country	Electrolyser power (MW)	Electrolyser type	Methanation type	Institutions	References
CO_2 methanation/ Biogas upgrading	RENERG2	2013–2016	Villigen	CH	0.100	PEM	Chemical	PSI, EMPA, ETH Zurich, ZHAW, EPFL	Competence Center Energy and Mobility (CCEM) (2020), Elber et al. (2018)
CO_2 methanation	Pilot- und Demonstrationsanlage Power-to-Methane HSR	2014–2017	Rapperswil	CH	0.025	n.a.	Chemical	SR-IET, HSR, Audi, Climeworks, Erdgas Obersee, Erdgas Regio	Friedl (2017), Friedl et al. (2015)
Biogas upgrading	Solothurn Plant (STORE&GO)	2015–2019	Solothurn	CH	0.350	PEM	Biological	Electrochaea, DVGW, E.ON, Regio Energie, EII Spa, HSR, PoliTo, JKU Linz-EI, RUG, Atmostat, CEA, Climeworks, DBI-GUT, BFP, EDI, EMPA, EPFL-IPESE, EPFL-MER, EPFL-CEN, Energy Valley	DVGW Research Centre at Engler-Bunte-Institute of Karlsruhe Institute of Technology (KIT) (2020), STORE&GO Switzerland (2019)
CO_2 methanation	Dietikon plant	2019-n.a.	Dietikon	CH	2.500	PEM	Biological	Limeco, Eniwa, Energie Zürichsee Linth, St. Galler Stadtwerke, EWB, GW Dietikon, GW Schlieren, SWL Energie, Industrielle Betriebe Interlaken, Swisspower, Schmack BioEnergie, Siemens, TBF	Limeco (2019a, 2019b)

CO_2 methanation	Alpha-plant	2009	Stuttgart	DE	0.025	ALK	Chemical	ETOGAS, ZSW	ETOGAS Gmbh (2013)
CO_2 methanation	Alpha-plant	2010–2011	Werlte	DE	0.025	ALK	Chemical	ETOGAS, ZSW, EWE Biogas	Specht (2012)
CO_2 methanation	CO2RRECT	2010–2013	Niederaußem	DE	0.300	PEM	Chemical	Bayer Technology Services, Bayer MaterialScience, RWE Power, Siemens, Invite, CAT Catalytic Center, RWTH Aachen, MPI Magdeburg, FHI Berlin, LIKAT Rostock, RUB, TU Dortmund, TU Dresden, Universität Stuttgart, KIT, TU Darmstadt	dena - Deutsche Energie-Agentur (2013), Karlsruhe Institute of Technology (2013), Kraftwerk Forschung (2013)
Biogas upgrading	Alpha-plant	2011	Morbach	DE	0.025	ALK	Chemical	ETOGAS, ZSW, Fraunhofer IWES, Juwi	Specht (2012), Zentrum Sonnenenergie-Wasserstoff-Forschung Baden-Württemberg (2013)
Biogas upgrading	Alpha-plant	2012	Bad Hersfeld	DE	0.025	ALK	Chemical	ETOGAS, ZSW, Fraunhofer IWES, HBFZ	Reuter and Schmid (2012)
Biogas upgrading	Power-to-Gas im Eucolino	2012	Schwandorf	DE	0.108	n.a.	Biological	Microbenergy, Schmack	Agricola (2013), Benjaminsson, Benjaminsson, and Rudberg (2013)
CO_2 methanation	Power to Gas 250	2012–2014	Stuttgart	DE	0.250	ALK	Chemical	ETOGAS, ZSW, Fraunhofer IWES	Zuberbühler and Specht (2012)

(*Continued*)

Table 9.1 (Continued)

Process type	Project/ plant name	Year	Location	Country	Electrolyser power (MW)	Electrolyser type	Methanation type	Institutions	References
Biogas upgrading	Kommunale Kläranlagen als Energiespeich	2012–2014	Emden	DE	0.312	ALK	Chemical	Stadtwerke Emden, BEE, Thalen Consult, IBIS Umwelttechnik, GA-Group	Schröder (2013)
CO_2 methanation	Audi e-gas	2013	Werlte	DE	6.000	ALK	Chemical	ETOGAS, ZSW, Fraunhofer IWES, EWE Biogas, Audi	Otten (2014), Waldstein (2015)
Biogas upgrading	Mikrobielle Methanisierung	2013	Schwandorf	DE	0.275	n.a.	Biological	Microbenergy, Schmack, FENES, ZVKSW	Ostbayerische Technische Hochschule Regensburg (2013)
Biogas upgrading	BioPower2Gas	2013–2016	Allendorf	DE	0.300	PEM	Biological	Microbenergy, Schmack, Carbotech, EnergieNetz, EAM EnergiePlus, CUBE, DBFZ, IdE	IEA Bioenergy (2018), IdE Institut dezentrale Energietechnologien GmbH (2020), The Viessmann Group (2015)
CO_2 methanation	HELMETH	2014–2017	Karlsruhe	DE	0.008	SOEC	Biological	Sunfire, KIT, PoliTo, ERIC, TS-Torino, NTUA, DVGW	Fuel Cell and Hydrogen Joint Undertaking (2014), European Commission (2018b), Karlsruher Institut für Technologie (KIT) (2020)

CO_2 methanation	WindGas Falkenhagen (STORE&GO)	2015–2019	Falkenhagen	DE	2.000	ALK	Chemical	Electrochaea, DVGW, E.ON, Regio Energie, EII Spa, HSR, PoliTo, JKU Linz-EI, RUG, Atmostat, CEA, Climeworks, DBI-GUT, BFP, EDI, EMPA, EPFL-IPESE, EPFL-MER, EPFL-CEN, Energy Valley	DVGW Research Centre at Engler-Bunte-Institute of Karlsruhe Institute of Technology (KIT) (2020), Hüser (2018)
Biogas upgrading	Energiepark Pirmasens-Winzeln	2015-n.a.	Pirmasens	DE	1.800	ALK	Biological	PFI	PFI (2015)
Biogas upgrading	Alpha-plant	2016–2019	Bad Hersfeld	DE	0.050	ALK	Chemical	Fraunhofer IEE, HBFZ, Landwirtschaftszentrum Eichhof	Fraunhofer-Institut für Energiewirtschaft und Energiesystemtechnik (2016)
CO_2 methanation	Exytron Zero-Emission-Wohnpark	2016-n.a.	Alzey	DE	0.063	ALK	Chemical	EXYTRON, Deutsche Reihenhaus, e-rp	EXYTRON (2017, 2020)
Biogas upgrading	INFINITY 1	2017-n.a.	Pfaffenhofen a. d. Ilm	DE	1.000	PEM	Biological	Bürgerenergie Pfaffenhofen, Stadtwerke Pfaffenhofen, Electrochaea, Max Streicher Anlagentechnik, H-TEC Systems, Audi Stiglmayr Paffenhofen	Electrochaea (2017)
Biogas upgrading	ORBIT (2nd site)	2020-n.a.	Ibbenbüren	DE	0.001	n.a.	Biological	OTH Regensburg, Universität Regensburg, Electrochaea, MicroPyros Viessmann, Westnetz, DVGW	Friedrich-Alexander-Universität Erlangen-Nürnberg (2020)

(*Continued*)

Table 9.1 (Continued)

Process type	Project/ plant name	Year	Location	Country	Electrolyser power (MW)	Electrolyser type	Methanation type	Institutions	References
Biogas upgrading	Pre-commercial	2011–2013	Foulum	DK	0.250	PEM	Biological	Electrochaea, E.ON, Energie 360°, EWZ, NEAS, AU	Adamsen (2013), Hofstetter (2013)
Biogas upgrading	El-opgraderet biogas	2013–2016	Foulum	DK	0.050	SOEC	Chemical	Haldor Topsoe, Ea, AU, PlanEnergi, HMN Naturgas, NGF, EnergiMidt, DGC, Cemtec, Xergi	Energiforskning.dk (2015a), Haldor Topsøe A/S (2018)
Biogas upgrading	P2G-BioCat project	2014–2016	Avedøre	DK	1.200	ALK	Biological	Electrochaea, Hydrogenics, Audi, NEAS, HMN Gashandel, Biofos, Insero	Electrochaea (2016), Electrochaea.dk ApS (2014)
Biogas upgrading	MeGa-stoRE 2–Optimising and Upscaling	2015–2018	Heden	DK	0.070	ALK	Chemical	AU, NGF, DTU, Elplatek, GreenHydrogen	Danmarks Tekniske Universitet Mekanik (2019), Energiforskning.dk (2015b)
Biogas upgrading	RENOVAGAS	2014–2016	Jerez de la Frontera	ES	0.015	ALK	Chemical	Enagas, ICP-CSIC, CNH2, FCC-Aqualia, GasNatural Fenosa, Tecnalia, Abengoa	Centro Nacional del Hidrogeno (2018), Sánchez (2015)
n.r.	Jupiter 1000	2014-n.a.	Fos-sur-Mer	FR	1.000	ALK	Chemical	GRTgaz, ATMOSTAT, CEA, CNR, Leroux&Lotz, Marseille Fos Port, McPhy, RTE, TGIF	GRTgaz (2018)

Biogas upgrading	Power-to-Gas Hungary	2016-n.a.	n.r.	HU	10.000	n.a.	Biological	Electrochaea, MVM, University of Chicago	Bio-based news (2016), Power-to-Gas Hungary Kft. (2020)
CO_2 methanation	STORE&GO	2015–2019	Troia	IT	0.200	ALK	Chemical	Electrochaea, DVGW, E.ON, Regio Energie, EII Spa, HSR, PoliTo, JKU Linz-EI, RUG, Atmostat, CEA, Climeworks, DBI-GUT, BFP, EDI, EMPA, EPFL-IPESE, EPFL-MER, EPFL-CEN, Energy Valley	STORE&GO (2019)
CO_2 methanation	Technical assumptions, technology demonstration and results P2G project	2013–2015	Rozenburg	NL	0.008	PEM	Chemical	DNV GL, TKI Gas, Stedin, Rotterdam Council, Ressort Wonen	DNV GL Oil & Gas (2015), Holstein et al. (2015)
Biogas upgrading	W2P2G	2014-n.a.	Wijster	NL	0.400	n.a.	Biological	Attero, Gasunie, Audi, Ponooc, Tennet, Provincie Drenthe, Pon	Kwak (2020), Rozzi et al. (2020)
Syngas upgrading	Energy Valley Delfzijl	2016-n.a.	Delfzijl	NL	12.000	PEM	Chemical	Torrgas, Siemens, Stedin, Gasunie, A.Hak, Hanzehogeschool/EnTranCe, Groningen Seaports and Energy Valley	European North Sea Energy Alliance (2015), Noordelijke Innovation Board (2007)
Syngas upgrading	DemoSNG	2011–2015	Köpig	SE	0.006	PEM	Chemical	DVGW-EBI, KIT, KTH, Cortus, Gas Natural Fenosa	Buchholz and Reimert (2014), Mesfun et al. (2017)

(*Continued*)

Table 9.1 (Continued)

Process type	Project/ plant name	Year	Location	Country	Electrolyser power (MW)	Electrolyser type	Methanation type	Institutions	References
n.r.	CymruH_2Wales project	2008	Baglan near Swansea	UK	0.049	n.a.	n.r.	University of South Wales, IT Power	Thema et al. (2019), University of South Wales - Sustainable Environment Research Centre (2018)
CO_2 methanation	CO_2 recycling via reaction with hydrogen	2009	Reno	US	0.005	n.a.	Chemical	DRI, RCO2 AS	Hoekman et al. (2010)
CO_2 methanation	P2G Solar Energy Storage RD&D	2014–2016	Golden	US	0.250	n.a.	Biological	NREL, SoCalGas	Rasberry (2015), Reed (2015)

ALK, alkaline; *PEM*, proton exchange membrane; *SOEC*, solid oxide electrolytic cell.

each, and exploits the CO_2 coming from a nearby anaerobic digester to produce 325 $Nm^3\,h^{-1}$ of SNG, mainly intended for mobility purposes (Otten, 2014). The claimed efficiency without heat utilisation is 54%.

Within the project STORE&GO three Power-to-Methane demonstration facilities were built and operated in three different European countries: Germany, Italy and Switzerland. The first two were based on CO_2 catalytic methanation. The plant in Falkenhagen (DE) comprises 2 MW of electrolysers (six alkaline modules of 330 kW each) producing hydrogen that is combined with the CO_2 from a bioethanol plant to produce 1400 m^3 SNG per day, with a methanation efficiency of 85% (STORE&GO, 2020). Within the same project, the other demo facility with catalytic methanation was installed in Troia (Southern Italy). With a capacity of 200 kW of alkaline electrolysers, it captures CO_2 from air and comprises an SNG liquefaction system, which makes the plant easily operable in stand-alone mode and eases the product transport. The system complexity implies, as main drawback, a lower efficiency, ranging around 29% during operation, with an upper potential of 46% after system optimisation and heat integration.

Still in the MW range, in France the first industrial demonstrator was built in 2018 in Fos-sur-Mer within the project Jupiter 1000. The facility is composed of 1 MW electrolysers (0.5 MW alkaline and 0.5 MW proton exchange membrane electrolysers) and a CO_2 capture plant from a nearby industrial site, for a methane production of 25 $m^3\,h^{-1}$ to be injected into the gas grid.

In the hundred kilowatts scale, in Emden the feasibility of using CO_2 from a wastewater treatment plant to an electrolyser to supply the city was first investigated. The facility comprises a 312 alkaline electrolyser and an SNG production of 59 $m^3\,h^{-1}$ (Bailera et al., 2017).

A 300 kW test facility was built in 2013 at the Niederaußem power plant within the CO2RRECT project. The plant combined the scrubbed CO_2 from the lignite-fired power plant in Niederaußem with the hydrogen produced from the electrolysis, using up to 5 $Sm^3\,h^{-1}$ of H_2 and up to 1 $Sm^3\,h^{-1}$ of CO_2 (Kraftwerk Forschung, 2013).

In 2012 the Power-to-Gas 250 project, the world's largest plant of its kind at that time, was realized by the partnership of ETOGAS, ZSW, Fraunhofer IWES. The facility was built in Stuttgart (Germany) and it aimed at testing different reactor concepts (plate or tubular reactor) to be implemented in the 6 MW facility in Werlte (Specht and Zuberbühler, 2013; Zuberbühler and Specht, 2012). The maximum output was 300 $m^3\,day^{-1}$ of SNG, with 99% purity thanks to recirculation and a final membrane separation step.

Completed in 2017, the RENERG2 project aimed at improving technologies and concepts in Power-to-Methane applications: development of a new anode catalyst for both proton exchange membrane electrolysers and co-electrolysers, increase of the knowhow on CO_2 methanation from gasification derived gas, experimental and numerical investigation of the early flame phase of premixed methane/air and hydrogen enriched methane/air combustion, utilisation of the hydrogen/methane mixtures on mobile and stationary applications, besides completing a feasibility analysis of the Power-to-Gas technology in the future Swiss low voltage grid

(Competence Center Energy and Mobility (CCEM), 2020; Elber, Hamburger, and Lutz, 2018).

Passing to the decades kW scale, the German start-up EXYTRON together with Leibniz Institute for Catalysis (LIKAT) and the University of Rostock proposed a decentralised concept of storing renewable energy to SNG at the property itself to be used for the building's energy needs (Ikäheimo et al., 2020). The concept was first tested in Rostock (Germany), then the so-called "EXYTRON Zero-Emission-Wohnpark" was built in Alzey (Germany) (EXYTRON, 2017). The facility comprises 63 kW of electrolysis and is able to supply the energy needs of 37 residential units. The CO_2 coming from the SNG combustion is stored to be re-employed for SNG synthesis; also the surplus heat generated by the exothermic reaction is used for water and space heating (Carbon Commentary, 2017).

In 2015 in Denmark the MeGa-stoRE project started to build and test a demonstration plant capable to clean biogas from impurities and convert the CO_2-part of the biogas to CH_4 in a catalytic reactor (Danmarks Tekniske Universitet Mekanik, 2019). The project was articulated in two phases. In the first phase (MeGa-StoRE 1) the technical proof-of-concept of the cleaning process and the Sabatier reactor on the proposed Power-to-Gas technology was proved (Aarhus University, 2015). The plant was installed at Lemvig Biogas with a capacity of 1 $Nm^3\,h^{-1}$ (Yde, 2016). In the second step (MeGa-StoRE 2 – Optimising and Upscaling) a novel cleaning technology to clean the biogas from impurities was developed, a 16 $Nm^3\,h^{-1}$ alkaline electrolyser was developed by downscaling a bigger prototype by GreenHydrogen and the methanation reactor developed in the first phase was upscaled up to 10 $Nm^3\,h^{-1}$ SNG (Danmarks Tekniske Universitet Mekanik, 2019; Energiforskning.dk, 2015b).

In the years 2013–17 in Foulum (Denmark) a demonstration facility was installed under the project "El-opgraderet biogas" with the aim of proving the efficient upgrading of biogas to pipeline quality by catalytic conversion of the CO_2 content in the biogas by reaction with hydrogen produced by a 50 kW solid oxide electrolyser cell. The plant capacity was around 10 $Nm^3\,h^{-1}$ of methane with 98% purity.

From 2014 to 2017 the first Power-to-Methane plant in Switzerland was built and operated at the Institute for Energy Technology of the Hochschule für Technik Rapperswil (HSR-IET) in Rapperswil (IET-Institut fur Energietechnik, 2020). The required hydrogen was produced by a 25 kW alkaline electrolyser, supplied by 50 m^2 PV panels. The CO_2 is captured from air with Climeworks technology and reacts with hydrogen in a fixed bed reactor on a nickel catalyst previously developed by ETOGAS, for an overall methane production of 1 $m^3\,h^{-1}$ (Friedl, 2017; Friedl et al., 2015).

In this regard, it is worth mentioning the Alpha-plant, which comprises several small scale facilities born from the collaboration between Audi, ETOGAS, the Centre for Solar Power and Hydrogen Research Baden-Württemberg (ZSW), the Fraunhofer Institute for Wind Energy and Energy System Technology (Fraunhofer IWES) and EWE Biogas GmbH & Co. KG. The first plant was built in Stuttgart in 2009 and was constituted by a 25 kW alkaline electrolyser and an air purification system to capture CO_2 from air. The second plant was built in Werlte (2010–11) to test the Power-to-Gas concept with CO_2 from biogas purification with pressure swing absorption technology. The third alpha-plant was built in Morbach in 2011, to prove the feasibility of operating a Power-to-Gas plant with raw biogas, which was finally tested on a longer period in the fourth alpha plant in Hersfeld in 2012. All these steps led then to the Power-to-Gas 250 project, which has been described above.

In the few kW scale, remarkable projects are the RENOVAGAS project in Jerez de la Frontera (Spain) (Centro Nacional del Hidrogeno, 2018; Sánchez, 2015), the DNV GL plant in Rozenburg (Netherlands) (DNV GL Oil & Gas, 2015; Holstein et al., 2015), the DemoSNG project in Köpig (Sweden) (Buchholz and Reimert, 2014; Mesfun et al., 2017) and the DRI plant in Reno (USA) (Hoekman et al., 2010).

It is worth mentioning that the biggest Power-to-Gas facility (12 MW) is under construction in Delfzijl (Netherlands), where the syngas from the Torrgas biocoal-gasifier is upgraded with hydrogen coming from electrolysis and delivered to nearby industries. Thanks to the synergy realized in the Energy Valley, the oxygen co-produced with the hydrogen by the electrolysis process will be employed in the gasification process of Torrgas (European North Sea Energy Alliance, 2015).

9.1.2 Biological methanation

During the years, in analogy to what happened to catalytic methanation, biological methanation has undergone a scaling up process.

The world's first grid-scale power-to-gas plant (with an estimated power consumption of 10 MW) is under construction in Hungary by the Hungarian company MVM and the German Electrochaea (Bio-based news, 2016). The employed technology will be based on biological methanation with methanogenic archaea, with a proprietary biocatalyst developed by Electrochaea that is already employed in other plants. The contract has been signed in 2016, but no further information is available yet.

Still in the MW power range, the companies Limeco, microbEnergy and Schmack Biogas partnered to build a 2.5 MW Power-to-Methane plant in Dietikon (Switzerland). The feedstock will be sewage gas from a nearby wastewater treatment plant, and hydrogen will be produced by proton exchange membrane electrolysers supplied by Siemens, which exploits electricity coming from a waste recycling plant (The Viessmann Group, 2019).

In 2019 the company PFI Germany upgraded the existing Felsalbe sewage plant with a 1.8 MW electrolyser, for an overall hydrogen production of $400\ m^3\ h^{-1}$ and an oxygen generation of $200\ m^3\ h^{-1}$ used for aeration purposes in activation tanks (PFI, 2015). The methanation reactor is a pressurised trickle bed reactor developed by PFI together with the Schmitt Company, in countercurrent configuration which enhances mass transfer, hence the contact between the gaseous substrate and the microorganisms. The plant has a capacity of $100\ m^3\ h^{-1}$ of biogas, which is fed into the Gas Grid of Pirmansens Utility Company (PFI Germany, 2016).

Another remarkable project in the MW scale was the BioCat project. The aim was to design, engineer, and construct a commercial-scale (1 MW) power-to-gas facility to produce grid-quality methane under intermittent operations and demonstrate its capabilities to provide energy storage services to the Danish energy system (Electrochaea, 2016). The plant was built at the BIOFOS wastewater treatment centre in Avedøre (Denmark) and it was operated for 8 months in 2016. During this time, the system used 42,193 Nm^3 biogas, 170 m^3 water and 708,215 kWh electricity for system operations to produce 129,290 Nm^3 hydrogen for methanation of ~16,000 Nm^3 of CO_2 from the biogas and making available ~15,000 Nm^3 renewable methane and 85,000 kWh heat for use at Avedøre's facility. The system could operate from 30 $Nm^3\,h^{-1}$ as hydrogen input and 19 $Nm^3\,h^{-1}$ biogas throughput, and up to three times those flow rates. The reactor and biological catalyst in BioCat were able to produce grid quality gas – greater than 97% CH_4, <2% H_2, <1% CO_2, and <5ppm H_2S in the product gas, requiring only an additional drying to remove water vapour step before grid injection (Electrochaea.dk ApS, 2014).

Still in the MW scale, in 2017 the project "Infinity 1" was launched in the city of Pfaffenhofen a.d. Ilm (Germany). Hydrogen is generated by 1 MW electrolyser and is added to the local sewage treatment plant to generate biomethane through the Electrochaea's technology (Electrochaea, 2017). The plant was planned to be finished by 2020, but no further information has been released so far.

In the W2P2G project, a Dutch consortium led by Attero aims at integrating municipal waste incineration and organic waste digestion plants with a 400 kW electrolyser and a methanator to increase the biomethane production. In the process, the methanation, incineration and digestor heat flows are also recovered (Kwak, 2020). The project was launched in 2016 but limited information is available.

Within the already introduced project STORE&GO, a pilot facility based on the biological methanation process was built in Solothurn (Switzerland). The facility comprised a methanator where methane was produced by archaea from hydrogen produced by a 350 kW electrolyser and carbon dioxide from a wastewater treatment plant. The facility was integrated with the municipal gas distribution network and was able to reach a product gas quality of more than 99% of volume of CH_4. The overall Power-to-Gas efficiency was 73%, increased to 76% when recovering the heat from the low temperature electrolyser.

In the years 2013–16 a demonstration plant was built in Allendorf (Germany) within the project BioPower2Gas. The plant was constituted by a 300 kW proton exchange membrane electrolyser, a 5 m^3 biological methanation reactor and it was connected to the national gas grid (Heller, 2016).

Other smaller pilot and demonstration facilities have been built since 2011 to test the main features of biological methanation. In order of decreasing plant size, we recall the projects: Mikrobielle Methanisierung (Ostbayerische Technische Hochschule Regensburg, 2013) and Power-to-Gas im Eucolino (Agricola, 2013) in Schwandorf (Germany), a pre-commercial plant (Adamsen, 2013; Hofstetter, 2013) in Foulum (Denmark), P2G Solar Energy Storage RD&D (Rasberry, 2015) in Golden (USA) and ORBIT (2nd site) (Friedrich-Alexander-Universität Erlangen-Nürnberg, 2020) in Ibbenbüren (Germany).

For further details, some exhaustive and detailed reviews have been written on this topic by Bailera et al. (2017), Blanco and Faaij (2018), Buttler and Spliethoff (2017), Götz et al. (2016), Thema, Bauer, and Sterner (2019), Wulf, Linßen, and Zapp (2018).

9.2 Power-to-Methanol projects

This section presents 8 Power-to-Methanol pilot and demonstration plants worldwide (see Table 9.2).

The George Olah Renewable Methanol plant was the first industrial scale plant for fuel production from CO_2. It was commissioned in 2011 in Svartsengi (Iceland), and it is fully operational since 2012. The required carbon dioxide is captured from an adjacent geothermal power plant, and it undergoes a purification process and compression before being mixed with the hydrogen coming from a 6 MW alkaline electrolyser that uses Iceland's grid electricity. After catalytic conversion, the final product is purified and water is removed in a distillation column that exploits the geothermal steam (Carbon Recycling International (CRI), 2018). The concept novelty relies on proving and testing, for the first time at an industrial scale, the direct CO_2 conversion to methanol, in comparison with the analogous, more widespread process starting from syngas. The main advantages of the CO_2-to-methanol process are the higher reaction selectivity which results in fewer byproducts and the milder reaction conditions due to the less exothermic reaction. On the other hand, the main disadvantages are the lower reactivity of CO_2 mixtures if compared to CO-rich syngas and a higher water consumption (Marlin, Sarron, and Sigurbjörnsson, 2018). The plant achieves an annual production capacity of 4000 tonnes of methanol per year. Within the CirclEnergy project context the plant was also tasted while operating under variable electric load and both commercial assessment and market analysis were realized. They showed a rapidly growing market potential, limited competition and the financial viability of the project (European Commission, 2018a).

The largest Power-to-Methanol facility is under construction in Delfzijl industrial park in the Netherlands within the project Djewels. The aim is to demonstrate the operational readiness of a 20 MW electrolyser for the production of renewable methanol by taking the technology from TRL7 to TRL8. Six European partners are involved and the project is supposed to last five years (Nouryon Industrial Chemicals B.V., 2020). Being the project still under development, not much information has been released yet.

Another project in the MW scale is the Power2Met project. The goal is to develop, design and test a pilot plant for a complete standardized, and modular Power-to-Methanol process that utilises CO_2 from biogas and makes it react with green hydrogen in a catalytic process (Aalborg Universitet, 2019). In the first phase, a downscaled pilot plant is established at Aalborg University producing biomethanol from biogas CO_2, using electricity from the grid for hydrogen production. The electrolysis system is composed of pressurised (35 bar) alkaline stacks with a

Table 9.2 Power-to-Methanol projects, listed alphabetically by country code and chronologically within the same country code.

Project/plant name	Year	Location	Country	Electrolyser power (MW)	Electrolyser type	Process type	Institutions	References
Power-to-methanol project	2020-n.a.	Antwerp	BE	2.500–5.000	n.a.	Captured CO_2 to methanol	Inovyn, Engie, Oiltanking, Indaver, Fluxys, Flemish Environmental Holding Company, Port of Antwerp	Inovyn (2020), Port of Antwerp (2020)
Shunli plant	2019-n.a.	Anyang	CN	n.a.	n.a.	CO_2 from coke to methanol	Henan Schuncheng Group, CRI	Carbon Recycling International (CRI) (2018)
MefCO_2	2016–2019	Niederaussem	DE	0.600	PEM	CO_2 captured from a coal-fired power plant	i-deals, NIC Slovenia, MHPS, Cardiff Catalysis Institute, CRI, DIME UniGe, Hydrogenics, University of Duisburg-Essen, RWE	Bowker (2019), Carbon Recycling International (CRI) (2018), European Commission (2016)
Carbon2Chem	2018-n.a.	Duisburg	DE	2.000	ALK	Steel mill gases to methanol	Thyssenkrupp, MPI-CEC, FhG-Umsicht, Siemens, ZBT Duisburg, Akzo Nobel, Clariant, FhG-ISE, RUB, Linde, Evonik, Covestro, BASF, Volkswagen, KIT, RWTH Aachen, TU Kaiserlautern	Büker et al. (2018), BusinessEurope (2018)

Power2Met	2019–2021	Aalborg	DK	10.000	ALK	Biogas CO_2 from biogas plant	Green Hydrogen Systems, REintegrate, Aalborg University, Hydrogen Valley, E.ON, NGF Nature Energy, Drivkraft Danmark, Rockwool, Process Engineering, Holtec Automatic-Nord, Lillegaarden EL	Aalborg Universitet (2019), Baden (2020), Green Hydrogen Systems (2020)
George Olah Renewable Methanol Plant	2011	SvartsengI	IS	6.000	ALK	CO_2 captured from geothermal off-gas	CRI	Bergins et al. (2015), Carbon Recycling International (CRI) (2018), Marlin et al. (2018)
Djewels	2019–2025	Delfzijl	NL	20.000	n.a.	n.a.	Nouryon, McPhy, BioMCN, DeNora, Hinicio	Nouryon Industrial Chemicals B.V. (2020), Recharge (2020)
FReSMe	2017-n.a.	Luleå	SE	n.a.	n.a.	Steel (residual blast furnace) gases to methanol	i-deals, ECN, SWERIM, CRI, NIC, PoliMi, SSAB, Stena Rederi, Kisuma, Tata Steel, Array Industries	Carbon Recycling International (CRI) (2018), European Commission (2017)

ALK, alkaline.

nominal power of 375 kW. The pilot facility achieves a daily methanol production capacity of 1000 litres per day. Starting in 2022, the second phase of the project will aim at bringing the technology at market level by building a scalable facility, with a capacity of 10 MW, based on methanol reactor modules and electrolyser modules. The companies Green Hydrogen Systems and REIntegrate will be in charge of the plant building, the latter being responsible for the methanol reactor and distillation system. CO_2 will be separated from E.On's biogas plant and the electricity supplied by wind turbines and PV plants (Baden, 2020).

In 2020 the building of the first industrial-scale demonstration Power-to-Methanol plant in Belgium was announced (Inovyn, 2020). The facility, located in the Port of Antwerp, will comprise of 2.5–5 MW of electrolysers and a CO_2 capture plant exploiting the CO_2 from not yet specified plants. The demonstration unit will produce up to 8000 tonnes of renewable methanol per year (Port of Antwerp, 2020). The produced methanol will be used by the chemical companies within the Port of Antwerp cluster (Inovyn, 2020).

The Carbon2Chem project was launched in 2016 by a consortium made up of 17 partners. The plant was built in Duisburg (Germany) and has been operational since 2018. The project aims at using emissions from steel production as a raw material for chemicals synthesis (BusinessEurope, 2018). In the Duisburg plant, blast furnace gases from the steel production are purified and used for the methanol synthesis. The facility comprises a 2 MW alkaline electrolyser, a novel sorptive and catalytic gas purification system and a pressure swing adsorption purification system for coke oven gas. Since the main challenge is to ensure appropriate gas composition of the synthesis gas, pilot plant tests will be carried out to assess the influence of fluctuations in the steel mill gas supply. In parallel, plant dynamic simulations, economic and life cycle analysis are performed (Büker et al., 2018).

Finally, the MefCO2 project was carried out in the years 2016–19 by a European consortium made up of nine partners. The main goal of the project was to demonstrate the economic feasibility of valorizing captured CO_2 by using it to produce methanol. The pilot facility was built in Niedeaussem (Germany) and was able to produce 1 tonne of methanol per day. It comprises a 40 m high capture unit, with a capture capacity of $7.2\ t\ day^{-1}$ and a separation efficiency of 90% of the processed flue gases. After being compressed and conditioned, the CO_2 is supplied to the methanol reactor, owned by CRI. The electrolyser, manufactured by Hydrogenics, is a 600 kW proton exchange membrane electrolyser that is able to produce $120\ m^3\ H_2\ h^{-1}$ at a purity of 99.99%. The methanol is finally stored in a $26\ m^3$ tank.

Besides the aforementioned plants, it is worth mentioning other projects that are still under development or with still limited information available, but that are contributing to widen the international knowhow on Power-to-Methanol technologies.

The FReSMe project integrates two previous EU funded projects: the STEPWISE (focused on CO_2 capture from blast furnace gas) (European Commission, 2020) and the $MefCO_2$, which we described above. Besides the previous technologies improvement, optimisation and integration in a pilot plant in

Lulea (Sweden), the project includes the end-user demonstration using the produced methanol to supply a ferry (line Gothenburg-Kiel) (European Commission, 2017). The pilot plant will run for three months, using up to 800 $m^3 h^{-1}$ of blast furnace gas to produce up to 50 kg h^{-1} of methanol.

Finally, CRI is also involved in the design and commissioning of the Shanli CO_2-to-Methanol plant in Anyang City (Henan Province, China), where a plant owned by the Chinese chemical company Henen Schuncheng Group converts coal into metallurgical coke, then used for steel manufacturing. The Power-to-Methanol facility will produce 110,000 tonne of methanol per year from coke oven gas, therefore becoming the world's largest production of fuel from CO_2, and the commissioning is planned for late 2021. Differently from the other plants described in this chapter, no electrolyser is included yet, the necessary hydrogen will have a fossil provenance. Even if not a proper Power-to-Methanol plant, the plant retrofitting with the CRI's Emission-to-Liquid facility will contribute to a significant reduction in the carbon footprint of steel production.

9.3 Power-to-Ammonia Projects

Differently from the Power-to-Methane and the Power-to-Methanol projects, the Power-to-Ammonia projects are aimed at a refurbishment of existing ammonia production plants by a progressive substitution of the fossil-derived hydrogen with electrolytic hydrogen. The core industrial procedure for the synthesis is the Haber-Bosch technology. This section reviews seven Power-to-Ammonia plants worldwide (see Table 9.3).

The first successful demonstration of Power-to-Ammonia plant using renewable hydrogen and power generation from gas turbines fuelled by synthesised ammonia was announced in Koriyama (Japan) in 2018, by JGC Corporation and the National Institute of Advanced Industrial Science and Technology. The project started in 2014 and the demonstration plant was put into operation by 2018. Besides successfully testing a new ruthenium catalyst capable of an efficient ammonia synthesis at low temperatures and low pressures, the consortium tested the ammonia synthesis from hydrogen generated from solar panels, therefore subjected to load changing operation. The plant reached an ammonia production of 20 kg per day and the ammonia fuelled gas turbine has a power generation of 47 kW (Fujimura et al., 2019; Yamagami and Fujiwara, 2018).

In the same period, Siemens put into operation the so-called "Green ammonia demonstrator" at the Rutherford Appleton Laboratory in Oxford (UK) (Ammonia Energy Association, 2018). The pilot plant, with a limited capacity of 30 kg of ammonia per day, has the aim to provide insights on the business case for ammonia as a flexible energy storage vector. For this purpose, the facility comprises also a storage and a "combustion asset", to demonstrate the full process round trip (Ammonia Industry, 2018a). Unfortunately, limited information on the project is available.

Table 9.3 Power-to-Ammonia plants, listed alphabetically by country code and chronologically within the same country code.

Project/plant name	Year	Location	Country	Electrolyser power (MW)	Electricity source for electrolysis	Institutions	References
Pilbara plant	2017–2020	Pilbara	AU	50–60	100 MW solar/natural gas	Yara International, ASA, ENGIE	Ammonia Industry (2018b, 2019a), Ammonia Energy Association (2019)
Port Lincoln plant	2018-n.a.	Port Lincoln	AU	30	Wind/solar	H2U, ThyssenKrupp	Ammonia Industry (2018b, 2018c)
Moranbah plant	2019-n.a.	Moranbah	AU	160	210 MW solar	QNP, Neoen, Worley	Australian Renewable Energy Agency (2019, 2020)
Prillex plant	2019-n.a.	Antofagasta	CL	47	62 MW solar	ENEAX, ENGIE	Ammonia Energy Association (2019), Enaex (2019)
Koriyama plant	2014–2018	Koriyama	JP	n.a.	Wind/solar	JGC, AIST, National Institute of Technology, JGC Catalysts, Chemicals Ltd.	Ammonia Energy Association (2018), Ammonia Industry (2018a, 2019b), Yamagami and Fujiwara (2018)
Kapuni project	2019-n.a.	Kapuni	NZ	n.a.	16 MW wind turbines	Ballance Agri-Nutrients, Hiringa Energy	Ammonia Energy Association (2019), Ballance Agri-Nutrients (2019)
Oxford plant	2018-n.a.	Oxford	UK	n.a.	n.a.	Siemens, University of Oxford, STFC, Cardiff University	Ammonia Energy Association (2018), Ammonia Industry (2018a), The Guardian (2018)

The world's largest green ammonia plant powered by renewable hydrogen is planned to be built in Queensland (Australia). The project aims at refurbishing the existing Moranbah ammonia plant with up to 160 MW electrolyser and a 210 MW solar farm co-located at Moranbah for hydrogen production. The plant currently achieves an annual capacity of 360,000 tonnes of ammonium nitrate to be used in the mining industry, and it employs natural gas-derived hydrogen (Australian Renewable Energy Agency, 2020). The estimated production capacity of the refurbished plant is 60,000 tonnes of ammonia per year. The broad project goal is to investigate the feasibility and economic profitability of renewable hydrogen on large scale for domestic use and for export, while reducing the ammonia production environmental footprint (Brown, 2020).

In the decade MW range scale, a consortium made up of Yara International, Engie and ASA started a revamping of Yara Pilbara ammonia production plant (Australia). The project comprises the installation of 200 hectare PV panels for a capacity of 100 MW, coupled with 50–60 MW of electrolyser for hydrogen production. This revamping allows the production of 80 tonnes of green ammonia per day, just a small fraction (less than 4%) of the total annual capacity (840,000 tonnes per year).

In 2019 the companies ENAEX and ENGIE signed a partnership for a feasibility study of a pilot plant for green ammonia production. The project aims at decarbonising the existing ammonium nitrate production plant (Prillex) owned by ENAEX and located in Mejillones (Chile) (Ammonia Energy Association, 2019; Enaex, 2019). The pilot plant will comprise of 62 MW of solar and 47 MW of electrolyser, for a capacity of 20,000 tonnes of ammonia per year (Brown, 2020).

In 2018 a demonstrator project was announced to be built in Port Lincoln (South Australia), by the two companies H2U and Thyssenkrupp Industrial Solutions. The facility will comprise a 30 MW electrolyser system and an ammonia production facility of 50 tonnes per day. The produced green ammonia is expected to be sold to the local agricultural market, as fertilizer. In parallel, the techno-economic viability of the renewable ammonia chain and the role of ammonia in Australia's energy future will be investigated (Ammonia Industry, 2018c, 2018b).

In 2019 the first green ammonia plant in New Zealand was announced, carried out by the companies Ballance Agri-Nutrients and Hiringa Energy. The project will take place at Ballance's Kapuni ammonia-urea plant and will act as a starting point to develop a green hydrogen market in New Zealand to fuel heavy transport (Ballance Agri-Nutrients, 2019). The hydrogen will be produced by four wind turbines with a total capacity of 16 MW, which will decarbonise around 2% of the plant total production (Ammonia Energy Association, 2019).

Besides the aforementioned ones, other large scale green ammonia projects are planned or announced, in Norway, Denmark, Australia, Chile (Brown, 2020), with yet limited information available. This fact, though, confirms the significant interest on Power-to-Ammonia not only to decarbonise the fertilizer industry, but also to synthetise green ammonia as an energy storage and energy vector, potentially addressed to other end-use (power generation, mobility) and to speed the hydrogen economy up.

9.4 Statistics of projects as for number and electrolyser size

Power-to-Methane has been the most developed and investigated technology with a total of 36 investigated projects since 2009 (Fig. 9.1). Within this technology, a higher number of plants are reported for the chemical pathway (21) with respect to the biological one (15). Power-to-Methanol and Power-to-Ammonia show a similar number of developed projects, with eight and seven commissioned or planned plants, respectively.

Even if Power-to-Methane has been deeply investigated and records the higher number of projects, Power-to-Methanol and Power-to-Ammonia show a drastically higher plant average and installed capacity than the Power-to-Methane (Figs. 9.2 and 9.3). Regarding the Power-to-Methane technology, the total installed capacity of the chemical route (22.55 MW) is slightly higher than the installed capacity via

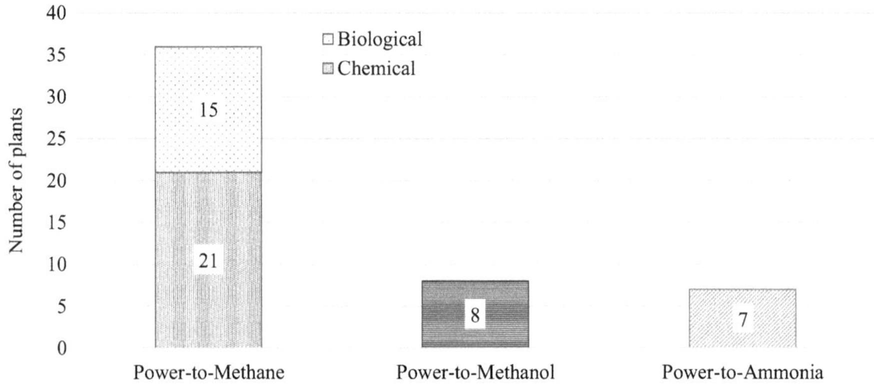

Figure 9.1 Number of Power-to-X pilot and demonstration plants.

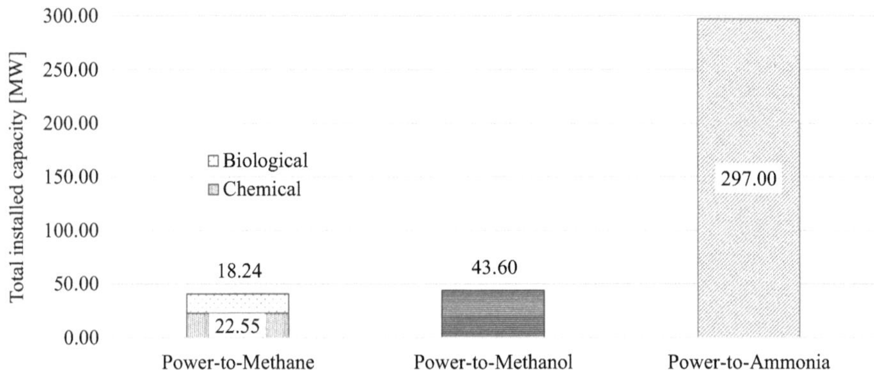

Figure 9.2 Total installed capacity [MW].

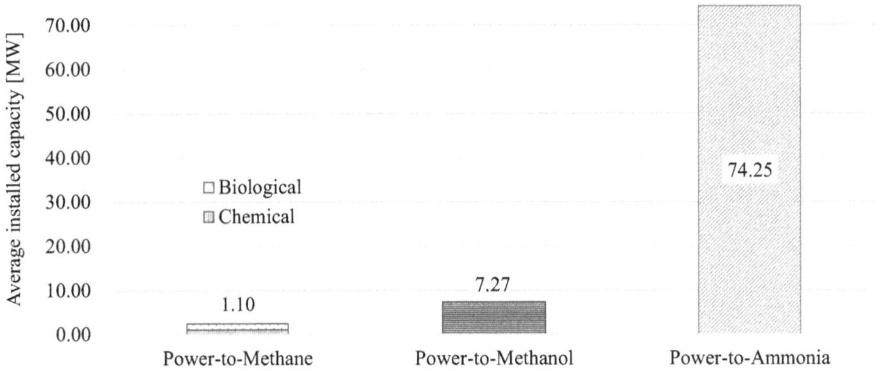

Figure 9.3 Average installed capacity [MW].

the biological route (18.24 MW), even though its average capacity is lower (1.02 MW) than the biological route (1.30 MW). As a matter of fact, as detailed in Section 9.1, both Power-to-Methane pathways have undergone a progressive upscaling. The Power-to-Methanol total and average installed capacity of 43.60 MW and 7.27 MW, namely. The Power-to-Ammonia records higher values by one order of magnitude on both indicators (297.00 and 74.25 MW, respectively).

9.5 Project geographic distribution

As for number of plants, Europe holds the leadership of Power-to-X facilities with a total of 43, followed by Oceania (4) and America (3). On the other hand, Oceania leads in terms of installed capacity, with 250 MW installed or planned, followed by Europe (84.19 MW) and America (47.26 MW).

Germany holds the leadership of Power-to-X plants in terms of number of projects, with 20 projects (Fig. 9.4). Other active countries are Denmark, Netherlands and Switzerland, with 5, 4 and 4 facilities respectively. This distribution is different when considering the overall installed capacity: the country reporting the highest installed capacity is Australia (250 MW), before Chile (47.00 MW), the Netherlands (22.41 MW) and Germany (15.17 MW). This is ascribable to the large size of the Power-to-Ammonia plants that have been installed up to now if compared to the Power-to-Methane ones.

It must be remarked that the installed capacity displayed in Fig. 9.5 does not comprise the facilities where the electrolyser power is not declared, such as the FReSme plant (Sweden), the Koriyama plant (Japan), the Oxford plant (UK) and the Kapuni plant (New Zealand).

Overall, is it clear that in the last decades there has been an increasing interest in Power-to-X plants worldwide for their potential of storing renewable energy into hydrogen vectors while decarbonising many sectors of society. This has led to an increasing number of projects worldwide to investigate new synthesis routes

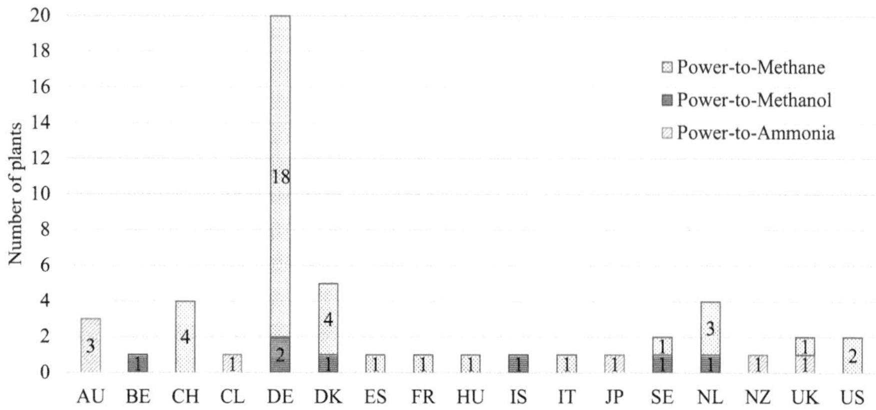

Figure 9.4 Distribution of Power-to-X plants worldwide in terms of number of projects.

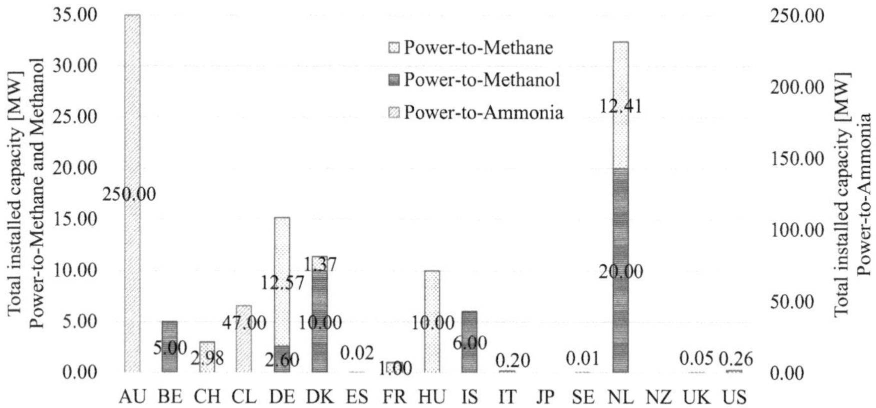

Figure 9.5 Distribution of Power-to-X plant worldwide in terms of installed capacity [MW].

(especially in the case of Power-to-Methane and Power-to-Methanol) or to decarbonise the existing plants (as in the case of Power-to-Ammonia). Further efforts are still necessary to optimise the pathways and make them competitive with the conventional ones and make the hydrogen economy take off.

References

Aalborg Universitet (2019). Power2Met – Renewable Energy to Green Methanol.

Aarhus University (2015). Final report MeGa-stoRE Project no. 12006.

Adamsen, A.P.S. (2013). 'Three innovative methanation technologies', Workshop for the promotion of biomethane in Denmark. Aarhus University, Department of Engineering.

Agricola, A.-C. (2013). 'Die dena-Strategieplattform Power to Gas', in BMBVS-Fachworkshop 'Power-to-Gas: Beitrag zum nachhaltigen Verkehr'. Berlin, Germany.
Ammonia Energy Association (2018). Green ammonia demonstration plants now operational, in Oxford and Fukushima.
Ammonia Energy Association (2019). Green Ammonia Plants in Chile, Australia, New Zealand.
Ammonia Industry (2018a). Green ammonia pilot plants now running, in Oxford and Fukushima.
Ammonia Industry (2018b). Renewable ammonia demonstration plant announced in South Australia.
Ammonia Industry (2018c). ThyssenKrupp's "green hydrogen and renewable ammonia value chain".
Ammonia Industry (2019a). Ammonia plant revamp to decarbonize: Yara Pilbara.
Ammonia Industry (2019b). JGC Corporation demonstrates "world's first" carbon-free ammonia energy cycle.
Australian Renewable Energy Agency (2019). Queensland green ammonia plant could use renewable hydrogen.
Australian Renewable Energy Agency (2020). Renewable hydrogen could power Moranbah ammonia facility.
Baden, N. (2020). 'Power2Met webinar'.
Bailera, M., et al. (2017). Power to gas projects review: lab, pilot and demo plants for storing renewable energy and CO_2. *Renewable and Sustainable Energy Reviews*, *69*, 292–312. Available from https://doi.org/10.1016/j.rser.2016.11.130, Elsevier.
Bailera, M. et al. (2020). Energy Storage - Hybridization of Power-to-Gas Technology and Carbon Capture.
Ballance Agri-Nutrients (2019). Kapuni 'green' hydrogen project seen as catalyst for NZ market.
Benjaminsson, G., Benjaminsson, J. and Rudberg, R.B. (2013). 'Power-to-Gas - A technical review (El-till-Gas-System, ekonomi och teknik) SGC Rapport 2013:284'.
Bergins, C., et al. (2015). Power to methanol solutions for flexible and sustainable operations in power and process industries. *POWER-GEN Europe*, *2015*, 1–18.
Bio-based news (2016). Munich-based clean-tech startup Electrochaea and Hungarian utility MVM establish power-to-gas joint venture.
Blanco, H., and Faaij, A. (2018). A review at the role of storage in energy systems with a focus on power to gas and long-term storage. *Renewable and Sustainable Energy Reviews*, *81*(1), 1049–1086. Available from https://doi.org/10.1016/j.rser.2017.07.062, Elsevier Ltd.
Bowker, M. (2019). Methanol synthesis from CO_2 hydrogenation. *ChemCatChem*, *11*(17), 4238–4246. Available from https://doi.org/10.1002/cctc.201900401.
Brown, T. (2020). 'The forces driving industrial scale-up of Green Ammonia pilot plants and their implications for the fertiliser industry', in. IFS webinar, p. 154.
Buchholz, D. and Reimert, R. (2014). 'An optimized SNG production route', in Wold Bioenergy 2014 Conference & Exhibition on Biomass for Energy. Jönköping (Sweden).
Büker, K. et al. (2018). 'Carbon2Chem ® – A Cross-Industrial Network', in Challenges for Petrochemicals and Fuels: Integration of Value Chains and Energy Transition DGMK Conference October 10 – 12, 2018. Berlin, Germany, pp. 25–32.
BusinessEurope (2018). The Carbon2Chem® project.
Buttler, A., and Spliethoff, H. (2017). Current status of water electrolysis for energy storage, grid balancing and sector coupling via power-to-gas and power-to-liquids: a review.

Renewable and Sustainable Energy Reviews. Available from https://doi.org/10.1016/j.rser.2017.09.003. (February).
Carbon Commentary (2017). Exytron: the world's first 'power-to-gas' system with integrated CO_2 collection and reuse.
Carbon Recycling International (CRI) (2018). Carbon Recycling Internation Projects.
Centro Nacional del Hidrogeno (2018). RENOVAGAS - Getting synthetic natural gas from renewable sources.
Competence Center Energy and Mobility (CCEM) (2020). RENewable enERgies in future enerGy supply (RENERG2).
Danmarks Tekniske Universitet Mekanik (2019). Final report - MeGa-stoRE 2- Phase 2.
dena - Deutsche Energie-Agentur (2013) CO_2RRECT.
DNV GL Oil & Gas (2015). Power-to-Gas project in Rozenburg, The Netherlands.
DVGW Research Centre at Engler-Bunte-Institute of Karlsruhe Institute of Technology (KIT) (2020). STORE&GO.
Elber, U., Hamburger, A. and Lutz, P. (2018). 'CCEM_Final Activity Report_2006–2017', p. 208.
Electrochaea (2016). BioCat Project-Power-to-Gas via biological catalysis.
Electrochaea (2017). 'Electrochaea realisiert Power-to-Gas- Anlage für ein nachhaltiges Pfaffenhofen'.
Electrochaea.dk ApS (2014). Power-to-Gas via Biological Catalysis (P2G-Biocat) Final report.
Enaex (2019). ENAEX & ENGIE: On the path to a smooth zero carbon transition in the Chilean Mining Sector.
Energiforskning.dk (2015a). El upgraded biogas.
Energiforskning.dk (2015b). MeGa-stoRE, Optimering og opskalering.
ETOGAS Gmbh (2013). 'Power to Gas: Smart energy conversion and storage'.
European Commission (2016). MefCO2 - Methanol fuel from CO_2.
European Commission (2017). FReSMe - From Residual Steel Gases to Methanol.
European Commission (2018a). CIRCLENERGY-Production of renewable methanol from captured emissions and renewable energy sources, for its utilisation for clean fuel production and green consumer goods.
European Commission (2018b). Final Report Summary - HELMETH (Integrated High-Temperature Electrolysis and Methanation for Effective Power to Gas Conversion).
European Commission (2020). STEPWISE SEWGS Technology Platform for cost effective CO2 reduction the in the Iron and Steel Industry.
European North Sea Energy Alliance (2015). Power2Gas – energy system of the future.
EXYTRON (2017). 'Das Projekt Alzey in Rheinland-Pfalz. Der weltweit erste ZERO-Emission-Wohnpark'.
EXYTRON (2020). The principle of Smart Energy Technology/Zero Emission Technology.
Fraunhofer-Institut für Energiewirtschaft und Energiesystemtechnik (2016). Direktmethanisierung von Biogas im Technikumsmaßstab: 50 kWel Power-to-Gas-Anlage.
Friedl, M. (2017). Pilot- und Demonstrationsanlage Power-to-Methane HSR.
Friedl, M.J. et al. (2015). 'Pilot and Demonstration Plant Power-to-Methane, HSR', in Heat & Electricity Storage 2nd Symposium. Villingen (Switzerland).
Friedrich-Alexander-Universität Erlangen-Nürnberg (2020). BMWi-Projekt ORBIT – bmwiOptimierung eines Rieselbett-Bioreaktors für die dynamische mikrobielle Biosynthese von Methan mit Archaeen in Power-to-Gas-Anlagen, Teilvorhaben: Auslegung, Konstruktion und Simulation eines Rieselbettreaktors für die biologische.

Fuel Cell and Hydrogen Joint Undertaking (2014). PROGRAMME REVIEW Report 2014. https://doi.org/10.1016/S1158-1360(08)72717-8.

Fujimura, Y. et al. (2019). 'Demonstration and optimization of Green Ammonia production operation responding to fluctuating hydrogen production from renewable energy', Ammonia Energy Topical Conference, 12(November).

Götz, M., et al. (2016). Renewable power-to-gas: a technological and economic review. *Renewable Energy*, *85*, 1371–1390. Available from https://doi.org/10.1016/j.renene.2015.07.066.

Green Hydrogen Systems (2020). First eMethanol plant in Denmark produces liquid electricity.

GRTgaz (2018). Jupiter 1000 - Premier démonstrateur industriel français de POWER TO GAS.

Haldor Topsøe A/S (2018). Electrical upgrading of biogas.

Hashimoto, K. et al. (2002). 'Materials for global carbon dioxide recycling', 44, pp. 371–386.

Hashimoto, K. et al. (2014). 'The production of renewable energy in the form of methane using electrolytic hydrogen generation', pp. 1–9.

Heller, T. (2016). 'BioPower2Gas – Power-to-Gas with Biological Methanation', (MicrobEnergy GmbH).

Hoekman, S. K., et al. (2010). CO_2 recycling by reaction with renewably-generated hydrogen. *International Journal of Greenhouse Gas Control*, *4*, 44–50. Available from https://doi.org/10.1016/j.ijggc.2009.09.012.

Hofstetter, D. (2013). 'Biomethane Production via Power-to-Gas', UK Biomethane Day, (May).

Holstein, J. et al. (2015). 'Experiences with one year operation of power-to-gas in Rozenburg, NL', in 2nd International Conference on Renewable Energy Gas Technology. Barcelona, Spain.

Hüser, T. (2018). Power-to-Gas-Anlage um Methanisierung erweitert.

IdE Institut dezentrale Energietechnologien GmbH (2020). BioPower2Gas.

IEA Bioenergy (2018). 'BioPower2Gas in Germany'.

IET-Institut fur Energietechnik (2020). Power-to-Methane Demonstrationsanlage der Schweiz.

Ikäheimo, J. et al. (2020). 'Target context and requirements for conversion module deployment in the smart grid', (773839).

Inovyn (2020). INOVYN plays role in ambitious 'Power to Methanol' project in Antwerp.

Karlsruhe Institute of Technology (2013). CO_2 Reaction Using Regenerative Energies and Catalytic Technologies (CO_2RRECT).

Karlsruher Institut für Technologie (KIT). (2020). HELMETH Integrated High-Temperature ELectrolysis and METHanation for Effective Power to Gas Conversion.

Kraftwerk Forschung (2013). CO_2RRECT: Storing electricity while simultaneously capturing CO_2.

Kwak, M. (2020). Attero "Power(ed) to (by) Gas" - The Waste to Power to Gas W2P2G project.

Limeco (2019a). Jetzt geben wir gemeinsam Gas für eine saubere Energiezukunft. Mit der ersten industriellen Power-to-Gas-Anlage der Schweiz.

Limeco (2019b). Leuchtturmprojekt bei Limeco: Gemeinden sagen Ja zur schweizweit ersten industriellen Power-to-Gas-Anlage.

Marlin, D. S., Sarron, E., and Sigurbjörnsson, Ó. (2018). Process advantages of direct CO_2 to methanol synthesis. *Frontiers in Chemistry*, *6*(September), 1–8. Available from https://doi.org/10.3389/fchem.2018.00446.
Mesfun, S. et al. (2017). BIO-SNG production by means of biomass gasification combined with MCEC technique.
Noordelijke Innovation Board (2007). 'The Green Hydrogen Economy in the Northern Netherlands', European Parliament Workshop, pp. 1–8.
Nouryon Industrial Chemicals B.V. (2020). Djewels - Pioneering the European renewable eFuels market.
Ostbayerische Technische Hochschule Regensburg (2013). HS.R und MicrobEnergy setzen Power-to-Gas an Kläranlage Schwandorf um.
Otten, R. (2014). 'Audi- The first industrial PtG plant - Audi e-gas as driver for the energy turnaround'.
PFI (2015). Biorefinery at Pirmasens-Winzeln Energy Park.
PFI Germany (2016). 'Biomethane for the Gas Grid of Municipal Utility Companies', pp. 1–3.
Port of Antwerp (2020). 'Power-to-methanol demonstration project'.
Power-to-Gas Hungary Kft. (2020). Power-to-Gas Hungary.
Rasberry, T. (2015). SoCalGas Comments on June 1, 2015 Workshop and Support of AB 1257 Report.
Recharge (2020). Europe's largest green-hydrogen project awarded €11m grant by EU body.
Reed, J. (2015). 'Hydrogen Energy Storage Activities', DOE Hydrogen and Fuel Cell Technical Advisory Committee Meeting.
Reuter, A. and Schmid, J. (2012). Fraunhofer IWES Annual Report 2011/2012.
Rozzi, E., et al. (2020). Green synthetic fuels: Renewable routes for the conversion of non-fossil feedstocks into gaseous fuels and their end uses. *Energies*, *13*(2). Available from https://doi.org/10.3390/en13020420.
Sánchez, M. (2015). 'RENOVAGAS-Proceso de Generación de Gas Natural Renovable', in Jornada PTEHPC. Grupo de Trabajo Técnico de Almacenamiento y Distribución de Hidrógeno. Huelva.
Schröder, H. (2013). "Power to gas": Versuchsanlage in Planung, Ostfriesen-Zeitung.
Specht, M. (2012). 'Power-to-Gas - Speicherinnovation für erneuerbare Energien: Ergebnisse'. Berlin, Germany.
Specht, M. and Zuberbühler, U. (2013). 'Power-to-Gas: Technology and System Operation Results', in Elements of a Greenhouse Gas Neutral Society. Berlin, p. 46.
STORE&GO (2019). The STORE&GO demonstration site at Troia, Italia.
STORE&GO (2020). The STORE&GO demonstration site at Falkenhagen, Germany.
STORE&GO Switzerland (2019). 'For the first time in Switzerland, renewable methane is produced using an ancient microorganism', pp. 31–32.
The Guardian (2018). Siemens pilots the use of ammonia for green energy storage.
The Viessmann Group (2015). 'First commercial PtG-plant with biological methanation goes live'. Hannover Messe.
The Viessmann Group (2019). Green light for first industrial power-to-gas plant.
Thema, M., Bauer, F., and Sterner, M. (2019). Power-to-gas: electrolysis and methanation status review. *Renewable and Sustainable Energy Reviews*, *112*(January), 775–787. Available from https://doi.org/10.1016/j.rser.2019.06.030, Elsevier Ltd.
University of South Wales - Sustainable Environment Research Centre (2018). Cymru H_2 Wales.

Waldstein, G. (2015). 'Catalytic methanation – the Audie e-gas project as an example of industrialized technology for power to gas', in 2nd International Conference on Renewable Energy Gas Technology, pp. 1–4.

Wulf, C., Linßen, J., and Zapp, P. (2018). Review of power-to-gas projects in Europe. *Energy Procedia, 155*, 367–378. Available from https://doi.org/10.1016/j.egypro.2018.11.041, Elsevier B.V.

Yamagami, A. and Fujiwara, H. (2018). 'World's First Successful Ammonia Synthesis Using Renewable Energy-Based Hydrogen and Power Generation', PR & IR Department, JGC Corporation, p. 3.

Yde, L. (2016). 'MeGa-stoRE', in Partnerskabet for Brint og Brændselsceller. AU Foulum.

Zentrum Sonnenenergie-Wasserstoff-Forschung Baden-Württemberg (2013). Ergebnisse // Results ZSW 2018.

Zuberbühler, U. and Specht, M. (2012). 'Power-to-Gas: Construction and start-up of a 250 kW research plant', in Proceedings of the 7th International Renewable Energy Storage Conference. Berlin, Germany.

Power-to-fuel potential market

10

Daniele Candelaresi and Giuseppe Spazzafumo
Department of Civil and Mechanical Engineering, University of Cassino and Southern Lazio, Cassino, Italy

In the present chapter, a rough idea of the potential penetration of power to fuel in the market is tried to be given, from a perspective of complete replacement of fossil fuels with hydrogen and/or synthetic carbon-neutral fuels. Starting from an analysis of world energy consumption, the amounts of hydrogen, methane, methanol, dimethyl ether, ammonia, urea and formic acid necessary for a complete energy transition were calculated, with a focus on their application in various sectors, such as industrial, transport, building and electricity generation sectors.

Currently the energy market is still heavily reliant on fossil fuels. In 2017 the World Total Primary Energy Supply (TPES) amounted to 13972 Mtoe (million tonnes of oil equivalent). Of these, 27.1% was derived from coal, 32.6% from oil and 22.2% from natural gas (NG), for a total of 81.3% (equal to 11359.2 Mtoe) of energy demand met by fossil fuels (IEA, 2019a). The remaining part is supplied by nuclear (4.9%), hydroelectric (2.5%), biofuels and waste (9.5%) and other renewable sources such as geothermal, wind and solar (1.8%) (Fig. 10.1).

Renewable energy sources (RES) are growing strongly, but they still cover a minimal part of the global energy demand. This is linked both to intrinsic problems of renewable energy sources, already described in Chapter 1 (e.g. storage), and to the difficulty in the electrification of some end uses. By means of power-to-fuel it would therefore be possible to circumvent these obstacles, allowing the gradual, and potentially full, replacement of fossil fuels with renewable-based options. To make it clear how this replacement can be made, it is first necessary to understand how and where fossil fuels are used today.

To date, coal (world production: 3789 Mtoe) is mainly used in the energy transformation sector inside electricity plants or cogenerated heat and power (CHP) plants for power generation, but also in blast furnaces and coke ovens, while main final uses of coal in its form, known as Total Final Consumption (TFC), are in iron and steel production, nonmetallic minerals industry, chemical and petrochemical industry, other nonspecified industries and residential sector (IEA, 2019c). Main uses of coal are shown in Table 10.1.

Almost all crude oil production (4560 Mtoe) is used inside refineries to obtain oil products (world production: 4510 Mtoe). These oil products are mainly middle distillates (1403 Mt, million tonnes), motor gasoline (1006 Mt), fuel oil (434.7 Mt), LPG/ethane/naphtha (414 Mt), aviation fuels (314.6 Mt) and other products such as lubricants, white spirit, bitumen, paraffin waxes and others (567 Mt) (IEA, 2019a).

Power to Fuel. DOI: https://doi.org/10.1016/B978-0-12-822813-5.00009-6

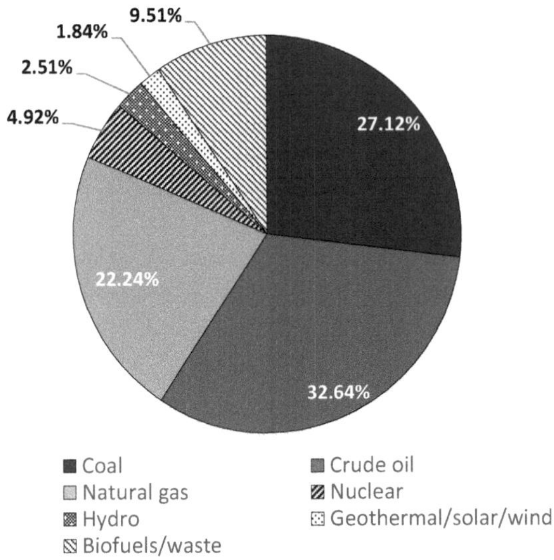

Figure 10.1 World total primary energy supply (TPES) in 2017.
Source: Data from IEA (2019a), Key World Energy Statistics, International Energy Agency. Paris: IEA. Available at: https://www.iea.org/reports/key-world-energy-statistics-2020.

Table 10.1 Main uses of coal.

	Mtoe	% (Coal TPES)	% (Coal TFC)
Energy transformation sector			
Electricity plants and CHP plants	2351	62.05%	-
Blast furnaces and coke ovens	282.2	7.45%	-
Others	135.8	3.58%	-
Total Energy transformation sector	*2769*	*73.08%*	-
Final uses			
Iron and steel production	313	8.26%	30.7%
Nonmetallic minerals industry	217	5.73%	21.3%
Chemical and petrochemical industry	108.46	2.86%	10.6%
Others nonspecified industries	77.25	2.03%	7.6%
Residential sector	75.55	2.00%	7.4%
Others	228.74	6.04%	22.4%
Coal Total Final Consumption	*1020*	*26.92%*	*100%*

TFC, Total final consumption; *TPES*, total primary energy supply.
Source: Data from IEA (2019c), World Energy Balances, International Energy Agency. Paris: IEA. Available at: https://www.iea.org/reports/world-energy-balances-overview.

In the energy transformation industry, oil products are used in relatively small amounts, mainly for energy industry's own use and in electricity plants, while almost all oil products that exit from refineries find employment in final consumptions within various sectors.

The largest share of oil products consumptions belongs to the transportation sector (2588 Mtoe), dominated by road transport and to a lesser extent by navigation and aviation. In addition to transport, not considering the nonenergy uses of oil, this is mainly consumed in the industrial (mainly chemical and petrochemical, nonmetallic minerals, construction, mining, and quarrying), residential, commercial and public services sectors and in the agricultural/forestry sector (IEA, 2019c). Main uses of oil derivatives, both in energy transformation sector and in final energy uses, are shown in Table 10.2.

Table 10.2 Main uses of oil derivatives.

	Mtoe	% (Oil TPES)	% (Oil TFC)
Energy transformation sector			
Energy industry own use	201.6	4.47%	-
Electricity plants and CHP plants	175.59[a]	3.89%	-
Others	162.63	3.61%	-
Total Energy transformation sector	*539.82*	*11.97%*	-
Final uses			
Road transport	1960.37	43.46%	49.38%
Navigation	271.35	6.02%	6.83%
Aviation	323.41	7.17%	8.15%
Rail	28.84	0.64%	0.73%
Nonenergy uses	633.51	14.05%	15.96%
Industrial sector	314.98	6.98%	7.93%
Residential sector	214.76	4.76%	5.40%
Commercial and public services sector	84.66	1.88%	2.13%
Agricultural/forestry sector	107.02	2.37%	2.69%
Others	31.28	0.69%	0.79%
Oil Total final consumption	*3970.18*	*88.03%*	*100%*

TFC, Total final consumption; *TPES*, total primary energy supply.
[a]Another 40.19 Mtoe from crude oil are used for electricity generation.
Source: Data from IEA (2019c), World Energy Balances, International Energy Agency. Paris: IEA. Available at: https://www.iea.org/reports/world-energy-balances-overview.

Table 10.3 Main natural gas (NG) uses.

	Mtoe	% (NG TPES)	% (NG TFC)
Energy transformation sector			
Electricity plants and CHP plants	1199.62	38.61%	-
Energy industry own use	290.77	9.36%	-
Heat plants	64.05	2.06%	-
Others	50.02	1.61%	-
Total Energy transformation sector	*1604.46*	*51.64%*	-
Final uses			
Industrial sector	567.60	18.27%	37.78%
Residential sector	440.58	14.18%	29.32%
Commercial and public services sector	190.31	6.13%	12.67%
Transport sector	104.71	3.37%	6.97%
Nonenergy uses	185.96	5.99%	12.38%
Others	13.17	0.42%	0.88%
Natural gas Total final consumption	*1502.34*	*48.36%*	*100%*

TFC, Total final consumption; *TPES*, total primary energy supply.
Data from IEA (2019c), World Energy Balances, International Energy Agency. Paris: IEA. Available at: https://www.iea.org/reports/world-energy-balances-overview.

World production of NG (3106.8 Mtoe) is consumed roughly half in the energy transformation industry (power plants and CHP plants, heat plants and energy industry's own use) and the other half as end uses in various sectors (Table 10.3).

With respect to TFC, NG is largely used in all industries, in residential sector, commercial and public services and finally in transport sector for road transportation and pipeline transport. The remaining part is for nonenergy uses (IEA, 2019c).

A first important opportunity for the power-to-fuel technique to penetrate the market would be its use to achieve the complete decarbonisation of the hydrogen currently produced, since hydrogen is already used in a variety of applications, especially industrial ones. The decarbonisation of currently used hydrogen would therefore allow, indirectly, a reduction in the carbon footprint of those products obtained from hydrogen. According to International Energy Agency (IEA, 2019b), the demand for hydrogen in the world is continuously growing: in 2018 it reached 70 million tonnes (Mt) in its pure form plus 45 Mt mixed with other gases. Hydrogen in its pure form is required mainly for specific applications such as refineries, ammonia production and use in fuel cells. Hydrogen mixed with other gases (e.g. in syngas), is mainly used in methanol production, in steel production in direct reduction of iron ore processes and other applications where it is required as a fuel or a feedstock. All this hydrogen is almost entirely produced from fossil fuels: 75% from NG (mainly by means of steam methane reforming) and 23% from coal (via coal gasification process), while electrolysis only accounts for 2% of today's global hydrogen production. Being produced from fossil fuels, this hydrogen is responsible for the annual emission of 830 Mt_{CO2}/year. If all the hydrogen required today was produced by electrolysis, using electricity from RES, many sectors could therefore be decarbonised such as the production of ammonia or, partially, refineries and methanol production. Producing all of today's dedicated hydrogen (70 Mt_{H2}) from electricity would result in an electricity demand of 3600 TWh, more than the total annual electricity generation of European Union (IEA, 2019b). In this sense it is possible to start having a first idea of the potential of power-to-fuel (115 Mt_{H2}/year), without taking into account the growth in demand expected in the coming years. In addition, electrolytic hydrogen could be used to produce other hydrogen-based substances, further increasing the range of possibilities. If one thought then, to replace each fossil fuel almost entirely with electrolytic hydrogen or synthetic carbo-neutral fuels, very important opportunities would appear. Present and potential uses of hydrogen and hydrogen-based chemicals will therefore be discussed below.

10.1 Industry

Hydrogen use today is dominated by industrial applications. In both pure and mixed forms, hydrogen is primarily used for oil refining (33%), ammonia production (27%), methanol production (11%), and steel production via direct reduction of iron ore (3%) (IEA, 2019b). Another future potential industrial use of hydrogen, currently still unexplored, would consist in generation of high-temperature heat.

10.1.1 Refineries

In refineries hydrogen is used as a feedstock, reagent or energy source, accounting for about 38 Mt_{H2}/year. This hydrogen demand is met for about one-third by on-site refineries by-products (e.g. from catalytic naphtha reforming), while about two-thirds are met by dedicated on-site production or merchant supply. Hydrogen in refineries is mainly used to remove impurities from crude oil, such as sulphur, and to upgrade heavier crude, through two processes: hydrotreatment and hydrocracking (IEA, 2019b).

Hydrotreatment is the process by which sulphur is removed, together with other impurities, by crude oil to obtain low-sulphur diesel fuel for instance, and to meet regulation standards regarding a variety of fuels. This process is often simply referred to as "desulphurisation" and it is largely hydrogen consuming. Due to growing concern about air quality, increasing regulatory pressure is expected to further reduce the sulphur content of fuels, thereby increasing the demand for hydrogen by refineries.

Hydrocracking is a process used to upgrade heavy residual oils into other oil products or lighter fractions, as they present a higher market value. Since the demand for heavy residual oil is decreasing while demand for light and middle distillate products is growing, the hydrogen need for hydrocracking is also growing. In addition, hydrogen is used by refineries for a variety of other minor processes such as upgrade of oil sands and hydrotreating of biofuels.

Under current trends, overall hydrogen demand in refineries is expected to grow by 7%, up to 41 Mt_{H2}/year in 2030 (IEA, 2019b). If an evolution in line with Paris Agreement is instead considered, the hydrogen demand by refineries would decline, as a consequence of oil demand decline. The replacement of this fossil-based hydrogen, with hydrogen produced by power-to-fuel would provide a potential early market for hydrogen from cleaner pathways, which could lower the emissions intensity of transport fuels.

In a long-term perspective, however, if all the oil used to produce fuels were replaced with hydrogen produced via electrolysis, about 3572 Mtoe/year coming from fossil fuels would be avoided, corresponding (with the same energy content) to a hydrogen production of 1246 Mt_{H2}/year.

10.1.2 Chemical sector

Power-to-fuel would represent an important opportunity to reduce levels of carbon emissions derived from the chemical sector since, also in this case, the vast majority of hydrogen required is produced from fossil fuels. The chemical sector output consists of a large variety of chemicals such as plastics, fertilisers, explosives and solvents. Anyway, seven primary chemicals can be considered: ammonia and methanol mostly, followed to a lesser extent by other "elementary bricks" used to obtain higher value chemicals, namely ethylene, propylene, benzene, toluene and xylene. These seven chemicals account roughly for around two-thirds of the chemical sector's energy consumption and feedstocks. In particular, ammonia production (175 Mt/year), which requires 31 Mt_{H2}/year, and methanol production (97 Mt/year), requiring 12 Mt_{H2}/year, represent the second and third hydrogen consuming sectors. In 2018, around 270 Mtoe/year of fossil fuels were used to produce the hydrogen needed for these two products, mainly from NG (IEA, 2019b). Only Asia-Pacific region, namely China, presents a large share of hydrogen production from coal.

A large part of ammonia (around 80%) is used for production of fertilisers such as urea and ammonium nitrate, while the remaining part is used for the manufacturing of explosives, synthetic fibres, pharmaceuticals and other chemicals for industrial applications (Market Research Future, 2020; Ye *et al.*, 2017).

On the other hand, methanol is used as a basis for the production of several other chemicals like dimethyl ether (DME), formic acid, formaldehyde, methyl methacrylate and various solvents (Grand View Research, 2019). In some regions, methanol also find application in the methanol-to-gasoline process. Two other processes, still in the demonstration phase, namely methanol-to-olefins and methanol-to-aromatics could increase the demand for methanol, for the production of plastics and other high added value chemicals (Tian *et al.*, 2015; Wang *et al.*, 2014).

Demand for hydrogen for ammonia and methanol production is set to increase from today's 44 MtH_2/year to 57 Mt_{H2}/year by 2030 and to 65 Mt_{H2}/year by 2050, without taking into account the possibility of their use as hydrogen carriers or as synthetic renewable fuels (IEA, 2019b).

Obviously, to completely de-fossilise ammonia and methanol production, it is not enough to de-fossilise only hydrogen production: considering the current demand level, ammonia would need also 144 Mt_{N2} with low environmental impact (e.g. separated from air using RES power), urea would need 112 Mt_{CO2} and methanol of around 133 Mt_{CO2} captured from air or from biogenic sources.

10.1.3 Iron and steel production

Direct reduction of iron-electric arc furnace (DRI-EAF) is a method for producing steel from iron ore, using a mixture of hydrogen and carbon monoxide as a reducing agent (Hsieh, 1979; Krüger *et al.*, 2020). This process today accounts for a hydrogen demand of 4 Mt_{H2}/year, corresponding to a 7% of the global primary steel production (1809 Mt of steel in 2018). Around 75% of this dedicated hydrogen production comes from NG, while the remaining 25% is derived from coal. Like other sectors, also the iron and steel sector produces hydrogen as a by-product, often in mixed form with other gases (e.g. coke oven gas). A part of this by-product hydrogen (9 Mt_{H2}/year) is internally consumed within the sector or burned inside blast furnace (basic oxygen furnace process) and another part (5 Mt_{H2}/year) is distributed for use in other sectors (IEA, 2019b).

Hydrogen produced via electrolysis could be therefore an interesting opportunity of decarbonisation for the existing steel production processes. Furthermore, in order to reduce steel production emissions, many efforts are being made to promote the development of novel production processes, still in the demonstration phase, consisting of a particular DRI-EAF that use only hydrogen as the key reducing agent. First commercial scale plants are expected for 2030 (HYBRIT Project, 2019). On a current trends basis, the global steel demand is set to increase by around 6% by 2030. Hydrogen demand from the iron and steel sector would increase from 4 to 8 Mt_{H2}/year by 2030, accordingly with the growth of the DRI-EAF process, from 7% to 14% of primary steel production. In the hypothesis of a 100% steel production by means of DRI-EAF process by 2050, hydrogen demand could theoretically reach 62 Mt_{H2}/year. In case the hydrogen breakthrough iron-making technology (HYBRIT) concept would satisfy 100% of primary steel demand by 2050, the hydrogen demand would instead potentially be equal to 67 Mt_{H2}/year (IEA, 2019b).

10.1.4 High-temperature heat generation

High temperature heat production for use in industrial applications could become another important source of hydrogen demand in the coming years, although there are currently no applications in this regard. In the industrial sector, heat is a utility required for a wide range of different processes, such as chemical reactions, gasification, drying, melting, etc. Heat can be used directly, as in a furnace or an oven, or indirectly by means of a hot fluid such as steam, air, pressurised-water and so on. The classification of heat for industrial uses is given on the basis of temperature ranges: low temperature (<100°C), medium temperature (100°C–400°C) and high temperature (>400°C). The current demand for high-temperature heat in industry is about 1280 Mtoe/year, of which 370 Mtoe/year (29%) are not consumed in the chemical and iron/steel sectors. Roughly half of this high-temperature heat (185 Mtoe) is consumed inside cement production plants. To date, almost all industrial high-temperature heat is derived from fossil fuels (around 65% from coal, 20% NG and 10% from oil). A small part of high-temperature heat is also produced by using biomass or waste. Electricity is widely used in industries to produce high-temperature heat, such as in the production of carbon fibre or in the electric arc furnaces, however electricity is often produced using fossil fuels. The level of demand for high-temperature heat is expected to grow, from 370 to about 400 Mtoe/year by 2030, which could potentially be provided by 130–140 Mt_{H2}/year (IEA, 2019b) or 335 Mt of renewable synthetic methane. Similarly, it is possible to calculate, with the same energy content, the amount of renewable synthetic fuels needed to satisfy the high-temperature heat demand.

10.2 Transport

Although in recent years there has been a rapid technological evolution to allow the electrification of some segments of the transport sector, fuels still remain a reliable option for the majority of transport modes. Potentially, all types of transport could be run on hydrogen or synthetic renewable fuels thus enabling emission reduction. In particular, hydrogen has been a subject of interest for many years as a potential clean transport fuel, since it can be produced from water and could emit only water as its main waste product, especially if is used in a fuel cell. Strictly speaking, when hydrogen is burned with air in internal combustion engines or turbines, it produces nitrogen oxides as a secondary emission with respect to water (Verhelst and Wallner, 2009). However, one of the main advantages of hydrogen is that it does not emit carbon in its direct use. For all applications in transport, hydrogen presents a main technical difficulty linked to the volume occupied: considering the energy content on a mass basis, hydrogen contains 120 MJ/kg (around three times more energy per kg than gasoline or diesel), but its energy content on a volume basis is very low, around 10.7 MJ/m^3 at standard conditions. Typically the storage pressures used for transport applications are 350 bar and 700 bar (energy density of about 4700–4900 MJ/m^3, but still around seven times lower than that of conventional fuels). To further increase the energy density, hydrogen can be stored in liquid form, but this requires large amounts of energy for liquefaction. Furthermore, many other technologies for efficient, safe and compact hydrogen storage are under study. In the meantime, the use of other renewable fuels produced through power-to-fuel could allow to circumvent this obstacle, using already consolidated technologies while reducing emissions at the same time. All synthetic fuels presented in this book have a certain range of potential applications in transport, in particular methane, methanol and dimethyl ether (DME) have direct uses also in internal combustion engines, while ammonia, urea and formic acid could primarily play a role as hydrogen carriers. Some of these fuels could also be used directly in specific fuel cells. In the following paragraphs the various possibilities will be analysed case by case. In addition to those considered in this book, there are a number of other liquid fuels suitable for specific applications (e.g. synthetic kerosene or jet fuel in aviation).

10.2.1 Cars

To date, very small amounts of hydrogen are used for automotive purposes, mostly in demonstration projects or, to a limited extent, in niche market present in some regions of the planet such as California, Japan and Germany. In 2018, approximately 11,200 hydrogen cars were on the road, with a growth of 56% compared to 2017, equal to 4000 new cars sold (IEA, 2019b). Although they have been growing rapidly in recent years, these numbers are still very small when compared to the stock of battery electric cars in the same year (5.1 million) and to that of global cars (more than 1 billion). In the automotive sector, hydrogen can be used in fuel cell electric vehicles (FCEVs) or in internal combustion engines (ICEs).

FCEVs have the advantage of having no tailpipe emissions (like battery electric vehicles) with exception for water, therefore they would reduce local air pollution and also global carbon dioxide emissions if 'green hydrogen' is used. Furthermore, fuel cells have an efficiency of about 55%–60%, more than double that the average efficiency of a conventional oil-fuelled ICE, drastically reducing fuel consumption. Due to their characteristics, proton exchange membrane fuel cells (PEMFC) type are mainly used in the automotive sector. When hydrogen is used in ICEs instead, tailpipe emissions mainly consists of water, variable amounts of nitrogen oxides (depending on the operating point of the engine) and very small amounts of carbon monoxide and particulate matter, resulting from the partial combustion of the lubricant oil. Also in this case a reduction in local air pollution would be obtained, if compared to that deriving from traditional gasoline or diesel vehicles. Furthermore, thanks to improved and faster combustion, the use of hydrogen increases the efficiency of the engine compared to a traditional vehicle, although not to the levels of efficiency obtainable with a fuel cell.

Another possibility, explored in the field of research, consists in the use of hydrogen mixtures with a traditional fuel (diesel, gasoline or NG) to feed an ICE. In this case, the advantage lies in reducing emissions and increasing efficiency, as hydrogen improves combustion characteristics. It also makes possible to use modest amounts of hydrogen, thus representing a possible temporary solution, while waiting for a greater diffusion of fuel cells.

Currently the energy demand for road transport amounts to 1960.4 Mtoe of oil plus 43.8 Mtoe of NG, and 60.7% of this demand comes from light-duty passenger vehicles (1216.5 Mtoe) (EIA, 2019; IEA, 2019c). Assuming an average efficiency of 25% for current vehicles, 55% for FCEVs and 40% for hydrogen ICE vehicles, if all cars currently on the road would be replaced with FCEVs, roughly 192 Mt_{H2}/year would be required, while hydrogen-powered ICE cars would require around 265 Mt_{H2}/year.

One of the main obstacles to the spread of hydrogen as a fuel for transport lies in the scarce diffusion of refuelling points. On the contrary, NG can count on an already very extensive distribution network, especially in some countries (Argentina, India, Italy, for example).

Renewable synthetic methane could replace the NG used today to power ICEs, referred to as compressed NG (CNG) vehicles. If all CNG used today for road transport (43.8 Mtoe including also buses and trucks) would be replaced by renewable methane (with the same efficiency as current vehicles), 36.67 Mt_{CH4}/year would be needed. Instead, to completely replace oil and gas used only from cars, and fuel the entire global car fleet with renewable methane, 1018 Mt_{CH4}/year would be needed. Similarly, renewable methanol (2821 Mt/year) and DME (1774 Mt/year) could also be used to feed ICEs, with characteristics similar to those of gasoline and diesel. In particular, DME is very suitable as a possible substitute for diesel fuel. Unfortunately, the calorific value of these two fuels is lower than that of traditional fuels: methanol has an energy content per kg equal to about 40% of that of gasoline, while DME has a lower heating value of 28.7 MJ/kg, equal to approximately 65% of that of diesel fuel. Finally, ammonia (2731 Mt/year) could also be used in ICEs, but due to the low flame propagation speed (5−13 cm/s), it is not optimal for use in automotive engines, however it may burn more rapidly if used in mixture with other fuels (Valera-Medina *et al.*, 2018). Ammonia, urea and formic acid could instead be used as hydrogen carriers, even stored on board, releasing the hydrogen contained to be used for example in a hydrogen-powered fuel cell.

10.2.2 Trucks and buses

Regarding medium-duty and heavy-duty road transport, the same technologies already discussed for cars can be adopted. In this case, however, the power required by the propulsion system is greater, while there are less stringent limitations regarding the volume that can be occupied by the storage system.

In this sense, fuel cell electric buses and trucks can take advantage of large hydrogen tanks and use lower storage pressures, usually 350 bar. Compared to battery electric buses and trucks, fuel cell ones have faster refuelling times (a few minutes versus a few hours), greater autonomy (km that can be travelled between a refuelling and another) and less weight of the energy storage system. Also in this case one of the obstacles to the diffusion of these systems, lies in the lack of a capillary supply infrastructure.

Despite this, buses and trucks are seen as possible forerunners for the use of hydrogen, thanks to their mission characteristics: buses can refuel in their depot, run their public transport service for a certain time and then return back to the depot to refuel again. Therefore, by operating in hub-and-spoke missions, they can circumvent the problem of lack of refuelling points on the road. Trucks, on the other hand, are intended for long-distance journeys, especially long-haul trucks. They can therefore easily be used to reach two refuelling points even very far from each other without the need for intermediate refuelling, spending most of the time on highways, which could be more easily equipped with hydrogen fuelling stations.

As for cars, these vehicles can also use different technologies and alternative fuels. In some countries, CNG buses are quite common, and could easily run on renewable synthetic methane. The same applies to CNG trucks for logistical use, although less diffused. Higher-class trucks for heavier transports need monofuel CNG engines specially designed to meet the required power characteristics. IVECO and other companies are developing engines of this type and are carrying out research in the field of methane-hydrogen and diesel-hydrogen mixtures (IVECO, 2020; Vavra, Bortel, and Takats, 2019). Also in this case, renewable DME lends itself well, due to its characteristics, to replace diesel fuel (Towoju and Dare, 2017). Ammonia and urea are already commonly transported on-board on diesel trucks, albeit in small amounts, to ensure the proper functioning of the exhaust gas after-treatment systems, known as selective catalytic reduction (SCR). On a technological level, therefore, it would be quite simple to transport ammonia or urea as hydrogen carriers, to then use the latter inside hydrogen fuel cells.

The demand for energy from buses is equal to approximately 5.3% of the energy dedicated to road transport, or around 107.2 Mtoe, of which 104.9 Mtoe of oil and 2.3 Mtoe of NG (EIA, 2019; IEA, 2019c). Assuming an average efficiency of 25% for current engines, 55% for FCEVs and 40% for hydrogen-fuelled ICEs, buses could be powered alternatively by 17 Mt of hydrogen in fuel cells or 23.4 Mt of hydrogen in ICEs. Nevertheless, renewable methane (89.75 Mt), methanol (248.71 Mt), DME (156.42 Mt) or ammonia (240.71 Mt) in ICE could be used alternatively, assuming engines with the same efficiency as that of current vehicles ones.

Energy demand by trucks corresponds to 31.7% of the energy dedicated to road transport, or approximately 636 Mtoe (622 oil + 14 NG) (EIA, 2019; IEA, 2019c). Using the same assumptions made for buses, trucks could be powered alternatively by hydrogen in fuel cell (101 Mt), hydrogen in ICE (138.7 Mt), renewable methane (532.3 Mt), methanol (1475 Mt), DME (927.7 Mt) or ammonia (1427.6 Mt) in ICE.

10.2.3 Trains

Rail transport is already widely electrified in many countries. Anyway, hydrogen could be used to meet decarbonisation targets in nonelectrified railways. Today, nonelectrified lines are mostly served by diesel-powered trains. Some of these lines could easily be electrified, at low cost and with minor technical complications, while for other lines the electrification costs are very high or construction feasibility is technically challenging. In these situations, hydrogen fuel cell trains could take over, replacing diesel-powered trains. Some Alstom demonstration projects have already been launched in Germany, with Coradia iLint train, while some countries have already planned to purchase tens of hydrogen trains for the next few years, like UK, France, and recently also Italy with an agreement between Snam and Alstom (Global Railway Review, 2020; Snam, 2020). Hydrogen fuel cell trains could be competitive especially for long distance movement of large trains combined with a low-frequency network utilisation. These two conditions are quite common for rail freight. The energy demand from nonelectrified trains, therefore fuelled with petroleum-derived fuels, is equal to 28.84 Mtoe/year (EIA, 2019; IEA, 2019c), that using the same assumptions on efficiency described above, could be satisfied alternatively by 4.57 Mt_{H2} in hydrogen fuel cells, or by renewable methane (24.14 Mt), methanol (66.90 Mt), DME (42 Mt), or ammonia (64.7 Mt) in ICE.

10.2.4 Ships

Ships are typically powered by large and slow two-stroke diesel engines. In the naval sector, growing environmental concerns are slowly translating into regulatory pressures, in particular on the acceptable sulphur content in the diesel fuel burned and on emissions of nitrogen oxides and sulphur oxides. Nevertheless, the International Maritime Organization (IMO) has set the goal to reduce the total annual greenhouse gas emission by at least 50% by 2050 compared to 2008 and, eventually, fully eliminate harmful emissions (International Maritime Organization, 2018). Limitations on CO_2 emissions instead have been set via the energy efficiency design index adopted by IMO. To achieve these goals, the global maritime industry has begun to consider carbon-free and sulphur-free fuels such as hydrogen and ammonia, or carbo-neutral synthetic fuels such as renewable methane or renewable methanol.

The application of hydrogen fuel cells or electric batteries has been demonstrated on small ships used for shorter routes like ferries, but it still appears to be temporally distant for trans-oceangoing vessels (MAN Energy Solutions, 2019). Over the years, Man Energy Solutions has tested and developed various marine engines capable of using alternative fuels, including NG and liquefied natural gas (LNG), liquefied petroleum gas (LPG) propane and butane, ethane, ethanol, methanol, and finally ammonia.

Hydrogen in gaseous or liquid form would also be a suitable fuel for marine ICEs. Compressed gaseous hydrogen entails major problems regarding the volume occupied by storage tanks, with consequent loss of cargo, while liquid hydrogen (at −253°C) has a higher volumetric energy density but requires the use of cryogenic tanks. Liquid hydrogen ships could also be used for hydrogen imports and exports, consuming a small part of it for the propulsion of the ship itself. As already happens for LNG, liquid hydrogen would then need to be brought back in gaseous form once on land. For some boats and ships, the metal hydrides option is very interesting. They are very compact and very safe, but for mobile applications they have the drawback of being very heavy. However, this is not a drawback for boats and ships which have a fixed ballast to provide stability since metal hydride tanks can replace such a ballast.

Also ammonia is a very promising fuel for ships: it constitutes a quite good energy storage solution since it has a higher volumetric energy density than liquid hydrogen, but it is less expensive and complex to transport and store. Used as a fuel inside an ICE, it does not emit carbon or sulphur. In the case of marine engines, which are much slower than automotive ones, the ammonia flame propagation speed is sufficiently high (Dimitriou and Javaid, 2020). Moreover, ammonia can count on well-established production, management and storage methods, therefore has the potential to enter the market relatively quickly. If 'green ammonia' is used, it is also possible to strongly reduce the carbon footprint. Finally, ammonia transported by ship could act not only as a fuel but also as an efficient liquid hydrogen carrier for hydrogen imports/exports. Another important characteristic of ammonia is related to safety on board, because ammonia is much less explosive than hydrogen. Australia has recently launched a strategy that plans to export large amounts of green ammonia by ship to Japan and China, where it would then be converted into hydrogen (ANT Energy Solutions, 2013; Asian Rehub, 2020).

Starting from an energy demand from ships of 271.3 Mtoe (IEA, 2019c) and using the same assumptions on efficiencies described above, these could be powered alternatively by hydrogen in fuel cell (43.03 Mt), liquid or gaseous hydrogen in ICE (59.17 Mt), renewable methane (227.13 Mt), methanol (629.41 Mt), DME (395.85 Mt) or ammonia (609.2 Mt) in ICE.

10.2.5 Aviation

Decarbonisation is a major challenge for aviation. The aviation sector is responsible for the emission of about 900 Mt_{CO2}/year. Despite the fact that the efficiency improvement targets are set by the International Civil Aviation Organization (ICAO) at a growth of 2% per year (ICAO, 2018), emissions are expected to more than double by 2050. The Air Transport Action Group (ATAG) has set a goal of a 50% reduction in CO_2 emissions by 2050 compared to 2005 levels, while the European Union has set a more ambitious target of carbo-neutrality by the same year. In addition to carbon dioxide, aircraft also emit carbon monoxide, unburned hydrocarbons, nitrogen oxides, soot, and water vapour, which create contrails and cirrus clouds (McKinsey and Company, Clean Sky 2 JU and Fuel Cell and Hydrogen 2 JU, 2020). Given these targets, it is urgent to start implementing decarbonisation measures, even in the short-term.

To date, aircraft mostly use jet-fuel or jet-propellant, a fuel derived from kerosene, within aircraft turbines or piston-based engines. In some cases this can also be derived from – or mixed with – gasoline or naphtha, depending on the type of aircraft or engine. Although revolutionary electrically-propelled aircraft have been proposed, such as those powered by photovoltaic cells, fuel cells, or ultracapacitors, gas turbines will remain the most reliable and economically competitive option for many years. This is also because gas turbines have an excellent ratio between power output and weight, combined with the high energy density of liquid fuels, which allows aircraft to travel long distances. In this sense, large commercial aircraft, especially those used for longer journeys, seem to have few alternatives to liquid fuel, at least for the short and medium term. Moreover there are rigorous safety procedures, which imposes stringent quality standards on the characteristics of the propellant fuel used. Considering that aircraft are often refuelled in different states, and that some states could have different jet fuel quality, it is required that these technical fuel specifications are harmonised (ICAO, 2018).

To face environmental problems even in the near term, the aviation industry is developing alternative Sustainable Aviation Fuels (SAFs). These can be of the "drop-in" type, that is kerosene-like fuels, which can be distributed with the same infrastructure and can be burned in the same aviation turbines already in use without any adaptation, while allowing emissions to be reduced. Drop-in SAFs therefore, represent a quick substitute for conventional jet fuel, completely interchangeable or mixable with it, and can be used 'as is' on currently flying aircraft. On the contrary, any 'non drop-in' SAF would involve safety concerns and major adaptations. To meet these characteristics, a certain number of drop-in SAF have been developed in recent years. Among these, biofuels are very promising, such as Hydroprocessed Esters and Fatty Acids (HEFA) that can be produced from biomass or waste and other advanced biofuels producible from crops, algae, non-food biomass, municipal wastes, cooking oil and agricultural residues. Another important option is given by the so-called 'synfuels' or 'electrofuels' that use power combined with Fischer-Tropsch process to produce a liquid drop-in SAF similar to kerosene. By combining electrolytic H_2 produced from RES and CO_2 captured from a nonfossil source, a carbon-neutral drop-in fuel can therefore be obtained.

The global energy demand from aircrafts amounts to 323.4 Mtoe (IEA, 2019c). Through power-to-fuel technologies (with the same assumptions on efficiencies described above), aircrafts could be powered alternatively by hydrogen in fuel cell (51.29 Mt), hydrogen in turbine (70.52 Mt, better if liquid but also in gaseous form), or renewable methane (270.7 Mt), methanol (750.17 Mt), DME (471.80 Mt) or ammonia (726 Mt) in turbine. All these fuels are potentially exploitable to power aircraft, but not being 'drop-in' type fuels, they are among the new propulsion technologies.

The combustion of hydrogen inside aeronautical turbines (with low NO_x emission) is feasible, but requires the development of dedicated turbines still under study and the resolution of storage systems and refuelling problems. Assuming these technical developments, H_2 propulsion would be initially best suited for commuter, regional, short-range, and medium-range aircraft. Even long-range aircrafts could be powered by hydrogen, but they would be subject to major design changes and to the evaluation of economic convenience by 2050. Already in the early 2000s Airbus together with other airlines, universities and research centres explored the use of liquid hydrogen in aircrafts as part of the Cryoplane European project, while in recent years other projects involving hydrogen aircrafts financed by large private investors have begun to flourish (CNN, 2020; European Commission, 2002).

Finally, for all new propulsion technologies, the year of entry-into-service must be taken into consideration. Conventional aircraft development cycles occur about every 15–20 years until a new aircraft platform is introduced, while older fleets retire. For short-range aircraft, which make up the bulk of emissions, the next window of opportunity is expected to be around 2030–2035 (McKinsey and Company, Clean Sky 2 JU and Fuel Cell and Hydrogen 2 JU, 2020).

10.3 Buildings

The global buildings sector alone accounts for 2848 Mtoe, corresponding to about 30% of TFC. Of this energy demand (electricity included), 72.4% comes from the residential sector and 27.5% from commercial and public services. About 2200 Mtoe of this energy is dedicated to heat provision, such as space heating, hot water production and cooking. Roughly, half of this is produced directly from fossil fuels: NG accounts for 630 Mtoe, oil products for 299 Mtoe and coal for 109 Mtoe (IEA, 2019b, 2019c). Most of the remaining heat production in buildings is met by electrical equipment as electric resistance radiators and induction cook-stoves, heat pumps but also by district heating. To date, this energy is also produced mostly (85%) from fossil fuels, even if indirectly. Besides, solid biomass for heating purposes is still very significant in developing countries. Overall, the buildings sector is responsible for a 28% of global CO_2 emissions related to energy uses.

With respect to heat provision, buildings present two main challenges, namely the reduction of heat demand (by improving the energy class of buildings, plants and equipment) and the decarbonisation of emissions connected to heat production. In addition to heat pumps, many alternative technologies such as geothermal energy or solar thermal have been proposed, but their diffusion depends on many factors in the case of buildings: location and type of building, ownership, customer preferences, perceived comfort, reliability, capital and operating costs, etc. It is therefore very likely that various technologies will coexist in the future. In this scenario, power-to-fuel offers many possibilities for decarbonisation, both in the near and in the long term. The first that can come to mind is the partial or total replacement of NG with renewable synthetic methane. This would make possible to continue to use the already existing distribution infrastructure as well as existing devices such as NG boilers. Also hydrogen arises as a suitable decarbonisation solution both in the near term, through injections in small amounts in the NG pipelines, and in the long term in a dedicated 100% hydrogen infrastructure.

10.3.1 Renewable methane

To date, the global methane distribution network can count on a large extension of approximately 3 million km of pipelines, to which are added an enormous underground storage capacity and international trade of LNG transported by ship. The global demand for NG today (already presented in paragraph 10.1) is approximately 3900 billion m^3, corresponding to 3106.8 Mtoe (IEA, 2019c). Replacing all this NG with biomethane would be impossible due to concerns about land use change and competition of energy with food crops. Even considering replacing with biomethane only the 630 Mtoe of NG used for heating buildings, it would mean facing a 90-fold increase in biomethane production in the European Union and a 20-fold increase with respect to current world production (IEA, 2019b). Power-to-methane could overcome these limitations by producing renewable synthetic methane. Compared to hydrogen, the disadvantage consists in a lower production efficiency and a higher cost, associated with the additional methanation process. This could probably lead to an increase in gas prices for final consumers, if compared to NG. Regarding the production of heat in buildings alone, power-to-methane could replace the 630 Mtoe of NG used today with 527 Mt_{CH4} of synthetic renewable methane, while the replacement of all fossil fuels used for direct heat provision in buildings (1038 Mtoe) would need 868 Mt_{CH4}.

10.3.2 Hydrogen use in buildings

Currently, hydrogen is not used in the buildings sector, except for some demonstration projects, which are trying to explore different possible future uses of the fuel. Some of these projects are experimenting hydrogen blending in the NG pipelines, other projects instead involve supply of pure hydrogen to various devices such as hydrogen boilers, fuel cells or burners. The current largest project regarding 100% hydrogen supply via dedicated hydrogen pipeline to buildings is the H21 North of England (H21, 2018). This project also demonstrated the feasibility of reusing the existing (polyethylene) pipeline network and it is testing hydrogen boilers. In addition, the largest demonstration projects concerning stationary fuel cells for residential and commercial buildings and cogeneration for residential use are ene.field in Europe and ENE-FARM in Japan (Nagashima, 2018; Nielsen and Prag, 2017). In the Japanese project, developed since 2009 by a consortium that includes Panasonic and other major energy suppliers/fuel cell manufacturers, NG or LPG are reformed locally to produce hydrogen to feed fuel cells, thus bypassing the problem of hydrogen distribution up to the building. In addition to generating electricity, hot water is also produced for domestic use, achieving a total declared energy efficiency of 97% (ENE-FARM, 2020). Regarding numbers of units, around 300,000 should be in operation by the end of 2020, while 5.3 million units are expected by 2050 (Nagashima, 2018). In Europe, on the other hand, more than 1000 micro-CHP units have been installed around 11 countries and an increase to 2800 units is expected in the coming years.

Coming to a future perspective, a multitude of factors, including population density and energy class of buildings, could influence the future demand for hydrogen by this sector. For example, condominiums, multi-family buildings and large commercial buildings represent good opportunities for the use of hydrogen. In the second case, buildings older than 25 years (the majority of buildings in Europe and United States) represent an energy-intensive market share that will continue to demand a large amount of heat for many years to come, that could be satisfied by hydrogen. A complete electrification of consumption by means of electric heat pumps is not adequate for these buildings: unless major improvements are made in building energy efficiency at the same time, it could lead to large seasonal imbalances in the demand for electricity. Anyway, there are many opportunities for hydrogen use in buildings, which can be classified into two main categories: hydrogen blending in existing NG network and direct use of pure hydrogen for heat (and/or electricity) production in buildings. A third option regards indirect use of hydrogen to heat or cool buildings by using centralised systems in neighbourhoods.

10.3.2.1 Blending

Thanks to the possibility of injecting hydrogen into existing NG distribution infrastructure, it would be possible to delay the significant investment in a dedicated hydrogen distribution network later in time, focusing first on increasing the production of hydrogen from RES. It is also possible to vary the mixing percentages and this would ensure a gradual energy transition as well as a progressive reduction in CO_2 emissions. At the same time, this possibility has some disadvantages. First of all an obstacle to be overcome consists in the harmonisation of the regulations regarding the percentage of hydrogen admissible within the gas network, not only at national level but also at the border level. Secondly, a hydrogen injection could lead to an increase in the cost of NG for consumers, which must be carefully evaluated.

Another problem is linked to the lower calorific value of hydrogen per unit of volume compared to NG. The main concerns would be a reduction in the ability to transport energy by pipelines, the need for consumers to use larger volumes to meet an energy requirement, the adjustment of energy metering by gas meters and gas supply for some industries. Finally, there are technical limitations on the percentage of hydrogen admissible in a mixture with NG. The upper limit of this amount is mainly dictated by the various users or equipment connected to the network, in particular by those with a lower tolerance level.

Some existing components already have a high hydrogen tolerance without any need for upgrades, for instance many European gas heating and cooking appliances can tolerate up to 23% hydrogen. The Ameland project, in the Netherlands, tested successfully some equipment for heat provision in buildings such as boilers, gas hobs and cooking appliances (Kippers *et al.*, 2011) with hydrogen mixtures with natural gas until 20%. In some cases, no problems were found until 30%. Other elements of the network such as gas meters and distribution and transmission networks also appear to have high levels of tolerance to hydrogen. Generally speaking, the most problematic elements from this point of view are compressors (10%), engines (5%), turbines (2%−5%) and some type of CNG tanks (2%), but in many cases it is possible to overcome these limitations with minor modifications which allow to increase their tolerance, for example, by changing seals in gas turbines (Altfeld 7 Pinchbeck, 2013; IEA, 2019b). These limit values are derived from legislative and/or technical limitations designed for generic cases, but there are components designed specifically for individual specific applications that can easily exceed them (e.g. some stationary ICE).

According to International Energy Agency, considering that the current demand for NG in terms of volume is equal to 3900 billion m^3, a replacement of 3% of this

NG with an equal volume of hydrogen would need about 12 Mt_{H2}/year (IEA, 2019b). However, this approach would lead to a reduction in the energy delivered by gas. Instead, if a replacement of the energy that was originally supplied by NG with an equivalent amount of energy supplied by hydrogen is considered, a 3% hydrogen blending would require around 31.8 Mt_{H2}. Both of these numbers are very significant and could lead to a major boost to the diffusion of hydrogen, creating an economy of scale for electrolysers. Similarly, considering the possibility of increasing the hydrogen percentage at a later time, a 20% hydrogen blending in current global uses of natural gas would need 69 Mt_{H2} for a replacement for the same volume and 212 Mt_{H2} for a replacement for the same energy delivered. Considering only the natural gas used in buildings for the supply of heat, and a substitution approach for the same energy delivered, a 3% hydrogen blending would bring about 6 Mt_{H2} in the final consumption of the buildings. Similarly a 20% hydrogen mixture would be equivalent to over 39 Mt_{H2} consumed in buildings.

10.3.2.2 Pure hydrogen

From a longer term perspective, if a dedicated 100% hydrogen distribution network were developed, it would be possible to meet the energy demand of buildings in different ways. Hydrogen boilers could provide the thermal energy necessary for space heating of buildings and for the production of hot water, without any CO_2 emissions. Another option is the cogeneration of heat and electricity using stationary fuel cells, as already seen for the ene.field and ENE-FARM projects. Finally, heating, cooling and electricity demand of buildings could be met by sending hydrogen to local district power plants for the cogeneration of electricity and heat and distribution of hot or cold flows through district energy networks. Potentially if hydrogen became cost competitive, it could replace all the NG dedicated to the production of heat for buildings with 220 Mt_{H2}, while to replace all the heat for buildings generated directly from fossil fuels would take 362 Mt_{H2}.

10.4 Power generation

In 2017, 25,606 TWh of electricity were generated: 21,605 TWh came from electricity plants and 4001 TWh from CHP plants. With respect to the total, 38.52% of electricity was generated from coal, around 3.29% from oil, 22.97% from NG, and 10.29% from nuclear power. The production of hydroelectric energy is equal to 15.94% of the total, while other renewables, albeit in strong growth, still play a rather marginal role: geothermal, wind and solar account globally for 6.64% and biofuels and waste for 2.33% (IEA, 2019a, 2019c). Fossil sources together to a total of 64.78%. Regarding the amount of primary energy from fossil sources used to generate this electricity, 2351 Mtoe of coal, 215 Mtoe of oil and 1200 Mtoe of NG were used (see Tables 10.1–10.3). It could be thought of directly replacing electricity generated from fossil sources with electricity generated from RES, however the nonprogrammability of RES constitutes an obstacle to their greater exploitation. To increase the possibilities of RES exploitation, an energy storage system is needed to decouple production from demand. One of the possibilities is to produce a synthetic fuel, easier to store, and then reconvert it in electricity when needed. This technique is known as power-to-power. Although this double conversion has significant efficiency losses, to date it represents one of the few viable and reliable alternatives to store energy on a large scale and for long periods (e.g. seasonal storage).

All the synthetic fuels presented in previous chapters offer various opportunities for decarbonising the electricity generation sector. In general, the greater the number of conversion steps, the greater the energy losses and the reduction in efficiency. From this point of view, hydrogen would be the most advantageous solution; however the possibility of using existing distribution infrastructures and more or less consolidated technologies, also increases the interest in more complex fuels such as synthetic methane and ammonia. Taking methane as an example, which is already one of the most widespread energy sources, it could be possible to replace NG burned today in power generation plants (1200 Mtoe), with 1004 Mt of renewable synthetic methane without making any change to the existing plants. If electricity generation from coal and oil were also totally replaced by synthetic methane, 1968 Mt_{CH4} and 180 Mt_{CH4} would be needed respectively, for a total of 3152 Mt_{CH4} to obtain a complete defossilisation of this sector using already proven technologies.

To date, hydrogen is not used for the generation of electricity, except in very few industrial sites where it is recovered as a waste product from nearby petrochemical or steel industries and then burned in a gas turbine to produce electricity, or in a few small-scale demonstration plants such as power plants with stationary fuel cells. For example, near Venice in Italy, inside the "Andrea Palladio" ENEL power plant, a 12 MW combined-cycle gas turbine fuelled with hydrogen is installed (Balestri *et al.*, 2007; ENEL, 2009). This turbine uses hydrogen recovered from the nearby petrochemical complex EniChem of Porto Marghera. In particular, the hydrogen comes from the unaccumulable surplus produced by the chloralkali process and emissions for 17,000 t_{CO2}/year are avoided thanks to this turbine. In other cases, however, the hydrogen recovered from nearby industrial sites is not in pure form, but mixed with other gases and is burned in internal combustion engines to produce electricity (and heat). Some of these stationary engines can tolerate hydrogen-rich gases up to $70\%_{vol.}$ today, while future gas engines could run even on 100% H_2 (Herdin *et al.*, 2007). Even some gas turbines can already burn mixtures containing high percentages of hydrogen, such as a 40 MW turbine installed in Korea, near a refinery, that was fuelled with a gaseous mixture containing up to 95% hydrogen for the past 20 years (Moliere, 1999). Finally, another application of hydrogen for electricity (and eventually heat) generation, today mostly at demonstration and small-scale, is in power plants with stationary fuel cells. The installed power worldwide to date is equivalent to 1.6 GW of stationary fuel cells or 363,000 units, anyway only 70 MW of these are fuelled directly with green hydrogen, while the majority uses NG reformed locally to obtain hydrogen to be used in fuel cells. Very few solid oxide fuel cells, currently mostly in demonstration projects, directly use NG as a fuel. Numerically, most of these fuel cells belong to the ENE-FARM project and can be classified as micro-cogeneration systems rather than as power plants, while few high-powered units, between 100 kW and 2.4 MW are around the world, such as in Japan, Korea and United States (Leo and Fuel Cell Energy, 2016; Toshiba, 2017). These high-power units can be combined in a modular way to create power plants, such as a 50 MW fuel cell power plant under construction in Korea (FuelCellsWorks, 2019). In addition to traditional power plants and distributed generation, other possible interesting applications for the generation of electricity using fuel cells consist of the supply of electricity in those situations where one cannot rely on the grid, namely portable generators, uninterruptible power supplies and off-grid power solutions. Ammonia is another promising synthetic fuel for power generation, as it can be used either as a hydrogen carrier or as an actual fuel, alone or in combination with other fuels, in a boiler, a turbine, an internal combustion engine or even in a fuel cell.

10.4.1 Renewable synthetic fuels in power generation

Many fossil fuel plants currently already in operation will remain so for many years to come. In this sense, it is essential to find a way to reduce as much as possible their emissions. Most of the synthetic fuels already present can be used in combination with traditional fuels in existing power plants with only minor plant changes.

For instance, hydrogen could be co-fired together with coal, oil or NG in a traditional boiler for steam generation, or blended with NG in different percentages and burned in a gas turbine. This technique is known as dual fuel combustion. The reduction of emissions in this case would be greater than the simple reduction due to the replacement of a part of fossil fuel, as the addition of hydrogen improves the characteristics and completeness of combustion. Similarly, hydrogen can also be mixed with oil or gas in internal combustion engines for stationary generation, as already discussed for engines for transport applications. Another option consists of the so-called attached-cycles: in this case hydrogen is not co-fired with the fossil fuel, but is burned with oxygen in a mixing superheater, to superheat the evolving fluid from within the fluid itself (Spazzafumo, 2003). The oxy-combustion of hydrogen in fact generates only water vapour and heat. Taking a traditional coal-fired steam plant as an example, it is possible to replace the traditional superheater with a mixing superheater of this type. On the one hand, this simplifies the plant layout, avoiding the necessity to return to the boiler to re-heat the steam, on the other hand it makes possible to reach much higher superheating temperatures, thus improving plant efficiency and performance with respect to a traditional steam plant.

Given its high hydrogen content, ammonia can also play similar roles like hydrogen and can be used as an energy storage and then decomposed to obtain hydrogen, or directly as ammonia in combination with fossil fuels. In this sense, the co-combustion of ammonia with coal seems very promising, since ammonia is currently cheaper and easier to store and manage than hydrogen. In 2017, an electric company in Japan (Chugoku Electric) has demonstrated with success to be able to burn ammonia together with coal at one of its coal plants of 120 MW power. The ammonia content used in the demonstration is 1% in terms of total energy content of the two fuels (IEA, 2019b; Muraki, 2018). Furthermore, IHI has also demonstrated in 2018, in another 10 MW combustion facility in Japan, that it is possible to increase the ammonia content mixed with coal up to 20% in energy terms (Power Magazine, 2020). In particular, the success of the experiment lies in the fact that there was no leak of unburned ammonia in the exhaust gases and that only minor changes are required to the traditional coal plant. According to the IEA estimates, if 20% of ammonia mixed with coal were used in all coal-fired power plants in the world by 2030, an ammonia demand of 670 Mt/year would be created, and this would require 120 Mt_{H2}/year to produce ammonia. The use of green ammonia together with coal could avoid the emission of a considerable amount of CO_2 into the atmosphere, around 1.2 Gt_{CO2}/year according to IEA estimates. The reduction in carbon emissions is accompanied by another side of the coin. Since ammonia contains nitrogen, its use as a fuel also lead to the formation of fuel-NO_x, causing concerns about the rise in total NO_x emissions. However, the 10 MW demonstration plant managed to keep these emissions within the usual limits and prevent unburned ammonia from escaping into the exhaust gases.

Regarding electricity generation based 100% on renewable synthetic fuels, there are still some technical challenges that make this option more likely for a medium-long term time horizon. Most of the gas turbines used today are already able to accept hydrogen levels from 3% to 5%, some can accept up to 30%, while in other cases percentages close to 100% can be reached depending on the design of the turbine (Goldmeer, 2018; Mitsubishi Power, 2019). Standard turbines capable of being fuelled entirely with hydrogen are expected by 2030 (Ditaranto, Heggset, and Berstad, 2020; EUTurbines, 2019).

Ammonia direct combustion in gas turbines has been already demonstrated in some micro gas turbines (powers lower than 300 kW) (Shiozawa, 2019), while larger gas turbines (> 2 MW power) present some technical hurdles that researchers are trying to solve, such as flame stability and speed or containment of nitrogen oxide emissions (Valera-Medina *et al.*, 2018). In addition to the direct combustion of ammonia, it is also possible to use it just as a long-term storage and to decompose it to obtain nitrogen and hydrogen, to burn only the latter in a gas turbine (van Wijk and Noordelijke Innovation Board, 2017).

Assuming a scenario in which hydrogen would replace fossil fuels to feed current power plants and an efficiency of 50% for a hydrogen plant, it would take 353 Mt_{H2} to completely replace only NG and 643 Mt_{H2} to completely replace coal and oil. Similarly (assuming an efficiency of 40% for an ammonia plant), 2839 Mt of ammonia would be needed to replace NG and 5167 Mt_{NH3} to replace coal and oil.

Furthermore, for hydrogen exists the possibility of carrying out advanced thermodynamic cycles which reckon on direct steam generation, by means of oxy-combustion of hydrogen in special burners/mixture superheaters, which mix the steam already present with that generated by hydrogen combustion. The steam thus obtained is made to expand in a steam turbine, but the thermodynamic efficiency of the cycle is higher with respect to a traditional Hirn cycle because it is possible to exceed the temperature limits, normally dictated by the boiler (Spazzafumo, 2003). To make the most of these cycles would require turbines and materials capable of withstanding high temperatures, which could be developed in years to come.

Synthetic fuels obtained through power to fuel, in particular hydrogen, synthetic methane, methanol, DME and ammonia, can also be used in internal combustion engines for stationary generation, such as those for cogeneration plants. An interesting application of them is also in off grid systems and portable generators which will be treated in the next paragraph. Fuel cells stacks can also be used to obtain a decarbonised and flexible electric energy system at the same time. These have several advantages, in fact they have a high efficiency (today around 50%−55%) and show little or no emissions, moreover they are modular and without moving mechanical parts, which makes construction and management very simple both for a centralised generation plant or a distributed generation system. To date, however, stationary fuel cells still have to face some weaknesses, such as high costs, a shorter lifespan and lower power than turbines. At the moment, therefore, fuel cells seem more suitable for distributed generation, but in the future a normal building could house several modular fuel cell stacks for a zero-emission power plant.

Furthermore, fuel cells, depending on the type of cell, can be fuelled not only with hydrogen, but also with methane, methanol, ammonia, urea and formic acid. If the current generation of electricity from fossil sources in thermoelectric power plants were replaced with synthetic fuels in fuel cells (considering an efficiency of 50%), it would alternatively take 996 Mt_{H2} to totally replace fossil fuels, or 2388 Mt_{CH4}, 6618 Mt of methanol, 6405 Mt_{NH3}, 13,127 Mt of urea or 40,493 Mt of formic acid.

10.4.2 Back-up and off-grid power

Some applications require an uninterrupted supply of electricity even in the event of a blackout, In this case the constant supply of electricity can be guaranteed through uninterruptible power supplies that are activated only when needed.

Moreover there are isolated or mobile applications. In this case power generators are required. Nowadays, most generators for back-up electricity supply or for off-grid applications are internal combustion engines, often fuelled by diesel. Also hybrid systems with diesel generator and batteries or photovoltaic panels and batteries are used, while fuel cells, still not very widespread today, could represent in the near future the optimal solution to reduce emissions in this sector.

The most promising fuel cells for this type of applications currently are those fuelled by hydrogen, methanol or ammonia, directly or after reforming. These systems are used for example for data centres, telecommunication towers, hospitals or rural villages and small islands. Many telecom towers, especially in developing countries, are powered by diesel generators, as they are located in hardly accessible places where the electricity network is absent. Some telecom base stations in South Africa and Kenya are already powered by fuel cells today in the context of experimental projects (Ballard and GSMA, 2013; GenCell, 2018; GSMA, 2019).

Hospitals and clinics, as well as data centres and some banks need protection from blackouts. In this case, dedicated generators or uninterruptible power supplies with fuel cells could be used. In South Africa, a clinic that needs a continuous supply of electricity to refrigerate vaccines and medicines has relied on an uninterruptible power supply with stationary fuel cells since 2015 (Fuel Cells Bulletin, 2006, 2015).

Even in case of emergencies where it is not possible to rely on the electricity grid, such as field hospitals, relief in the event of hurricanes or earthquakes, mobile generators with fuel cells could be used, but also in outdoor events or concerts.

Another case of stationary generation demand may come from nonelectrified rural villages. Some trial project in South Africa have provided electricity to small rural villages using methanol fuel cells with appropriate methanol tanks and batteries (Fuel Cells Bulletin, 2013; ESI Africa's Power Journal, 2014).

10.4.3 Long term and large scale energy storage

Energy storage on a seasonal scale can make possible to balance seasonal variations both in terms of electricity demand and supply. Especially in an energy system strongly or totally based on RES, it will be necessary to store huge amounts of energy. Trying to explore some alternatives, to date, batteries are not yet a fully mature technology, they are very expensive for large plants, suffer from self-discharge over time and, moreover, a huge number of batteries and occupation of large spaces would be required for large-scale storage. This technology nowadays would also require a significant exploitation of rare earths and critical materials for batteries production, such as lithium and cobalt (Yao *et al.*, 2016). In a market with high shares of renewables, hydroelectric storage basins (or pumped-hydro), on the contrary now fully mature, will no longer be enough to absorb the grid energy fluctuations, while the construction of new artificial basins, however linked to strictly geomorphological problems, would involve evident impacts in natural environments. Although there are various other alternatives for energy storage, such as flow or solid-state batteries, flywheels and accumulation of compressed air, the best viable solution is the power to fuel technology which does not suffer from losses of stored energy along the time.

One of the advantages of the power to methane technology is the possibility of energy exchange between the electricity grid and the NG network which can be seen as a huge storage basin for renewable sources. As is already the case with the NG infrastructure, synthetic gases can also be stored for entire seasons in underground caves or salt caverns. Salt caverns, thanks to their characteristics of good tightness and low gas contamination, appear very promising in particular for the storage of hydrogen. In addition, other underground storage systems are currently being studied, such as depleted oil or NG fields and pore storage. Ammonia is another suitable synthetic fuel for long-term and large-scale energy storage, since large steel refrigerated tanks are already commonly used in the fertiliser industry for storing liquid ammonia.

Nomenclature

CHP Cogenerated Heat and Power
CNG Compressed Natural Gas
DME DiMethyl Ether
DRI-EAF Direct Reduction of Iron-Electric Arc Furnace

FCEV	Fuel Cell Electric Vehicle
ICE	Internal Combustion Engine
IMO	International Maritime Organization
LNG	Liquefied Natural Gas
LPG	Liquefied Petroleum Gas
NG	Natural Gas
PEMFC	Proton Exchange Membrane Fuel Cell
RES	Renewable Energy Sources
SAF	Sustainable Aviation Fuels
SCR	Selective Catalytic Reduction
TFC	Total Final Consumption
TPES	Total Primary Energy Supply

References

Altfeld, K., and Pinchbeck, D. (2013). Admissible hydrogen concentrations in natural gas systems. *Gas for Energy, March/2013*, 1–16. Available at: https://www.gas-for-energy.com.

ANT Energy Solutions (2013) 'Renewable Hydrogen Market Report'.

Asian Rehub (2020) *Asian Renewable Energy Hub Project*. Available at: https://asianrehub.com/ (Accessed: 22 December 2020).

Balestri, M. *et al.* (2007) 'Enel's Fusina hydrogen-fed power generation plant', *2007 International Conference on Clean Electrical Power, ICCEP '07*, pp. 456–463. https://doi.org/10.1109/ICCEP.2007.384254.

Ballard and GSMA (2013) *Fuel Cell Systems for Telecom Backup Power*. Available at: https://www.gsma.com/mobilefordevelopment/wp-content/uploads/2013/06/Ballard-Power-Systems.pdf.

CNN (2020) *Bill Gates, Amazon and British Airways are backing a hydrogen plane startup*. Available at: https://edition.cnn.com/2020/12/16/business/zeroavia-hydrogen-planes/index.html (Accessed: 22 December 2020).

Dimitriou, P., and Javaid, R. (2020). A review of ammonia as a compression ignition engine fuel. *International Journal of Hydrogen Energy*, *45*(11), 7098–7118. Available from https://doi.org/10.1016/j.ijhydene.2019.12.209, Elsevier Ltd.

Ditaranto, M., Heggset, T., and Berstad, D. (2020). Concept of hydrogen fired gas turbine cycle with exhaust gas recirculation: assessment of process performance. *Energy*, *192*, 116646. Available from https://doi.org/10.1016/j.energy.2019.116646, Elsevier Ltd.

EIA (2019) *International Energy Outlook 2019 - with projections to 2050*. Washington, DC 20585: U.S. Energy Information Administration. Available at: https://www.eia.gov/outlooks/ieo/pdf/ieo2019.pdf.

ENE-FARM (2020) *Panasonic Launches New 'ENE-FARM' Product, a Fuel Cell for Condominiums - FuelCellsWorks*. Available at: https://fuelcellsworks.com/news/panasonic-launches-new-ene-farm-product-a-fuel-cell-for-condominiums/ (Accessed: 30 November 2020).

ENEL (2009) ENEL: first hydrogen-fuelled power now on-line in Venice. Available at: https://corporate.enel.it/en/media/press/d/2009/08/enel-first-hydrogen-fuelled-power-now-on-line-in-venice- (Accessed: 21 December 2020).

ESI Africa's Power Journal (2014) *Methanol Fuel Cells trialled to power rural South African towns*. Available at: https://www.esi-africa.com/top-stories/methanol-fuel-cells-trialled-to-power-rural-south-african-towns/ (Accessed: 23 December 2020).

European Commission (2002) *Liquid hydrogen fuelled aircraft - system analysis (CRYOPLANE project)*. Available at: https://cordis.europa.eu/project/id/G4RD-CT-2000-00192/it (Accessed: 22 December 2020).

EUTurbines (2019) *The gas turbine industry commitments to drive Europe's transition to a decarbonised energy mix (press release)*. Available at: https://powertheeu.eu/wp-content/themes/euturbines/dl/EUTurbines-press-release-on-the-Commitments.pdf.

Fuel Cells Bulletin. (2006). Intelligent energy fuel cells showcased in S. Africa. *Fuel Cells Bulletin, 2006*(4), 1. Available from https://doi.org/10.1016/S1464-2859(06)70995-6.

Fuel Cells Bulletin. (2013). SA supports Ballard, Anglo American fuel cells deployment. *Fuel Cells Bulletin, 2013*(7), 6. Available from https://doi.org/10.1016/s1464-2859(13)70255-4.

Fuel Cells Bulletin. (2015). South African clinic using fuel cell power to benefit TB patients. *Fuel Cells Bulletin, 2015*(10), 5. Available from https://doi.org/10.1016/s1464-2859(15)30308-4.

FuelCellsWorks (2019) *South Korea: Work Begins on World's Largest Hydrogen Fuel Cell Power Plant*. Available at: https://fuelcellsworks.com/news/south-korea-work-begins-on-worlds-largest-hydrogen-fuel-cell-power-plant-commissioned/ (Accessed: 23 December 2020).

GenCell (2018) *Adrian Kenya Selects New GenCell A5 Fuel Cell Solution to Provide Green, Off-Grid Power to 800 Telecom Base Stations*. Available at: https://www.gencellenergy.com/news/adrian-kenya-800-telecom-base-stations/ (Accessed: 23 December 2020).

Global Railway Review (2020) *Alstom awarded Germany contract for supply of 30 Coradia Lint trains*. Available at: https://www.globalrailwayreview.com/news/98798/alstom-germany-contract-coradia-lint-trains/ (Accessed: 22 December 2020).

Goldmeer, J. (2018) 'Fuel flexible gas turbines as enablers for a low or reduced carbon energy ecosystem', (May).

Grand View Research (2019) *Methanol Market Size, Share & Trends Analysis Report By Application (Formaldehyde, Acetic Acid, MTBE, DME, Fuel Blending, MTO, Biodiesel), By Region, And Segment Forecasts, 2019 - 2025*. Available at: https://www.grandviewresearch.com/industry-analysis/methanol-market (Accessed: 24 December 2020).

GSMA (2019) *Case Study: GenCell*. Available at: https://www.gsma.com/futurenetworks/wiki/case-study-gencell/ (Accessed: 23 December 2020).

H21 (2018) *H21 North of England*. Available at: https://www.h21.green/wp-content/uploads/2019/01/H21-NoE-PRINT-PDF-FINAL-1.pdf.

Herdin, G. *et al.* (2007) 'Hydrogen and hydrogen mixtures as fuel in stationary gas engines', *SAE Technical Papers*, (724). doi: 10.4271/2007-01-0012.

Hsieh, J. (1979) *U.S. Patent 4160663A - Method for the direct reduction of iron ore*.

HYBRIT Project (2019) *HYBRIT - Towards fossil-free steel*. Available at: http://www.hybritdevelopment.com/ (Accessed: 21 September 2020).

ICAO. (2018). *Sustainable Aviation Fuels Guide*. International Civil Aviation Organisation.

IEA. (2019a). Key World Energy Statistics, International Energy Agency. Paris: IEA Available at. Available from https://www.iea.org/reports/key-world-energy-statistics-2020.

IEA. (2019b). The Future of Hydrogen—Seizing today's opportunities—Report prepared for the G20, Japan, International Energy Agency. Paris: IEA Available at. Available from https://webstore.iea.org/download/direct/2803.

IEA. (2019c). World Energy Balances, International Energy Agency. Paris: IEA Available at. Available from https://www.iea.org/reports/world-energy-balances-overview.

International Maritime Organization (2018) *Adoption of Initial IMO Strategy on Reduction of GHG Emissions from Ships, Note by the International Maritime Organization to the UNFCCC Talanoa Dialogue*.

IVECO (2020) *Natural Gas Powered Vehicles*. Available at: https://www.iveco.com/en-us/press-room/kit/Pages/E3-NaturalGasPoweredVehicles.aspx (Accessed: 22 December 2020).

Kippers, M.J. *et al.* (2011) *Pilot project on hydrogen injection in natural gas on island of Ameland in The Netherlands, International Gas Union Research Conference Proceedings*. Apeldoorn, Netherlands. Available at: http://members.igu.org/old/IGUEvents/igrc/igrc2011/igrc-2011-proceedings-and-presentations/posterpaper-session1/P1-34_MathijsKippers.pdf.

Krüger, A., et al. (2020). Integration of water electrolysis for fossil-free steel production. *International Journal of Hydrogen Energy, 45*(55), 29966–29977. Available from https://doi.org/10.1016/j.ijhydene.2020.08.116.

Leo, A. and FuelCell Energy (2016) 'Stationary Fuel Cell Power Systems with Direct FuelCell Technology Tackle Growing Distributed Baseload Power Challenge'. Available at: http://www.fuelcellenergy.com/wp-content/uploads/2016/10/Fuel-Cell-Technology-White-Paper.pdf.

MAN Energy Solutions (2019) 'Engineering the future two-stroke green-ammonia engine'. Available at: https://marine.man-es.com/docs/librariesprovider6/test/engineering-the-future-two-stroke-green-ammonia-engine.pdf?sfvrsn = 7f4dca2_4.

Market Research Future (2020) *Ammonia Market by Size, Share, Type and Research Analysis – 2028*. Available at: https://www.marketresearchfuture.com/reports/ammonia-market-2405 (Accessed: 21 December 2020).

McKinsey and Company, Clean Sky 2 JU and Fuel Cell and Hydrogen 2 JU (2020) *Hydrogen-powered aviation. A fact-based study of hydrogen technology, economics, and climate impact by 2050, Publications Office of the European Union*. Belgium. https://doi.org/10.2843/766989.

Mitsubishi Power (2019) *Hydrogen Power Generation Handbook*. Available at: https://power.mhi.com/catalogue/pdf/hydrogen_en.pdf.

Moliere, M. (1999) 'Heavy duty gas turbines in petrochemical plants: Samsung's Daesan plant (Korea) beats fuel flexibility records with over 95% hydrogen in process gas', *Proceedings of the PowerGen Asia Conference*, (September).

Muraki, S. (2018) 'Energy Carriers - Becoming a New Energy Society', *Cross-ministerial Strategic Innovation Promotion Program (SIP)*.

Nagashima, M. (2018) *Japan's hydrogen strategy and its economic and geopolitical implications, Études de l'Ifri*. Ifri.

Nielsen, E.R. and Prag, C.B. (2017) *Learning points from demonstration of 1000 fuel cell based micro-CHP units - Summary of analysis from the ene.field project*. Available at: http://enefield.eu/wp-content/uploads/2017/10/ene.field-Summary-Report.pdf.

Power Magazine (2020) *JERA Planning to Shift Coal Power Fleet to 100% Ammonia*. Available at: https://www.powermag.com/jera-planning-to-shift-coal-power-fleet-to-100-ammonia/ (Accessed: 23 December 2020).

Shiozawa, B. (2019) '"SIP Energy Carriers" Updates and Establishment of Green Ammonia Consortium'.

Snam (2020) *Agreement between Alstom and Snam for the development of hydrogen trains in Italy*. Available at: https://www.snam.it/en/Media/Press-releases/Agreement_Alstom_and_Snam_hydrogen_trains_in_Italy.html (Accessed: 22 December 2020).

Spazzafumo, G. (2003). An Overview of Thermodynamic Cycles Based on Direct Stoichiometric Combustion of Hydrogen and Oxygen. *Journal of Scientific & Industrial Research, 62*(January–February), 71–80.

Tian, P., et al. (2015). Methanol to olefins (MTO): From fundamentals to commercialization. *ACS Catalysis, 5*(3), 1922–1938. Available from https://doi.org/10.1021/acscatal.5b00007.

Toshiba (2017) *Toshiba's 100kW Pure Hydrogen Fuel Cell System Realizing CO2 Free Electricity Generation is now Operating in Japan*. Available at: https://www.toshiba.co.jp/about/press/2017_04/pr2701.htm (Accessed: 23 December 2020).

Towoju, O. A., and Dare, A. (2017). Di-methyl ether (DME) as a substitute for diesel fuel in compression ignition engines. *International Journal Of Advanced Research in Engineering & Management (IJAREM)*, 39–47, March.

Valera-Medina, A., et al. (2018). Ammonia for power. *Progress in Energy and Combustion Science*, *69*, 63–102. Available from https://doi.org/10.1016/j.pecs.2018.07.001, Elsevier Ltd.

van Wijk, A. and Noordelijke Innovation Board (2017) *The Green Hydrogen Economy now in the Northern Netherlands*. Groningen, the Netherlands.

Vavra, J., Bortel, I., and Takats, M. (2019). *A dual fuel hydrogen—diesel compression ignition engine and its potential application in road transport. SAE Technical Paper*. SAE International. https://doi.org/10.4271/2019-01-0564.

Verhelst, S., and Wallner, T. (2009). Hydrogen-fueled internal combustion engines. *Progress in Energy and Combustion Science*, *35*(6), 490–527. Available from https://doi.org/10.1016/j.pecs.2009.08.001.

Wang, T., et al. (2014). *Conversion of methanol to aromatics in fluidized bed reactor. Catalysis Today* (233, pp. 8–13). Elsevier B.V. https://doi.org/10.1016/j.cattod.2014.02.007.

Yao, L., et al. (2016). Challenges and progresses of energy storage technology and its application in power systems. *Journal of Modern Power Systems and Clean Energy. Springer Berlin Heidelberg*, *4*(4), 519–528. Available from https://doi.org/10.1007/s40565-016-0248-x.

Ye, L., et al. (2017). Reaction: "green" ammonia production. *Chem*, *3*(5), 712–714.

Index

Note: Page numbers followed by "*f*" and "*t*" refer to figures and tables, respectively.

FSC®
www.fsc.org
MIX
Papier | Fördert gute Waldnutzung
FSC® C083411

Kolibri 360 GmbH
Ferdinand-Jühlke-Straße 7
99095 Erfurt, Deutschland
produktsicherheit@kolibri360.de